INTRODUCTION TO

PHYSICAL GEOLOGY

GEOLOGIC FEATURES AND EVENTS DISCUSSED IN TEXT

1. Lava plateau: Columbia River plateau, Washington (Chapter 5)
2. Eruption of El Paricutin: San Juan Parangaricutiro, Mexico (Chapter 5)
3. Caldera eruption: Yellowstone National Park, Wyoming (Chapter 5)
4. Caldera eruption: Long Valley, California (Chapter 5)
5. Sediment deposition and unconformities: Grand Canyon, Arizona (Chapter 9)
6. Earthquake damage: Mexico City (Chapter 10)
7. Earthquakes at a transform plate boundary: The Northridge earthquake, California (Chapter 10)
8. Earthquakes at subduction zones: The Pacific Northwest (Chapter 10)
9. Earthquakes at plate interiors: New Madrid, Missouri (Chapter 10)
10. Magnetic orientations of the oceanic crust: Reykjanes Ridge, Iceland (Chapter 11)
11. Seamounts and oceanic islands: Hawaiian Island-Emperor Seamoun Chain, Hawaii (Chapter 11)
12. Mountain building at a subduction zone: The Andes, western South America (Chapter 12)
13. Mountain building at a collision zone between two plates carrying continental crust: The Himalayas (Chapter 12)
14. Mass wasting triggered by an earthquake: Madison River slide, Montana (Chapter 13)

©1990 Tom Van Sant, Inc. / The GeoSphere Project
Santa Monica, California

Graham R. Thompson, PhD
University of Montana

Jonathan Turk, PhD

INTRODUCTION TO

PHYSICAL GEOLOGY

Saunders College Publishing

Harcourt Brace College Publishers

Fort Worth Philadelphia San Diego New York Orlando
Austin San Antonio Montreal London Sydney Tokyo

Vice President/Publisher: John Vondeling
Acquisitions Editor: Jennifer Bortel
Product Manager: Nick Agnew
Developmental Editor: Sarah Fitz-Hugh
Project Editor: Sally Kusch
Production Manager: Charlene Catlett Squibb
Art Director: Caroline McGowan

Cover: Kayaker under waterfall, Bio Bio River, Chile. (© Tony Stone Images/Harvey Kennan)

Frontispiece: Kayaker on the Seven Teacups, Chile. (© Tony Stone Images/Harvey Kennan)

Printed in the United States of America

Introduction to Physical Geology

0-03-024348-3

Library of Congress Catalog Card Number: 97-67673

7890123456   032   10 987654321

PREFACE

Every day we walk, bicycle, or drive across landscapes that were sculptured by geologic processes. Even people who live in cities see evidence of geological activity in parks and road cuts and in the shapes of river valleys, lake shores, or coastlines. Mountains, prairies, and glaciers are familiar through travel, books, and film. Every year volcanic eruptions and earthquakes create human tragedy somewhere on the planet. In addition, the water we drink, the climate we live in, and the air we breathe are affected by geological processes. This book will enable a student to see the Earth as a geologist does and to understand geologic processes that change and shape our planet.

▶ PHILOSOPHY AND APPROACH

Geologic change affects humans, and at the same time, humans extract and sometimes pollute geologic resources. Thus, earthquakes and volcanoes destroy cities; industry and commerce depend on fuels and minerals; floods and storms threaten cities along river banks and coastlines; pesticide and fertilizer runoff from farms leach into ground water. Difficult questions arise as we try to harmonize our needs with fundamental Earth processes. We explore these issues by examining scientific processes without advocating opinion or philosophy.

We emphasize the scientific method throughout the book. In many cases, we discuss the basic observations or experiments that lead to the development of a theory. In other instances, we describe conflicting viewpoints about a topic to emphasize the nature of scientific debate. Finally, case studies demonstrate how geologic processes change our world. Examples include: *The Northridge Earthquake of January 1994* in Chapter 10 and *Flood Control and the 1993 Mississippi River Floods* in Chapter 14.

Introduction to Physical Geology is a streamlined essentials text, a frank abridgement of our successful *Modern Physical Geology*. Our goal in producing *Introduction to Physical Geology* is to create a solid introductory text respectful of the economies of time and expense that constrain today's students and instructors. This book is comprehensive, with broad appeal to many different types of first-time geology students. Yet it is as direct and lean as an introductory text can be. For example, in the interest of brevity, we have decided to confine our discussion to the geology of this planet. (For expanded coverage of the other planets, please see our recently published second edition of *Modern Physical Geology*.)

▶ ORGANIZATION

Modern geologists explain many phenomena, such as earthquakes, volcanoes, and the origins of mountain ranges and ocean basins in terms of plate tectonics theory. For that reason, we introduce plate tectonics in Chapter 2. We then explain the theory in more detail and discuss the consequences of plate tectonics throughout the remainder of the book. This sequence corresponds to modern geological thought and teaches geology as an active, growing science built on observation, experimentation, and reasoning, not as a list of facts to be memorized.

To accommodate a wide range of course syllabi, we organized this book into a core text of four subject areas supplemented with additional material in "Focus On" boxes and a fifth area that deals with special topics. The subject areas and the chapters they comprise are as follows.

Chapters 1–3: The Earth: Its Origin, Internal Processes, and Minerals
Chapters 4–9: Rocks
Chapters 10–12: An Active Earth
Chapters 13–18: Surface Processes
Chapters 19–20: Special Topics in Geology

The first four subject areas provide a complete, concise introduction to geology. "Focus On" boxes and the two final chapters complement the core material and expose students to important recent developments in modern geology.

Introduction to Physical Geology incorporates many helpful suggestions from professors who have used other books we have written and from many reviewers. We have written the book with clarity and timeliness our chief concerns. In addition to the coverage of tectonic plate activity early in the book, we offer coverage of such important issues as lakes and modeling methods of ground-water remediation. We have carefully selected most case studies to illustrate specific concepts and pique student interest. We have written richly descriptive captions that explicate and amplify the excellent photographs and illustrations accompanying this text. The boxes that expand on the main text include discussion questions to encourage students to think critically about what they are studying.

► SPECIAL FEATURES

FOCUS ON BOXES

Interesting side topics are presented throughout the book in the form of "Focus On" boxes. Thus, the student is drawn into the chapter through reading geological anecdotes and interesting ideas that reinforce the main topics.

USE OF ANALOGIES

Many geological processes occur beneath the Earth's surface where they cannot be directly observed. Some occur so slowly that even if we could see them, the changes would be detectable only over many human lifespans. However, all of these processes can be clearly understood through the use of examples and analogies. This book presents familiar examples and analogies to explain geological processes that cannot be experienced directly.

CHAPTER REVIEW MATERIAL

Important words are highlighted in bold type in the text. We then list these **Key Words** at the end of each chapter for review purposes and provide a short **Summary** at the end of each chapter.

QUESTIONS

Two types of questions are included in the end-of-chapter material. **Review Questions** can be answered in a straightforward manner from the material in the text. On the other hand, **Discussion Questions** challenge students to apply what they have learned to the analysis of situations not directly presented in the text. Often there are a number of correct answers to these questions; they are intended to provoke thought and discussion among students.

GLOSSARY AND APPENDICES

A glossary of geological terms is provided at the end of the book. In almost all cases, we use definitions given in the *Dictionary of Geological Terms,* published by the American Geological Institute. We recommend this excellent and inexpensive paperback to the student who wants a comprehensive glossary of geology. In addition, appendices cover metric units and rock symbols.

▶ ANCILLARIES

This text is accompanied by an extensive set of support materials.

INVESTIGATIONS INTO PHYSICAL GEOLOGY: A LABORATORY MANUAL

This laboratory manual written by Jim Mazzullo at Texas A&M University is intended to be used in the laboratory classes for introductory physical geology and to complement **Introduction to Physical Geology.** The focus of this manual is its diverse exercises, which are designed to develop the observational and analytical skills of students as they study the structure, composition, and origin of the Earth's crust and the tectonic and surficial processes that shape the continents and the sea floor. These exercises are accompanied by reviews of the basic principles of physical geology, as well as by a rich collection of maps, photographs, block diagrams, tables, and other resources and references.

INSTRUCTOR'S MANUAL/TEST BANK

The instructor's manual, written by the authors, provides teaching goals, alternate sequences of topics, and answers to discussion questions. The test bank is an important tool that features multiple choice, true/false, and completion questions for each chapter in the text.

OVERHEAD TRANSPARENCIES

A set of transparencies featuring new figures and photos is available.

SAUNDERS GEOLOGY VIDEODISC

The videodisc includes over 2000 images from eight of Saunders' best-selling geology, earth science, and geography texts. It includes almost one hour of live-action footage from *Encyclopedia Britannica* archives, featuring landscapes and geophenomena, and animated segments that bring geo-processes to life.

EARTH SYSTEMS STUDENT TUTORIAL CD-ROM

The Earth Systems CD-ROM provides a clear and engaging tutorial that walks students through the fundamental geologic concepts of plate tectonics and surface processes. It contains dramatic video and animation clips of such phenomena as earthquakes, volcanoes, and hurricanes. It provides students with the next best thing to direct experience of the forces that shape our planet.

EARTH SYSTEMS MEDIA ACTIVE™ CD-ROM FOR INSTRUCTORS

This invaluable teaching aid for use with Lecture Presentation software includes images of all the book's illustrations and many of its photos.

Saunders College Publishing may provide complimentary instructional aids and supplements or supplement packages to those adopters qualified under our adoption policy. Please contact your sales representative for more information. If as an adopter or potential user you receive supplements you do not need, please return them to your sales representative or send them to: Attn: Returns Department, Troy Warehouse, 465 South Lincoln Drive, Troy, MO 63379.

► ACKNOWLEDGEMENTS

We have not worked alone. The manuscript has been extensively reviewed at several stages and the numerous careful criticisms have helped shape the book and ensure accuracy:

Special thanks are due Mark J. Camp, The University of Toledo; R. Douglas Elmore, University of Oklahoma; and B.F. Rowell, Kutztown University, who guided our hands in shaping this sleek and useful volume.

Additionally, we wish to thank the following colleagues for their invaluable contributions to this edition and the first edition of *Modern Physical Geology*.

REVIEWERS

David Alt, *University of Montana*
Estella Atekwana, *Western Michigan University*
Joan Baldwin, *El Camino College*
Phillip Banks, *Case Western Reserve University*
Robert E. Behling, *West Virginia University*
Donald M. Burt, *Arizona State University*
Gary Byerly, *Louisiana State University*
F. Howard Campbell III, *James Madison University*
Mario V. Caputo, *Mississippi State University*
Maryanne Cella, *University of Dayton*
Wang-Ping Chen, *University of Illinois at Urbana-Champaign*
Roger Cooper, *Lamar University*
Larry E. Davis, *Washington State University*
Don Fisher, *Pennsylvania State University*
Roberto Garza, *San Antonio College*
Ernest Gilmour, *Eastern Washington University*
Fred Goldstein, *Trenton State College*
Bryan Gregor, *Wright State University*
Andrew Hajash, *Texas A & M University*
Vicki Harder, *Texas A & M University*
Donald W. Hyndman, *University of Montana*
Edward Kantor, *Troy State University*
David T. King, *Auburn University*
Martin Kleinrock, *Vanderbilt University*
Jim Mazzullo, *Texas A & M University*
Kula Misra, *University of Tennessee, Knoxville*
Howard Mooers, *University of Minnesota*

John E. Mylroie, *Mississippi State University*
Rainer Newberry, *University of Alaska, Fairbanks*
John Osmond, *Florida State University*
James F. Petersen, *Southwest Texas State University*
Louis Pinto, *Monroe Community College*
Thomas Prather, *Western State College*
Jennifer S. Prouty, *Corpus Christi State University*
John Reid, *University of North Dakota*
Mary Jo Richardson, *Texas A & M University*
Bruce Rueger, *Colby College*
David Schwimmer, *Columbus College*
James W. Sears, *University of Montana*
Steven D. Sherriff, *University of Montana*
Steve Simpson, *Highland Community College*
William A. Smith, *Western Michigan University*
George D. Stanley, Jr., *University of Southern Mississippi*
Daniel A. Sundeen, *University of Southern Mississippi*
Lawrence Tanner, *Bloomsburg University*
Charles P. Thornton, *Pennsylvania State University*
Alfred Traverse, *Pennsylvania State University*
Willis Weight, *Montana Tech of the University of Montana*
Peter W. Whaley, *Murray State University*

Geology is a visual science. We can readily observe rocks and landforms on the Earth's surface. Although we cannot see internal processes in action, these events can be visualized through the artist's eye. George Kelvin has painted many of the superb illustrations in this book and he has been a pleasure to work with. We would also like to thank Larry Davis, Don Hyndman, Dewey Moore, Roberto Garza, and David Schwimmer for contributing excellent photographs. Thanks also to Duwayne Anderson at Texas A&M University for his help with some of the visual ancillaries. A special thanks to Dana Desonie for reading and editing the entire manuscript with care. She has contributed significantly to sharpening the writing style of the text.

We would never have been able to produce this book without professional support both here in Montana and at the offices of Saunders College Publishing. Thanks to Christine Seashore for logistic collaboration. Special thanks to John Vondeling, our Publisher. One of us, Jonathan Turk, has worked with John for over twenty-five years and has developed a long-lasting friendship and a superb professional relationship with him. Jennifer Bortel, Acquisitions Editor, Sarah Fitz-Hugh, Developmental Editor, Sally Kusch, Managing Editor, George Semple, Picture Developmental Editor, Caroline McGowan, Art Director, Charlene Squibb, Production Manager, and Marne Evans, Copy Editor, have all worked hard and efficiently to produce the finished project.

GRAHAM R. THOMPSON
University of Montana
Missoula, Montana

JONATHAN TURK
Darby, Montana

July 1997

CONTENTS OVERVIEW

CONTENTS

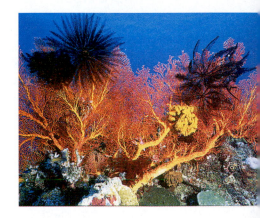

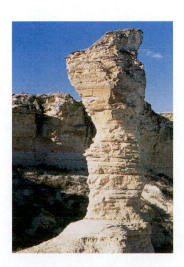

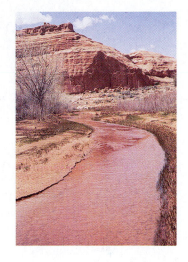

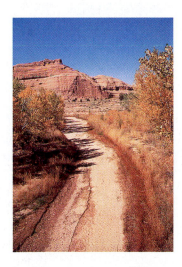

Geology and the Earth

Imagine walking on a rocky shore. You can see the pounding surf, hear stones clink together as waves recede, feel the wind blowing in your hair. But the cliffs don't move and the ground doesn't shake. Even though the Earth appears to be a firm foundation beneath your feet, it is a dynamic planet. Continents slowly shift position; mountains rise and then erode away. These motions escape casual observation because they are generally slow, although every year events such as volcanic eruptions and earthquakes remind us that geologic change can be rapid.

A storm-driven wave crashes against the Oregon coast. (H. Richard Johnston/Tony Stone)

► I.I THE SCIENCE OF GEOLOGY

Geology is the study of the Earth, including the materials that it is made of, the physical and chemical changes that occur on its surface and in its interior, and the history of the planet and its life forms.

THE EARTH AND ITS MATERIALS

The Earth's radius is about 6370 kilometers, nearly one and a half times the distance from New York to Los Angeles (Fig. 1–1). If you could drive a magical vehicle from the center of the Earth to the surface at 100 kilometers per hour, the journey would take more than two and a half days.

Most of the Earth is composed of **rocks**. Rock outcrops form some of our planet's most spectacular scenery: white chalk cliffs, pink sandstone arches, and the grey granite of Yosemite Valley. Rocks, in turn, are composed of **minerals** (Fig. 1–2). Although more than 3500 different minerals exist, fewer than a dozen are common. Geologists study the origins, properties, and compositions of both rocks and minerals.

Geologists also explore the Earth for the resources needed in our technological world: fossil fuels such as coal, petroleum, and natural gas; mineral resources such as metals; sand and gravel; and fertilizers. Some search for water in reservoirs beneath Earth's surface.

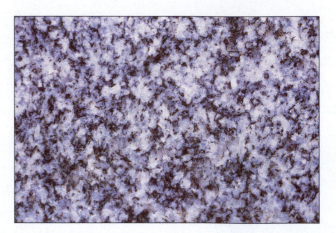

Figure 1–2 This granite rock is composed of different minerals, primarily quartz, feldspar, and hornblende. The mineral grains are a few millimeters in diameter.

INTERNAL PROCESSES

Processes that originate deep in the Earth's interior are called **internal processes**. These are the *driving forces* that raise mountains, cause earthquakes, and produce volcanic eruptions. Builders, engineers, and city planners might consult geologists, asking, "What is the probability that an earthquake or a volcanic eruption will damage our city? Is it safe to build skyscrapers, a dam, or a nuclear waste repository in the area?"

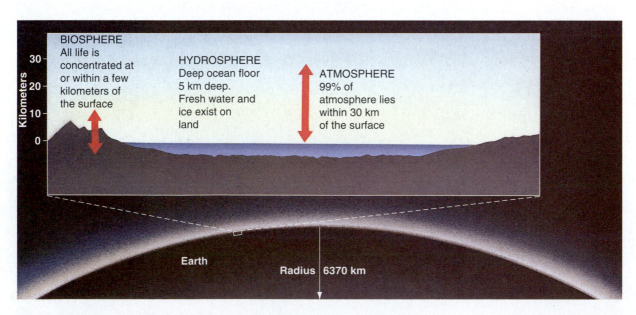

Figure 1–1 Most of the Earth is solid rock, surrounded by the hydrosphere, the biosphere, and the atmosphere.

SURFACE PROCESSES

Surface processes are all of those processes that sculpt the Earth's surface. Most surface processes are driven by water, although wind, ice, and gravity are also significant. The **hydrosphere** includes water in streams, wetlands, lakes, and oceans; in the atmosphere; and frozen in glaciers. It also includes ground water present in soil and rock to a depth of at least 2 kilometers.

Most of us have seen water running over the ground during a heavy rain. The flowing water dislodges tiny grains of soil and carries them downslope. If the rain continues, the water may erode tiny gullies into a hillside (Fig. 1–3). A gully may form in a single afternoon; over much longer times, the same process forms canyons and spacious river valleys. People build cities along rivers to take advantage of the flat land, fertile soil, and abundant water. But the erosion continues. Rivers wear away at their banks and bed and periodically flood adjacent land. Geologists seek to understand these processes and advise builders and planners to minimize loss of life and property.

The oceans cover more than 70 percent of our planet. Although oceanography is a separate scientific discipline, it overlaps with geology. Geologic processes form the ocean basins and alter their size and shape. Weathering and erosion of continents carry mud, sand, and salts to the sea. Earth is the only planet in the Solar System that has oceans. It is also the only planet that supports life. Oceanographers examine the oceans' influence on climate, the atmosphere, life, and the solid Earth.

THE ATMOSPHERE

The **atmosphere** is a mixture of gases, mostly nitrogen and oxygen (Fig. 1–4). It is held to the Earth by gravity and thins rapidly with altitude. Ninety-nine percent is concentrated within 30 kilometers of the Earth's surface, but a few traces remain even 10,000 kilometers above the surface. A brief look at our neighbors in space reminds us that the interactions among air, rock, and life affect atmospheric composition, temperature, and movement. The solid Earth, Venus, and Mars are approximately identical in composition. Yet the three planets have radically different atmospheres and climates. Today, the Venusian atmosphere is hot, acidic, and rich in carbon dioxide. The surface temperature is 450°C, as hot as the interior of a self-cleaning oven, and the atmospheric pressure is 90 times greater than that of the Earth. In contrast, Mars is frigid, with an atmospheric pressure only 0.006 that at the surface of the Earth. Venusian water has boiled off into space; almost all Martian water lies frozen in vast underground reservoirs.

Figure 1–3 Over long periods of time, running water can carve deep canyons, such as this tributary of Grand Canyon in the American southwest.

Figure 1–4 This storm cloud over Mt. Robson, British Columbia, is a visible portion of the Earth's atmosphere.

THE BIOSPHERE

The **biosphere** is the thin zone near the Earth's surface that is inhabited by life. It includes the uppermost solid Earth, the hydrosphere, and the lower parts of the atmosphere. Land plants grow on the Earth's surface, with roots penetrating at most a few meters into soil. Animals live on the surface, fly a kilometer or two above it, or burrow a few meters underground. Sea life also concentrates near the ocean surface, where sunlight is available. Some aquatic communities live on the deep sea floor, bacteria live in rock to depths of a few kilometers, and a few windblown microorganisms are found at heights of 10 kilometers or more. But even at these extremes, the biosphere is a very thin layer at the Earth's surface.

Paleontologists are geologists who study the evolution and history of life by examining fossils and other evidence preserved in rock and sediment. The study of past life shows us that the solid Earth, the atmosphere, the hydrosphere, and the biosphere are all interconnected. Internal processes such as volcanic eruptions and migrating continents have altered the Earth's climate and atmospheric composition. Life has altered the atmosphere. The atmosphere reacts with rocks.

▶ I.2   UNIFORMITARIANISM AND CATASTROPHISM

James Hutton was a gentleman farmer who lived in Scotland in the late 1700s. Although trained as a physician, he never practiced medicine and, instead, turned to geology. Hutton observed that a certain type of rock, called sandstone, is composed of sand grains cemented together (Fig. 1–5). He also noted that rocks slowly decompose into sand, and that streams carry sand into the lowlands. He inferred that sandstone is composed of sand grains that originated by the erosion of ancient cliffs and mountains.

Hutton tried to deduce how much time was required to form a thick bed of sandstone. He studied sand grains slowly breaking away from rock outcrops. He watched sand bouncing down streambeds. Finally he traveled to beaches and river deltas where sand was accumulating. Hutton concluded that the sequence of steps that he had observed must take a long time. He wrote that

> on us who saw these phenomena for the first time, the impression will not easily be forgotten. . . .
> We felt ourselves necessarily carried back to the time . . . when the sandstone before us was only beginning to be deposited, in the shape of sand and mud, from the waters of an ancient ocean. . . . The mind seemed to grow giddy by looking so far into the abyss of time.

Hutton's conclusions led him to formulate a principle now known as **uniformitarianism**. The principle states that geologic change occurs over long periods of time, by a sequence of almost imperceptible events. Hutton surmised that geologic processes operating today also operated in the past. Thus, scientists can explain events that occurred in the past by observing changes occurring today. Sometimes this idea is summarized in the statement "The present is the key to the past." For example, we can observe today each individual step that leads to the formation of sandstone. Even though it would take too long for us to watch a specific layer of sandstone form, we can infer that the processes occur slowly—step by step—over great periods of time.

If we measure current rates of geologic change, we must accept the idea that most rocks are much older than human history. Taking his reasoning one step further, Hutton deduced that our planet is very old. He was so overwhelmed by the magnitude of geological time that he wrote, "We find no vestige of a beginning, no prospect of an end."

William Whewell, another early geologist, agreed that the Earth is very old, but he argued that geologic change was sometimes rapid. He wrote that the geologic past may have "consisted of epochs of paroxysmal and catastrophic action, interposed between periods of comparative tranquility." Whewell was unable to give examples of such catastrophes. He argued that they happen so infrequently that none had occurred within human history.

Today, geologists know that both Hutton's uniformitarianism and Whewell's **catastrophism** are correct. Thus, over the great expanses of geologic time, **slow, uniform processes are significant, but improbable,**

Figure I–5   Sandstone cliffs rise above the Escalante River, Utah.

catastrophic events radically modify the path of slow change.

Gradual Change in Earth History

Within the past few decades, geologists have learned that continents creep across the Earth's surface at a rate of a few centimeters every year. Since the first steam engine was built 200 years ago, North America has migrated 8 meters westward, a distance a sprinter can run in 1 second. Thus continental motion is too slow to be observed except with sensitive instruments. However, if you could watch a time-lapse video of the past few hundred million years—only a small chunk of geologic time—you would see continents travel halfway around the Earth.

Catastrophic Change in Earth History

Chances are small that the river flowing through your city will flood this spring, but if you lived to be 100 years old, you would probably see a catastrophic flood. In fact, many residents of the Midwest saw such a flood in the summer of 1993, and California residents experienced one in January 1995 (Fig. 1–6).

When geologists study the 4.6 billion years of Earth history, they find abundant evidence of catastrophic events that are highly improbable in a human lifetime or even in human history. For example, giant meteorites have smashed into our planet, vaporizing enormous volumes of rock and spreading dense dust clouds over the sky. Similarly, huge volcanic eruptions have changed conditions for life across the globe. Geologists have suggested that these catastrophic events have driven millions of species into extinction.

▶ 1.3 GEOLOGIC TIME

During the Middle Ages, the intellectual climate in Europe was ruled by the clergy, who tried to explain natural history by a literal interpretation of the Bible. In the middle 1600s, Archbishop James Ussher calculated the Earth's age from the Book of Genesis in the Old Testament. He concluded that the moment of creation occurred at noon on October 23, 4004 B.C.

Hutton refuted this biblical logic and deduced that the Earth was infinitely old. Today, geologists estimate that the Earth is about 4.6 billion years old. In his book *Basin and Range*, about the geology of western North America, John McPhee offers us a metaphor for the magnitude of geologic time. If the history of the Earth were represented by the old English measure of a yard, the distance from the king's nose to the end of his outstretched hand, all of human history could be erased by a single stroke of a file on his middle fingernail.

THE GEOLOGIC TIME SCALE

Geologists have divided Earth history into units displayed in the **geologic time scale** (Table 1–1). The units are called eons, eras, periods, and epochs and are identi-

Figure 1–6 Torrential rains caused the Russian River in California to flood in January 1995. In this photograph, Tom Monaghan is salvaging a few possessions and wading across the second-story balcony, awaiting rescue. (Reuters/Bettmann)

Table I–I • THE GEOLOGIC TIME SCALE

TIME UNITS OF THE GEOLOGIC TIME SCALE

Eon	Era	Period	Epoch		DISTINCTIVE PLANTS AND ANIMALS
Phanerozoic Eon (*Phaneros* = "evident"; *Zoon* = "life")	Cenozoic Era	Quaternary	Recent or Holocene	"Age of Mammals"	Humans
			Pleistocene		Mammals develop and become dominant
		Tertiary — Neogene	Pliocene — 2		
			Miocene — 5		
			— 24		
		Tertiary — Paleogene	Oligocene — 37		
			Eocene — 58		Extinction of dinosaurs and many other species
			Paleocene — 66		
	Mesozoic Era	Cretaceous — 144		"Age of Reptiles"	First flowering plants, greatest development of dinosaurs
		Jurassic — 208			First birds and mammals, abundant dinosaurs
		Triassic — 245			First dinosaurs
	Paleozoic Era	Permian — 286		"Age of Amphibians"	Extinction of trilobites and many other marine animals
		Carboniferous — Pennsylvanian — 320			Great coal forests; abundant insects, first reptiles
		Carboniferous — Mississippian — 360			Large primitive trees
		Devonian — 408		"Age of Fishes"	First amphibians
		Silurian — 438			First land plant fossils
		Ordovician — 505		"Age of Marine Invertebrates"	First fish
		Cambrian — 538			First organisms with shells, trilobites dominant
	Proterozoic				First multicelled organisms
		2500	Sometimes collectively called Precambrian		
	Archean				First one-celled organisms
		3800			Approximate age of oldest rocks
	Hadean	4600±			Origin of the Earth

Time is given in millions of years (for example, 1000 stands for 1000 million, which is one billion). The table is *not* drawn to scale. We know relatively little about events that occurred during the early part of the Earth's history. Therefore, the first four billion years are given relatively little space on this chart, while the more recent Phanerozoic Eon, which spans only 538 million years, receives proportionally more space.

fied primarily by the types of life that existed at the various times. The two earliest eons, the Hadean and Archean, cover the first 2.5 billion years of Earth history. Life originated during Archean time. Living organisms then evolved and proliferated during the Proterozoic Eon (*protero* is from a Greek root meaning "earlier" or "before" and *zoon* is from the Greek word meaning "life"). However, most Proterozoic organisms had no hard parts such as shells and bones. Most were single celled, although some multicellular organisms existed. The Proterozoic Eon ended about 538 million years ago.

Then, within an astonishingly short time—perhaps as little as 5 million years—many new species evolved. These organisms were biologically more complex than their Proterozoic ancestors, and many had shells and skeletons. The most recent 13 percent of geologic time, from 538 million years ago to the present, is called the Phanerozoic Eon (*phaneros* is Greek for "evident"). The Phanerozoic Eon is subdivided into the Paleozoic Era ("ancient life"), the Mesozoic Era ("middle life"), and the Cenozoic Era ("recent life") (Fig. 1–7).

▶ 1.4   THE EARTH'S ORIGIN

THE EARLY SOLAR SYSTEM

No one can go back in time to view the formation of the Solar System and the Earth. Therefore, scientists will never be able to describe the sequence of events with certainty.

The hypothesis given here is based on calculations about the behavior of dust and gas in space and on observations of stars and dust clouds in our galaxy. Refer to the "Focus On" box on page 12 for a discussion of how scientists formulate a hypothesis.

The hypothesis states that about 5 billion years ago the matter that became our Solar System was an immense, diffuse, frozen cloud of dust and gas rotating slowly in space. This cloud formed from matter ejected from an exploding star. More than 99 percent of the cloud consisted of hydrogen and helium, the most abundant elements in the Universe. The temperature of this cloud was about −270°C. Small gravitational attractions among the dust and gas particles caused the cloud to condense into a sphere (Figs. 1–8a and 1–8b). As condensation continued, the cloud rotated more rapidly, and the sphere spread into a disk, as shown in Figure 1–8c. Some scientists have suggested that a nearby star exploded and the shock wave triggered the condensation.

More than 90 percent of the matter in the cloud collapsed toward the center of the disk under the influence of gravity, forming the **protosun**. Collisions among high-speed particles released heat within this early version of the Sun, but it was not a true star because it did not yet generate energy by nuclear fusion.

Heat from the protosun warmed the inner region of the disk. Then, after the gravitational collapse was nearly complete, the disk cooled. Gases in the outer part of the disk condensed to form small aggregates, much as snowflakes form when moist air cools in the Earth's

Figure 1–7   This 50-million-year-old fossil fish once swam in a huge landlocked lake that covered parts of Wyoming, Utah, and Colorado.

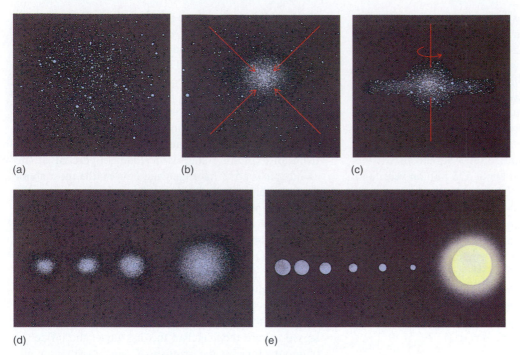

(a)

(b)

(c)

(d)

(e)

Figure 1–8 Formation of the Solar System. (a) The Solar System was originally a diffuse cloud of dust and gas. (b) This dust and gas began to coalesce due to gravity. (c) The shrinking mass began to rotate and formed a disk. (d) The mass broke up into a discrete protosun orbited by large protoplanets. (e) The Sun heated until fusion temperatures were reached. The heat from the Sun drove most of the hydrogen and helium away from the closest planets, leaving small, solid cores behind. The massive outer planets are still composed mostly of hydrogen and helium.

atmosphere. Over time, the aggregates stuck together as snowflakes sometimes do. As they increased in size and developed stronger gravitational forces, they attracted additional particles. This growth continued until a number of small rocky spheres, called **planetesimals**, formed, ranging from a few kilometers to about 100 km in diameter. The entire process, from the disk to the planetesimals, occurred quickly in geologic terms, over a period of 10,000 to 100,000 years. The planetesimals then coalesced to form a few large planets, including Earth.

At the same time that planets were forming, gravitational attraction pulled the gases in the protosun inward, creating extremely high pressure and temperature. The core became so hot that hydrogen nuclei combined to form the nucleus of the next heavier element, helium, in a process called **nuclear fusion**. Nuclear fusion releases vast amounts of energy. The onset of nuclear fusion marked the birth of the modern Sun, which still generates its energy by hydrogen fusion.

THE MODERN SOLAR SYSTEM

Heat from the Sun boiled most of the hydrogen, helium, and other light elements away from the inner Solar System. As a result, the four planets closest to the Sun—Mercury, Venus, Earth, and Mars—are now mainly rocky with metallic centers. These four are called the **terrestrial planets** because they are "Earthlike." In contrast, the four outer planets—Jupiter, Saturn, Uranus, and Neptune—are called the **Jovian planets** and are composed primarily of liquids and gases with small rocky and metallic cores (Fig. 1–9). Pluto, the outermost known planet, is anomalous. It is the smallest planet in the Solar System and is composed of rock mixed with frozen water and methane. Figure 1–10 is a schematic representation of the modern Solar System.

THE EVOLUTION OF THE MODERN EARTH

Scientists generally agree that the Earth formed by accretion of small particles, as discussed above. They also agree that the modern Earth is layered. The center is a dense, hot **core** composed mainly of iron and nickel. A thick **mantle**, composed mainly of solid rock, surrounds the core and contains 80 percent of the Earth's volume. The **crust** is a thin surface veneer, also composed of rock (Fig. 1–11).

(a)   (b)

Figure 1-9 (a) Mercury is a small planet close to the Sun. Consequently, most of the lighter elements have long since been boiled off into space, and today the surface is solid and rocky. (b) Jupiter, on the other hand, is composed mainly of gases and liquids, with a small solid core. This photograph shows its turbulent atmosphere. The scales in these two photographs are different. Jupiter is much larger than Mercury. (NASA)

Earth temperature and pressure increase gradually with depth. Ten meters below the surface, soil and rock are cool to the touch, but at a depth between about 100 kilometers and 350 kilometers, the mantle rock is so hot that one or two percent of it is melted, so that the entire mantle flows very slowly, like cold honey. This movement allows continents to move across the globe, ocean basins to open and close, mountain ranges to rise, volcanoes to erupt, and earthquakes to shake the planet. Rock is even hotter deeper in the mantle, but the intense pressure prevents it from melting. The outer core is composed of molten metal, but the inner core, which is as hot as the surface of the Sun, is under such intense pressure that it is solid. We will discuss these layers further in Chapters 2 and 10.

Although scientists agree that our planet is layered, they disagree on how the layering developed. Astronomers have detected both metallic and rocky meteorites in space, and many think that both metallic and rocky particles coalesced to form the planets. According to one hypothesis, as the Earth began to form, metallic particles initially accumulated to create the metallic core,

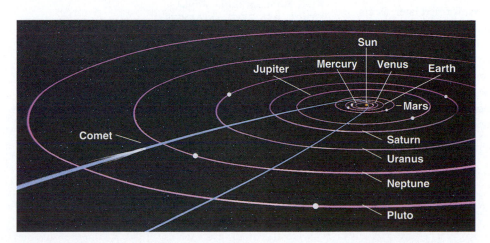

Figure 1-10 A schematic view of the Solar System.

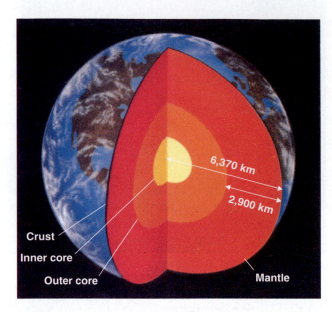

Figure I–II A schematic view of the interior of the Earth.

and then rocky particles collected around the core to form the rocky mantle. Thus, the Earth has always been layered.

An alternative hypothesis states that the rock and metal accumulated simultaneously during the initial coalescence, forming a homogeneous (non-layered) planet. The young Earth became hot as gravity pulled the small particles together and later as asteroids, comets, and planetesimals crashed into the surface. At the same time, radioactive decay heated the Earth's interior. Thus, our planet became so hot that all or most of it melted soon after it formed. Heavy molten iron and nickel gravitated toward the center and collected to form the core, while lighter materials floated toward the surface to form the mantle. In both hypotheses, the crust formed later, as discussed in Chapter 12.

How can we determine which of the two hypotheses is correct? By studying modern meteorites and lunar rocks, two geologists recently estimated that the core formed at least 62 million years after the Earth coalesced.[1] This interpretation supports the hypothesis that our planet was initially homogeneous and then separated

into the core and mantle at a later date. However, as discussed in "Focus On: Hypothesis, Theory, and Law," future research may change our views about a sequence of events that occurred so long ago.

▶ I.5 GEOLOGIC CHANGE AND THE ENVIRONMENT

The Earth's surface environment has changed frequently and dramatically during its long history. Atmospheric composition and climate have changed. Glaciers have covered huge portions of the continents and then melted to leave the land covered by tropical swamps or scorching deserts. Volcanic eruptions and meteorite impacts have occurred, and many scientists think that these events have caused global catastrophes that resulted in extinctions of large proportions of the Earth's species.

Primitive human-like species evolved in East Africa about 4 million years ago. Thus, *homo sapiens* and their immediate ancestors have lived on Earth for a mere 0.05 percent of its history. The Industrial Revolution began only 250 years ago. Yet within this minuscule slice of geologic history, humans have altered the surface of the planet. Today, farms cover vast areas that were recently forested or covered by natural prairies. People have paved large expanses of land, drained wetlands, dammed rivers, pumped ground water to the surface, and released pollutants into waterways and the atmosphere. Some of these changes have affected even the most remote regions of the Earth, including the Sahara desert, the Amazon rainforest, the central oceans, and the South Pole. Some scientists are concerned that these changes also threaten human well-being.

RISK ASSESSMENT AND COST–BENEFIT ANALYSIS

Many geologic processes put humans at risk. Volcanic eruptions, earthquakes, floods, and mudslides kill people and destroy cities. Human activities also create environmental hazards that jeopardize our health. Geologists and other geoscientists attempt to analyze the risks and costs of exposure to these hazards.

Risk assessment is the analysis of risk and the implementation of policy based on that analysis. Cost–benefit analysis compares the monetary expense of solving a problem with the monetary benefits of the solution.

Consider the following two examples of risk assessment and cost–benefit analysis.

[1]Der-Chuen Lee and Alex N. Halliday, "Hafnium-tungsten Chronometry and the Timing of Terrestrial Core Formation." *Science*, vol. 378, Dec. 21/28, 1995, p. 771.

California Earthquakes

About 85 percent of the people and industry in California are located close to the San Andreas fault zone, an active earthquake zone that parallels the Pacific coast from the Mexican border to Cape Mendocino, north of San Francisco. Earthquakes occur frequently throughout this region, causing property damage and loss of life. Geologists are unable to predict exactly when and where the next quake will strike, but they can identify high-risk regions with high probabilities of a devastating earthquake. Building codes are written accordingly.

Engineers can build structures that will withstand even the most severe earthquakes, but only at great expense. As a result, building codes represent a compromise between safety and cost. Construction requirements for nuclear power plants are stricter than those for bridges because if a nuclear power plant were to fail, thousands of people could die, whereas only a few would die if a bridge collapsed (Fig. 1–12).

In the late 1970s, the California state government authorized the Seismic Safety Commission to recommend upgrading state-owned structures to meet stricter engineering criteria. The commission evaluated soil and bedrock at probable earthquake zones and then recommended construction upgrades based on the probable number of lives saved per dollar. This cost–benefit analysis assures people that the state will not raise taxes a great deal to pay for costly reconstruction. The tradeoff is that in a large earthquake, some people will die when structures collapse.

Toxic Ground Water

At many mines and industrial sites, toxic chemicals have leaked into soil. Geologists measure how fast these compounds spread into ground water, and they ask, "Do the chemicals threaten drinking water resources and human health?" "Can the chemicals be contained or removed?"

To measure the risks associated with contaminated ground water, we must evaluate the toxicity of the pollutant. Because it is unethical to feed potentially toxic chemicals directly to humans, scientists may feed concentrated doses to laboratory rats. If the rats sicken or die, the scientists infer that the chemical may be toxic to humans in lower doses. However, a substance that is poisonous to rats may or may not be poisonous to humans. Scientists also question whether it is valid to extrapolate results of high-dose exposure to the effects of lower doses found in contaminated ground water.

Scientists also use epidemiological studies to measure the health hazard of a pollutant. For example, if the

Figure 1–12 Earthquake-resistant design allowed this bridge support to rupture during the 1994 Northridge, California, earthquake, but prevented total collapse of the bridge. (Earthquake Engineering Research Institute)

drinking water in a city is contaminated with a pesticide and a high proportion of people in the city develop an otherwise rare disease, then the scientists may infer that the pesticide caused the disease.

Because neither laboratory nor epidemiological studies can prove that low doses of a pollutant are harmful to humans, scientists are faced with a dilemma: "Should we spend money to clean up the pollutant?" Some argue that such expenditure is unnecessary until we can prove that the contaminant is harmful. Others invoke the precautionary principle, which says simply, "It is better to be safe than sorry." Proponents of the precautionary principle argue that people commonly act on the basis of incomplete proof. For example, if a mechanic told you that your brakes were faulty and likely to fail within the next 1000 miles, you would recognize this as an opinion, not a fact. Yet would you wait for proof that the brakes would fail or replace them now?

Pollution control is expensive. Water purification adds to the cost of manufactured goods. If pollutants do

On an afternoon field trip, you may find several different types of rocks or watch a river flow by. But you can never see the rocks or river as they existed in the past or as they will exist in the future. Yet a geologist might explain to you how the rocks formed millions or even a few billion years ago and might predict how the river valley will change in the future.

Scientists not only study events that they have never observed and never will observe, but they also study objects that can never be seen, touched, or felt. In this book we describe the center of the Earth 6370 kilometers beneath our feet, even though no one has ever visited it and no one ever will.

Much of science is built on inferences about events and objects outside the realm of direct experience. An inference is a conclusion based on thought and reason. How certain are we that a conclusion of this type is correct?

Scientists develop an understanding of the natural world according to a set of guidelines known as the scientific method, which involves three basic steps: (1) observation, (2) forming a hypothesis, and (3) testing the hypothesis and developing a theory.

Observation

All modern science is based on observation. Suppose that you observed an ocean current carrying and depositing sand. If you watched for some time, you would see that the sand accumulates slowly, layer by layer, on the beach. You might then visit Utah or Nevada and see cliffs of layered sandstone hundreds of meters high. Observations of this kind are the starting point of science.

Forming a Hypothesis

Simple observations are only a first step along the path to a theory. A scientist tries to organize observations to recognize patterns. You might note that the sand layers deposited along the coast look just like the layers of sand in the sandstone cliffs. Perhaps you would then infer that the thick layers of sandstone had been deposited in an ancient ocean. You might further conclude that, since the ocean deposits layers of sand slowly, the thick layers of sandstone must have accumulated over a long time.

If you were then to travel, you would observe that thick layers of sandstone are abundant all over the world. Because thick layers of sand accumulate so slowly, you might infer that a long time must have been required for all that sandstone to form. From these observations and inferences you might form the hypothesis that the Earth is old.

A hypothesis is a tentative explanation built on strong supporting evidence. Once a scientist or group of scientists proposes a hypothesis, others test it by comparison with observations and experiments. Thus, a hypothesis is a rough draft of a theory that is tested against observable facts. If it explains some of the facts but not all of them, it must be altered, or if it cannot be changed satisfactorily, it must be discarded and a new hypothesis developed.

Testing the Hypothesis and Forming a Theory

If a hypothesis explains new observations as they accumulate and is not substantively contradicted, it

---

escape, removal of the contaminant and restoration of the contaminated area may be even more costly. For example, if a pollutant has already escaped into ground water, it may be necessary to excavate thousands of cubic meters of soil, process the soil to remove the contaminant, and then return the soil to the excavated site.

However, pollution is also expensive. If a ground water contaminant causes people to sicken, the cost to society can be measured in terms of medical bills and loss of income resulting from missed work. Many contaminants damage structures, crops, and livestock. People in polluted areas also bear expense because tourism diminishes and land values are reduced when people no longer want to visit or live in a contaminated area. All of these costs are called externalities.

Cost–benefit analysis balances the cost of pollution control against the cost of externalities. Some people suggest that we should minimize the total cost even though this approach accepts significant pollution. Others argue that cost–benefit analysis is flawed because it ignores both the quality of life and the value of human life. How, they ask, can you place a dollar value on a life that ends early, or on the annoyance of a vile odor, a persistent cough, polluted streams, dirty air, or industrial noise? Such annoyances damage our sense of well-being.

People do not agree on an optimal level of pollution control or an acceptable level of pollution. There are no easy answers. In this textbook we will not offer solutions, but we will explain the scientific principles behind difficult questions.

becomes elevated to a **theory**. Theories differ widely in form and content, but all obey four fundamental criteria:

1. A theory must be based on a series of confirmed observations or experimental results.

2. A theory must explain all relevant observations or experimental results.

3. A theory must not contradict any relevant observations or other scientific principles.

4. A theory must be internally consistent. Thus, it must be built in a logical manner so that the conclusions do not contradict any of the original premises.

For example, the theory of plate tectonics states that the outer layer of the Earth is broken into a number of plates that move horizontally relative to one another. As you will see in later chapters, this theory is supported by many observations and seems to have no major inconsistencies.

Many theories can never be absolutely proven. For example, even though scientists are just about certain that their image of atomic structure is correct, no one has watched or ever will watch an individual electron travel in its orbit. Therefore, our interpretation of atomic structure is called atomic theory.

However, in some instances, a theory is elevated to a scientific law. A law is a statement of how events always occur under given conditions. It is considered to be factual and correct. A law is the most certain of scientific statements. For example, the law of gravity states that all objects are attracted to one another in direct proportion to their masses. We cannot conceive of any contradiction to this principle, and none has been observed. Hence, the principle is called a law.

Sharing Information

The final step in the scientific process is to share your observations and conclusions with other scientists and the general public. Typically, a scientist communicates with colleagues to discuss current research by phone, at annual meetings, or more recently, by electronic communications systems such as E-mail and Internet. When the scientist feels confident in his or her conclusions, he or she publishes them in a scientific journal. Colleagues review the material before it is published to ensure that the author has followed the scientific method, and, if the results are of general interest, the scientist may publish them in popular magazines or in newspapers. The authors of this text have read many scientific journals and now pass the information on to you, the student.

DISCUSSION QUESTION

Obtain a copy of a news article in a weekly news magazine. Underline the facts with one color pencil and the author's opinions with another. Did the author follow the rules for the scientific method in reaching his or her conclusions?

SUMMARY

Geology is the study of the Earth including the materials that it is made of, the physical and chemical changes that occur on its surface and in its interior, and the history of the planet and its life forms.

Most of the Earth is composed of **rocks**, and rocks are composed of **minerals**. **Internal processes** move continents and cause earthquakes and volcanoes; **surface processes** sculpt mountains and valleys. The **hydrosphere** consists of water in streams, lakes, and oceans; in the atmosphere; and frozen in glaciers. It also includes ground water that soaks soil and rock to a depth of 2 or 3 kilometers.

The **atmosphere** is a mixture of gases, mostly nitrogen and oxygen. Ninety-nine percent is concentrated in the first 30 kilometers, but a few traces remain even 10,000 kilometers above the Earth's surface. Organisms of the **biosphere**, including humans, affect and are affected by Earth's surface processes and the compositions of the hydrosphere and atmosphere. Paleontologists study the evolution and history of life from its beginning to the present.

The principle of **uniformitarianism** states that geologic change occurs over a long period of time by a sequence of almost imperceptible events. Thus, over the immense magnitude of geologic time, processes that occur too slowly or rarely to have an impact on our daily lives are important in Earth history. In contrast, **catastrophism** postulates that geologic change occurs mainly during infrequent catastrophic events. Today, geologists know that both uniformitarianism and catastrophism are correct.

The 4.6-billion-year history of the Earth is divided into **eons**, **eras**, **periods**, and **epochs**, which are based on the types of life that existed at various times.

The Solar System formed from dust and gases that rotated slowly in space. Within its center, the gases were pulled inward with enough velocity to initiate nuclear fusion and create the Sun. In the disk, planets formed from coalescing dust and gases. In the inner planets, most of the lighter elements escaped, but they are important components in the outer giants.

The modern Earth is made up of a dense **core** of iron and nickel, a rocky **mantle** of lower density, and a **crust** of yet lower density. One hypothesis states that both a core and mantle existed in the earliest Earth. An alternative hypothesis states that the Earth was initially homogeneous. The primordial planet was heated by energy from the original gravitational coalescence, by radioactive decay, and by bombardment from outer space. This heat caused all or most of the Earth to melt, and dense materials settled to the center to form the core, while less dense rock floated toward the surface to form the mantle.

Risk assessment is the analysis of risk of geologic and human-induced hazards and the implementation of policy based on that analysis. **Cost–benefit analysis** compares the monetary cost of solving a problem with the monetary benefits of the solution.

KEY WORDS

rocks *2*	atmosphere *3*	protosun *7*	Jovian planets *8*
minerals *2*	biosphere *4*	planetesimals *8*	core *8*
internal processes *2*	uniformitarianism *4*	nuclear fusion *8*	mantle *8*
surface processes *3*	catastrophism *4*	terrestrial planets *8*	crust *8*
hydrosphere *3*	geologic time scale *5*		

REVIEW QUESTIONS

1. Give a concise definition of geology.
2. Compare and contrast internal processes with surface processes.
3. List six types of reservoirs that collectively contain most of the Earth's water.
4. What is ground water? Where in the hydrosphere is it located?
5. What two gases comprise most of the Earth's atmosphere?
6. How thick is the Earth's atmosphere?
7. Compare and contrast uniformitarianism and catastrophism. Give an example of each type of geologic change.
8. How old is the Earth?

9. List the Earth's major eons in order of age. List the three eras that comprise the most recent eon.
10. Very briefly outline the formation of the Universe and the Solar System.
11. How did the Sun form? How is its composition different from that of the Earth? Explain the reasons for this difference.
12. Compare and contrast the properties of the terrestrial planets with those of the Jovian planets.
13. List the three major layers of the Earth. Which is the most dense, and which is the least dense?
14. Define cost–benefit analysis and risk assessment and give an example of how these policies are implemented.

DISCUSSION QUESTIONS

1. What would the Earth be like if it
   a. had no atmosphere?   b. had no water?
2. In what ways do organisms, including humans, change the Earth? What kinds of Earth processes are unaffected by humans and other organisms?

3. Redraw the geologic time scale with the size of each of the major eons proportional to its time span. How does your redrawn time scale compare with the one in Table 1–1? Speculate on why the time scale is drawn as it is.

4. Explain how the theory of the evolution of the Solar System explains the following observations: a. All the planets in the Solar System are orbiting in the same direction. b. All the planets in the Solar System except Pluto are orbiting in the same plane. c. The chemical composition of Mercury is similar to that of the Earth. d. The Sun is composed mainly of hydrogen and helium but also contains all the elements found on Earth. e. Venus has a solid surface, whereas Jupiter is mainly a mixture of gases and liquids with a small, solid core.

5. Jupiter is composed of solids such as rock, iron, and nickel; a vast amount of liquid hydrogen; and gases such as hydrogen, helium, ammonia, and methane. From your knowledge of the formation and structure of the Earth, which compounds do you predict would make up Jupiter's core, mantle, and outer shell?

6. The radioactive elements that are responsible for the heating of the Earth decompose very slowly, over a period of billions of years. How would the Earth be different if these elements decomposed much more rapidly— say, over a period of a few million years?

7. In Los Angeles, the risk of death per year from an automobile accident is 1 in 4000; the risk of death from an earthquake is about 1 in 50,000. Would you use these data to argue that additional reinforcement of bridges and buildings is unwarranted?

Plate Tectonics:
A First Look

About 1 million earthquakes shake the Earth each year; most are so weak that we do not feel them, but the strongest demolish cities and kill thousands of people. Most of us have seen televised coverage of volcanic eruptions blasting molten rock and ash into the sky, destroying villages and threatening cities. Over geologic time, mountain ranges rise and then erode away, continents migrate around the globe, and ocean basins open and close.

Before 1960, no single theory explained all of these manifestations of the active Earth. In the early 1960s, geologists developed the **plate tectonics theory**, which provides a single, unifying framework that explains earthquakes, volcanic eruptions, mountain building, moving continents, and many other geologic events. It also allows geologists to identify many geologic hazards before they affect humans.

Because plate tectonics theory is so important to modern geology, it provides a foundation for many of the following chapters of this book. We describe and explain the basic aspects of the theory in this chapter. In following chapters we use the theory to explain the active Earth.

India collided with southern Asia to raise the Himalayas, the Earth's highest mountain chain. (Tom Van Sant/Geosphere Project, Santa Monica Photo Science Library)

▶ 2.1   AN OVERVIEW OF PLATE TECTONICS

Like most great, unifying scientific ideas, the plate tectonics theory is simple. Briefly, it describes the Earth's outer layer, called the **lithosphere**, as a shell of hard, strong rock. This shell is broken into seven large (and several smaller) segments called **tectonic plates**. They are also called lithospheric plates, and the two terms are interchangeable (Fig. 2–1). The tectonic plates float on the layer below, called the **asthenosphere**. The asthenosphere, like the lithosphere, is rock. But the asthenosphere is so hot that 1 to 2 percent of it is melted. As a result, it is plastic, and weak. The lithospheric plates glide slowly over the asthenosphere like sheets of ice drifting across a pond (Fig. 2–2). Continents and ocean basins make up the upper parts of the plates. As a tectonic plate glides over the asthenosphere, the continents and oceans move with it.

Most of the Earth's major geological activity occurs at **plate boundaries**, the zones where tectonic plates meet and interact. Neighboring plates can move relative to one another in three different ways (Fig. 2–3). At a **divergent boundary**, two plates move apart, or separate. At a **convergent boundary**, two plates move toward each other, and at a **transform boundary**, they slide horizontally past each other. Table 2–1 summarizes characteristics and examples of each type of plate boundary. Plate interactions at these boundaries build mountain ranges and create earthquakes and volcanic eruptions.

▶ 2.2   THE EARTH'S LAYERS

The energy released by an earthquake travels through the Earth as waves. Geologists have found that earthquake waves abruptly change both speed and direction at certain depths as they pass through the Earth's interior. Chapter 10 describes how these abrupt changes reveal that the Earth is a layered planet. Figure 2–4 and Table 2–2 describe the layers.

THE CRUST

The **crust** is the outermost and thinnest layer. Because the crust is relatively cool, it consists of hard, strong rock. Crust beneath the oceans differs from that of continents. Oceanic crust is 5 to 10 kilometers thick and is composed mostly of a dark, dense rock called **basalt**. In contrast, the average thickness of continental crust is about 20 to 40 kilometers, although under mountain ranges it can be as much as 70 kilometers thick.

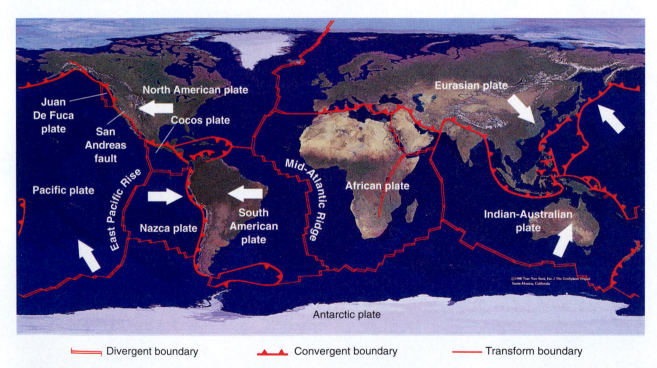

| ⚎ Divergent boundary | ▲▲ Convergent boundary | — Transform boundary |

Figure 2–1   The Earth's lithosphere is broken into seven large plates, separated by the red lines; they are called the African, Eurasian, Indian–Australian, Antarctic, Pacific, North American, and South American plates. A few of the smaller plates are also shown. White arrows indicate directions of plate movement and show that the plates move in different directions. The red lines also distinguish the three types of plate boundaries.   (Tom Van Sant, Geosphere Project)

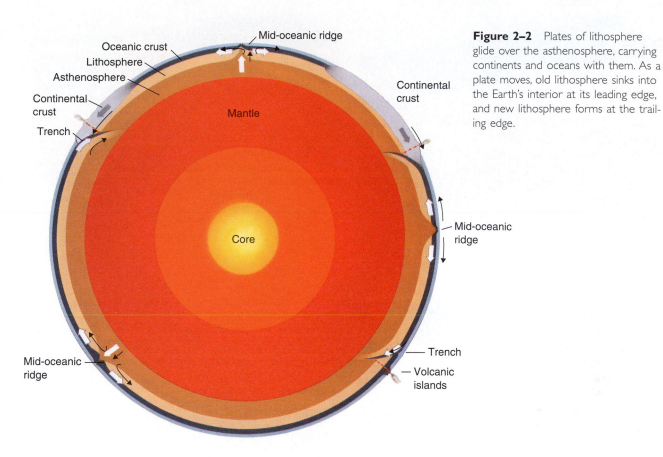

Figure 2–2 Plates of lithosphere glide over the asthenosphere, carrying continents and oceans with them. As a plate moves, old lithosphere sinks into the Earth's interior at its leading edge, and new lithosphere forms at the trailing edge.

Table 2–1 • CHARACTERISTICS AND EXAMPLES OF PLATE BOUNDARIES

TYPE OF BOUNDARY	TYPES OF PLATES INVOLVED	TOPOGRAPHY	GEOLOGIC EVENTS	MODERN EXAMPLES
Divergent	Ocean-ocean	Mid-oceanic ridge	Sea-floor spreading, shallow earthquakes, rising magma, volcanoes	Mid-Atlantic ridge
	Continent-continent	Rift valley	Continents torn apart, earthquakes, rising magma, volcanoes	East African rift
Convergent	Ocean-ocean	Island arcs and ocean trenches	Subduction, deep earthquakes, rising magma, volcanoes, deformation of rocks	Western Aleutians
	Ocean-continent	Mountains and ocean trenches	Subduction, deep earthquakes, rising magma, volcanoes, deformation of rocks	Andes
	Continent-continent	Mountains	Deep earthquakes, deformation of rocks	Himalayas
Transform	Ocean-ocean	Major offset of mid-oceanic ridge axis	Earthquakes	Offset of East Pacific rise in South Pacific
	Continent-continent	Small deformed mountain ranges, deformations along fault	Earthquakes, deformation of rocks	San Andreas fault

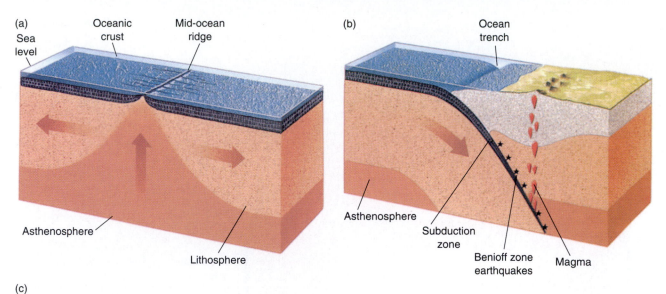

(a)
Sea level — Oceanic crust — Mid-ocean ridge

Asthenosphere

Lithosphere

(b)
Ocean trench

Asthenosphere

Subduction zone

Benioff zone earthquakes — Magma

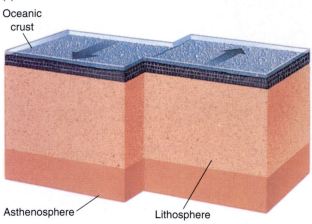

(c)
Oceanic crust

Asthenosphere

Lithosphere

Figure 2–3 Three types of boundaries separate the Earth's tectonic plates: (a) Two plates separate at a divergent boundary. New lithosphere forms as hot asthenosphere rises to fill the gap where the two plates spread apart. The lithosphere is relatively thin at this type of boundary. (b) Two plates converge at a convergent boundary. If one of the plates carries oceanic crust, the dense oceanic plate sinks into the mantle in a subduction zone. Here an oceanic plate is sinking beneath a less dense continental plate. Magma rises from the subduction zone, and a trench forms where the subducting plate sinks. The stars mark Benioff zone earthquakes that occur as the sinking plate slips past the opposite plate (described in Chapter 10). (c) At a transform plate boundary, rocks on opposite sides of the fracture slide horizontally past each other.

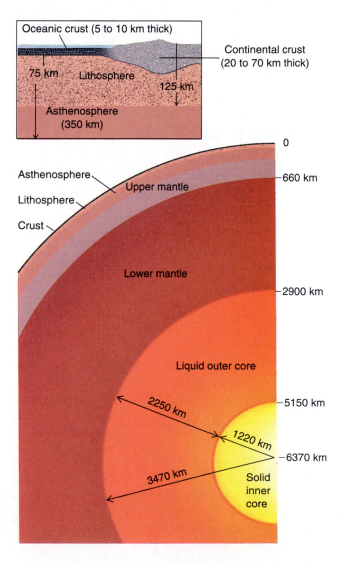

Oceanic crust (5 to 10 km thick)

Continental crust (20 to 70 km thick)

75 km   Lithosphere   125 km

Asthenosphere (350 km)

Asthenosphere — Upper mantle — 0
Lithosphere — 660 km
Crust

Lower mantle — 2900 km

Liquid outer core — 5150 km

2250 km   1220 km
3470 km   6370 km

Solid inner core

Figure 2–4 The Earth is a layered planet. The insert is drawn on an expanded scale to show near-surface layering.

Table 2–2 • THE LAYERS OF THE EARTH

	LAYER	COMPOSITION	DEPTH	PROPERTIES
Crust	Oceanic crust Continental crust	Basalt Granite	5–10 km 20–70 km	Cool, hard, and strong Cool, hard, and strong
Lithosphere	Lithosphere includes the crust and the uppermost portion of the mantle	Varies; the crust and the mantle have different compositions	75–125 km	Cool, hard, and strong
Mantle	Uppermost portion of the mantle included as part of the lithosphere	Entire mantle is ultramafic rock. Its mineralogy varies with depth		
	Asthenosphere		Extends to 350 km	Hot, weak, and plastic, 1% or 2% melted
	Remainder of upper mantle		Extends from 350 to 660 km	Hot, under great pressure, and mechanically strong
	Lower mantle		Extends from 660 to 2900 km	High pressure forms minerals different from those of the upper mantle
Core	Outer core	Iron and nickel	Extends from 2900 to 5150 km	Liquid
	Inner core	Iron and nickel	Extends from 5150 km to the center of the Earth	Solid

Continents are composed primarily of a light-colored, less dense rock called **granite**.

THE MANTLE

The **mantle** lies directly below the crust. It is almost 2900 kilometers thick and makes up 80 percent of the Earth's volume. Although the chemical composition may be similar throughout the mantle, Earth temperature and pressure increase with depth. These changes cause the strength of mantle rock to vary with depth, and thus they create layering within the mantle. The upper part of the mantle consists of two layers.

The Lithosphere

The uppermost mantle is relatively cool and consequently is hard, strong rock. In fact, its mechanical behavior is similar to that of the crust. The outer part of the Earth, including both the uppermost mantle and the crust, make up the lithosphere (Greek for "rock layer"). The lithosphere can be as thin as 10 kilometers where tectonic plates separate. However, in most regions, the lithosphere varies from about 75 kilometers thick beneath ocean basins to about 125 kilometers under the continents. A tectonic (or lithospheric) plate is a segment of the lithosphere.

The Asthenosphere

At a depth varying from about 75 to 125 kilometers, the strong, hard rock of the lithosphere gives way to the weak, plastic asthenosphere. This change in rock properties occurs over a vertical distance of only a few kilometers, and results from increasing temperature with depth. Although the temperature increases gradually, it crosses a threshold at which the rock is close to its melting point. As a result, 1 to 2 percent of the asthenosphere is liquid, and the asthenosphere is mechanically weak and plastic. Because it is plastic, the asthenosphere flows slowly, perhaps at a rate of a few centimeters per year. Two familiar examples of solid materials that flow are Silly Putty™ and hot road tar. However, both of these solids flow much more rapidly than the asthenosphere rock. The asthenosphere extends from the base of the lithosphere to a depth of about 350 kilometers. At the base of the asthenosphere, increasing pressure causes the mantle to become mechanically stronger, and it remains so all the way down to the core.

THE CORE

The **core** is the innermost of the Earth's layers. It is a sphere with a radius of about 3470 kilometers and is composed largely of iron and nickel. The outer core is molten because of the high temperature in that region. Near its center, the core's temperature is about 6000°C, as hot as the Sun's surface. The pressure is greater than 1 million times that of the Earth's atmosphere at sea level. The extreme pressure overwhelms the temperature effect and compresses the inner core to a solid.

To visualize the relative thickness of the Earth's layers, let us return to an analogy used in Chapter 1. Imagine that you could drive a magical vehicle at 100 kilometers per hour through the Earth, from its center to its surface. You would pass through the core in about 35 hours and the mantle in 29 hours. You would drive through oceanic crust in only 6 minutes, and most continental crust in about half an hour. When you arrived at the surface, you would have spent the last $3\frac{1}{2}$ hours traversing the entire asthenosphere and lithosphere.

▶ 2.3 PLATES AND PLATE TECTONICS

In most places, the lithosphere is less dense than the asthenosphere. Consequently, it floats on the asthenosphere much as ice floats on water. Figure 2–1 shows that the lithosphere is broken into seven large tectonic plates and several smaller ones. Think of the plates as irregularly shaped ice floes, packed tightly together floating on the sea. Ice floes drift over the sea surface and, in a similar way, tectonic plates drift horizontally over the asthenosphere. The plates move slowly, at rates ranging from less than 1 to about 16 centimeters per year (about as fast as a fingernail grows). Because the plates move in different directions, they bump and grind against their neighbors at plate boundaries.

The great forces generated at a plate boundary build mountain ranges and cause volcanic eruptions and earthquakes. These processes and events are called **tectonic** activity, from the ancient Greek word for "construction." Tectonic activity "constructs" mountain chains and ocean basins. In contrast to plate boundaries, the interior portion of a plate is usually tectonically quiet because it is far from the zones where two plates interact.

DIVERGENT PLATE BOUNDARIES

At a divergent plate boundary, also called a **spreading center** and a **rift zone**, two lithospheric plates spread apart (Fig. 2–5). The underlying asthenosphere then oozes upward to fill the gap between the separating plates. As the asthenosphere rises between separating plates, some of it melts to form molten rock called **magma**.[1] Most of the magma rises to the Earth's surface, where it

[1]It seems counterintuitive that the rising, cooling asthenosphere should melt to form magma, but the melting results from decreasing pressure rather than a temperature change. This process is discussed in Chapter 5.

cools to form new crust, the top layer of the lithosphere. Most of this activity occurs beneath the seas because most divergent plate boundaries lie in the ocean basins.

Both the asthenosphere and the lower lithosphere (the part beneath the crust) are parts of the mantle and thus have similar chemical compositions. The main difference between the two layers is one of mechanical strength. The hot asthenosphere is weak and plastic, but the cooler lithosphere is strong and hard. As the asthenosphere rises, it cools, gains mechanical strength, and, therefore, transforms into new lithosphere. In this way, new lithosphere continuously forms at a divergent boundary.

At a spreading center, the rising asthenosphere is hot, weak, and plastic. Only the upper 10 to 15 kilometers cools enough to gain the strength and hardness of lithosphere rock. As a result, the lithosphere, including the crust and the upper few kilometers of mantle rock, can be as little as 10 or 15 kilometers thick at a spreading center. But as the lithosphere spreads, it cools from the top downward. When the lithosphere cools, it becomes thicker because the boundary between the cool, strong rock of the lithosphere and the hot, weak asthenosphere migrates downward. Consequently, the thickness of the lithosphere increases as it moves away from the spreading center. Think of ice freezing on a pond. On a cold day, water under the ice freezes and the

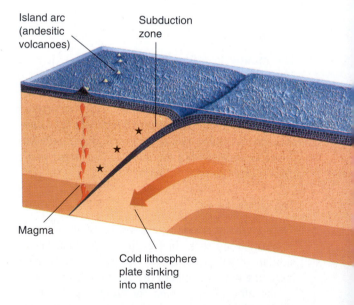

Figure 2–5 Lithospheric plates move away from a spreading center by gliding over the weak, plastic asthenosphere. In the center of the drawing, new lithosphere forms at a spreading center. At the sides of the drawing, old lithosphere sinks into the mantle at subduction zones.

ice becomes thicker. The lithosphere continues to thicken until it attains a steady state thickness of about 75 kilometers beneath an ocean basin, and as much as 125 kilometers beneath a continent.

The Mid-Oceanic Ridge: Rifting in the Oceans

A spreading center lies directly above the hot, rising asthenosphere. The newly formed lithosphere at an oceanic spreading center is hot and therefore of low density. Consequently, the sea floor at a spreading center floats to a high elevation, forming an undersea mountain chain called the **mid-oceanic ridge** (Fig. 2–6). But as lithosphere migrates away from the spreading center, it cools and becomes denser and thicker; as a result, it sinks. For this reason, the sea floor is high at the mid-oceanic ridge and lower away from the ridge. Thus, the average depth of the sea floor away from the mid-oceanic ridge is about 5 kilometers. The mid-oceanic ridge rises 2 to 3 kilometers above the surrounding sea floor and, thus, comes within 2 kilometers of the sea surface.

If you could place two bright red balls on the sea floor, one on each side of the ridge axis, and then watch them over millions of years, you would see the balls migrate away from the rift as the plates separated. The balls would also sink to greater depths as the hot rocks cooled (Fig. 2–7).

Oceanic rifts completely encircle the Earth, running around the globe like the seam on a baseball. As a result,

the mid-oceanic ridge system is the Earth's longest mountain chain. The basaltic magma that oozes onto the sea floor at the ridge creates approximately 6.5×10^{18} (6,500,000,000,000,000,000) tons of new oceanic crust each year. The mid-oceanic ridge system and other features of the sea floor are described further in Chapter 11.

Splitting Continents: Rifting in Continental Crust

A divergent plate boundary can rip a continent in half in a process called **continental rifting**. A **rift valley** develops in a continental rift zone because continental crust stretches, fractures, and sinks as it is pulled apart. Continental rifting is now taking place along a zone called the East African rift (see Fig. 2–1). If the rifting continues, eastern Africa will separate from the main portion of the continent, and a new ocean basin will open between the separating portions of Africa. The Rio Grande rift is a continental rift extending from southern Colorado to El Paso, Texas. It is unclear whether rifting is still taking place here or the process has ended.

CONVERGENT PLATE BOUNDARIES

At a convergent plate boundary, two lithospheric plates move toward each other. Convergence can occur (1) between a plate carrying oceanic crust and another carrying continental crust, (2) between two plates carrying oceanic crust, and (3) between two plates carrying continental

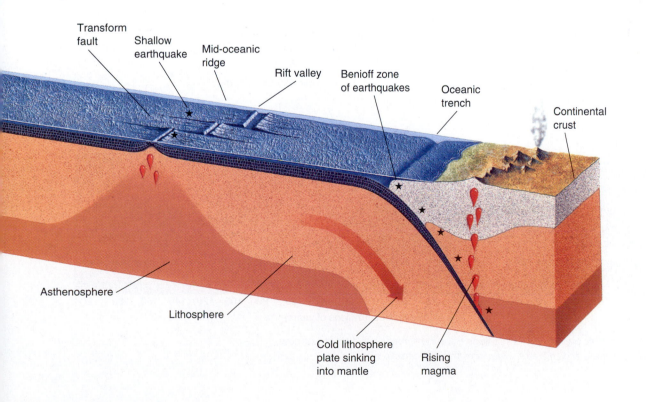

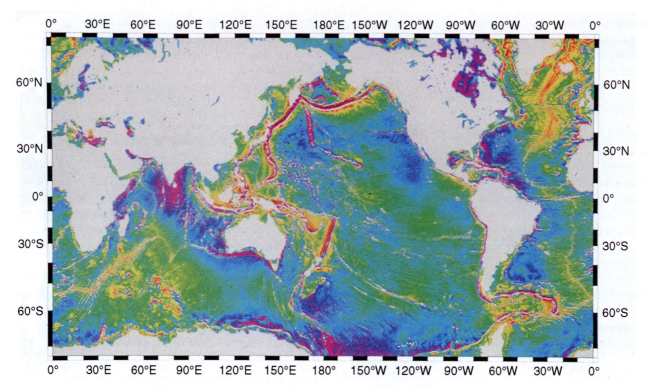

Figure 2–6 Sea floor topography is dominated by huge undersea mountain chains called mid-oceanic ridges and deep trenches called subduction zones. Mid-oceanic ridges form where tectonic plates separate, and subduction zones form where plates converge. The green areas represent the relatively level portion of the sea floor that lies about 5 kilometers underwater. The yellow-orange-red hues are mountains, primarily the mid-oceanic ridges. The blue-violet-magenta areas are trenches. (Scripps Institution of Oceanography, University of California, San Diego)

crust. Differences in density determine what happens where two plates converge. Think of a boat colliding with a floating log. The log is denser than the boat, so it sinks beneath the boat.

When two plates converge, the denser plate dives beneath the lighter one and sinks into the mantle. This process is called **subduction**. Generally, only oceanic lithosphere can sink into the mantle. Attempting to stuff

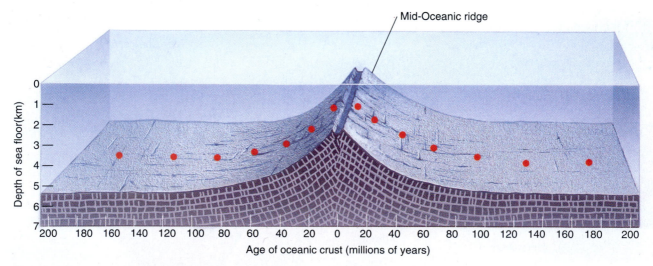

Figure 2–7 Red balls placed on the sea floor trace the spreading and sinking of new oceanic crust as it cools and migrates away from the mid-oceanic ridge.

a low-density continent down into the mantle would be like trying to flush a marshmallow down a toilet: It will not go because it is too light. In certain cases, however, small amounts of continental crust may sink into the mantle at a subduction zone. These cases are discussed in Chapter 12.

A **subduction zone** is a long, narrow belt where a lithospheric plate is sinking into the mantle. On a world-wide scale, the rate at which old lithosphere sinks into the mantle at subduction zones is equal to the rate at which new lithosphere forms at spreading centers. In this way, global balance is maintained between the creation of new lithosphere and the destruction of old lithosphere.

The oldest sea-floor rocks on Earth are only about 200 million years old because oceanic crust continuously recycles into the mantle at subduction zones. Rocks as old as 3.96 billion years are found on continents because subduction consumes little continental crust.

Convergence of Oceanic Crust with Continental Crust

When an oceanic plate converges with a continental plate, the denser oceanic plate sinks into the mantle beneath the edge of the continent. As a result, many subduction zones are located at continental margins. Today, oceanic plates are sinking beneath the western edge of South America; along the coasts of Oregon, Washington, and British Columbia; and at several other continental margins (see Fig. 2–1). We will return to this subject in Chapters 11 and 12.

Convergence of Two Plates Carrying Oceanic Crust

Recall that newly formed oceanic lithosphere is hot, thin, and light, but as it spreads away from the mid-oceanic ridge, it becomes older, cooler, thicker, and denser. Thus,

Figure 2–8 A collision between India and Asia formed the Himalayas. This figure shows Rushi Konka, eastern Tibet.

the density of oceanic lithosphere increases with its age. When two oceanic plates converge, the denser one sinks into the mantle. Oceanic subduction zones are common in the southwestern Pacific Ocean and are discussed in Chapter 11.

Convergence of Two Plates Carrying Continents

If two converging plates carry continents, neither can sink into the mantle because of their low densities. In this case, the two continents collide and crumple against each other, forming a huge mountain chain. The Himalayas, the Alps, and the Appalachians all formed as results of continental collisions (Fig. 2–8). These processes are discussed in Chapter 12.

TRANSFORM PLATE BOUNDARIES

A transform plate boundary forms where two plates slide horizontally past one another as they move in opposite directions (Fig. 2–3C). California's San Andreas fault is the transform boundary between the North American plate and the Pacific plate. This type of boundary can occur in both oceans and continents and is discussed in Chapters 10, 11, and 12.

▶ 2.4 THE ANATOMY OF A TECTONIC PLATE

The nature of a tectonic plate can be summarized as follows:

1. A plate is a segment of the lithosphere; thus, it includes the uppermost mantle and all of the overlying crust.

2. A single plate can carry both oceanic and continental crust. The average thickness of lithosphere covered by oceanic crust is 75 kilometers, whereas that of lithosphere covered by a continent is 125 kilometers (Fig. 2–9). Lithosphere may be as little as 10 to 15 kilometers thick at an oceanic spreading center.

3. A plate is composed of hard, mechanically strong rock.

4. A plate floats on the underlying hot, plastic asthenosphere and glides horizontally over it.

5. A plate behaves like a large slab of ice floating on a pond. It may flex slightly, as thin ice does when a skater goes by, allowing minor vertical movements. In general, however, each plate moves as a large, intact sheet of rock.

6. A plate margin is tectonically active. Earthquakes and volcanoes are common at plate boundaries. In contrast, the interior of a lithospheric plate is normally tectonically stable.

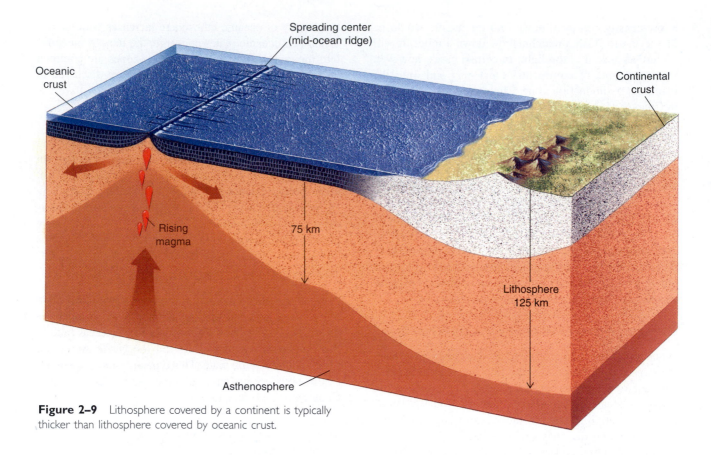

Figure 2–9 Lithosphere covered by a continent is typically thicker than lithosphere covered by oceanic crust.

7. Tectonic plates move at rates that vary from less than 1 to 16 centimeters per year.

▶ 2.5 CONSEQUENCES OF MOVING PLATES

As we mentioned previously, the plate tectonics theory provides a unifying explanation for earthquakes, volcanoes, mountain building, moving continents, and many other manifestations of the Earth's dynamic nature. In this section, we introduce some of these consequences of plate tectonics processes.

VOLCANOES

A volcanic eruption occurs where hot magma rises to the Earth's surface. Volcanic eruptions are common at both divergent and convergent plate boundaries. Three factors can melt rock to form magma and cause volcanic eruptions. The most obvious is rising temperature. However, hot rocks also melt to form magma if pressure decreases or if water is added to them. These magma-forming processes are discussed further in Chapter 4.

At a divergent boundary, hot asthenosphere rises to fill the gap left between the two separating plates (Fig. 2–7). Pressure decreases as the asthenosphere rises. As a result, portions of the asthenosphere melt to form huge quantities of basaltic magma, which erupts onto the Earth's surface. The mid-oceanic ridge is a submarine chain of volcanoes and lava flows formed at a divergent plate boundary. Volcanoes are also common in continental rifts, including the East African rift and the Rio Grande rift.

At a convergent plate boundary, cold, dense oceanic lithosphere dives into the asthenosphere. The sinking plate carries water-soaked mud and rock that once lay on the sea floor. As the sinking plate descends into the mantle, it becomes hotter. The heat drives off the water, which rises into the hot asthenosphere beneath the opposite plate. The water melts asthenosphere rock to form huge amounts of magma in a subduction zone. The magma then rises through the overlying lithosphere. Some solidifies within the crust, and some erupts from volcanoes on the Earth's surface. Volcanoes of this type are common in the Cascade Range of Oregon, Washington, and British Columbia; in western South America; and near most other subduction zones (Fig. 2–10).

EARTHQUAKES

Earthquakes are common at all three types of plate boundaries, but less common within the interior of a tectonic plate. Quakes concentrate at plate boundaries simply because those boundaries are zones of deep fractures in the lithosphere where one plate slips past another. The slippage is rarely smooth and continuous. Instead, the fractures may be locked up for months or for hundreds of years. Then, one plate suddenly slips a few centimeters or even a few meters past its neighbor. An earthquake is vibration in rock caused by these abrupt movements.

MOUNTAIN BUILDING

Many of the world's great mountain chains, including the Andes and parts of the mountains of western North America, formed at subduction zones. Several processes combine to build a mountain chain at a subduction zone. The great volume of magma rising into the crust thickens the crust, causing mountains to rise. Volcanic eruptions build chains of volcanoes. Additional crustal thickening may occur where two plates converge for the same reason that a mound of bread dough thickens when you compress it from both sides.

Great chains of volcanic mountains form at rift zones because the new, hot lithosphere floats to a high level, and large amounts of magma form in these zones. The mid-oceanic ridge, the East African rift, and the Rio Grande rift are examples of such mountain chains.

OCEANIC TRENCHES

An **oceanic trench** is a long, narrow trough in the sea floor that develops where a subducting plate sinks into the mantle (Figs. 2–3b and 2–5). To form the trough, the sinking plate drags the sea floor downward. A trench can form wherever subduction occurs—where oceanic crust sinks beneath the edge of a continent, or where it sinks beneath another oceanic plate. Trenches are the deepest parts of the ocean basins. The deepest point on Earth is in the Mariana trench in the southwestern Pacific Ocean, where the sea floor is as much as 10.9 kilometers below sea level (compared with the average sea-floor depth of about 5 kilometers).

MIGRATING CONTINENTS AND OCEANS

Continents migrate over the Earth's surface because they are integral parts of the moving lithospheric plates; they simply ride piggyback on the plates. Measurements of these movements show that North America is now moving away from Europe at about 2.5 centimeters per year, as the mid-Atlantic ridge continues to separate. South

Figure 2–10 Mount Hood in Oregon is a volcanic peak that lies near a convergent plate boundary.

America is drawing away from Africa at a rate of about 3.5 centimeters per year. As the Atlantic Ocean widens, the Pacific is shrinking at the same rate. Thus, as continents move, ocean basins open and close over geologic time.

▶ 2.6 THE SEARCH FOR A MECHANISM

Geologists have accumulated ample evidence that lithospheric plates move and can even measure how fast they move (Fig. 2–11). However, geologists do not agree on an explanation for why the plates move. Studies of the Earth's interior show that the mantle flows slowly beneath the lithosphere. Some geologists have suggested that this mantle flow drags the lithospheric plates along. Others suggest that another force moves the plates, and the movement of the plates causes the mantle to flow.

MANTLE CONVECTION

Convection occurs when a fluid is heated. For example, as a pot of soup is heated on a stove, the soup at the bottom of the pot becomes warm and expands. It then rises because it is less dense than the soup at the top. When the hot soup reaches the top of the pot, it flows along the surface until it cools and sinks (Fig. 2–12). The convection continues as long as the heat source persists. A similar process might cause convection in the Earth's mantle.

The mantle is heated internally by radioactive decay and from below by the hot core. Although the mantle is solid rock (except for small, partially melted zones in the asthenosphere), it is so hot that over geologic time it flows slowly. According to one hypothesis, hot rock rises from deep in the mantle to the base of the lithosphere.

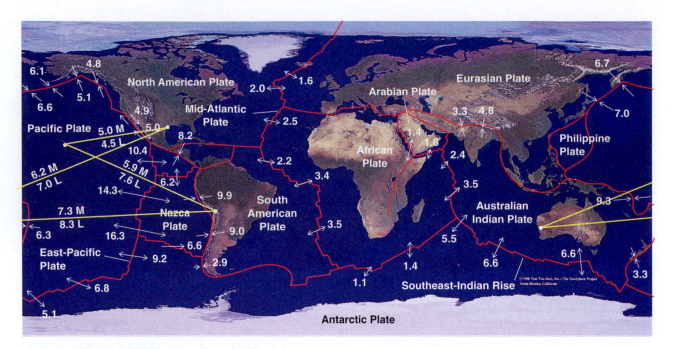

Figure 2–11 Plate velocities in centimeters per year. Numbers along the mid-oceanic ridge system indicate the rates at which two plates are separating, based on magnetic reversal patterns on the sea floor (discussed in Chapter 11). The arrows indicate the directions of plate motions. The yellow lines connect stations that measure present-day rates of plate motions with satellite laser ranging methods. The numbers followed by L are the present-day rates measured by laser. The numbers followed by M are the rates measured by magnetic reversal patterns. (Modified from NASA report, Geodynamics Branch, 1986. Tom Van Sant, Geosphere Project)

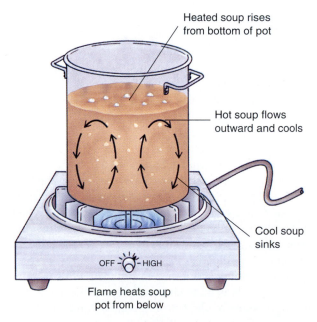

Figure 2–12 Soup convects when it is heated from the bottom of the pot.

At the same time, nearby parts of the cooler upper mantle sink. Thus, convection currents develop as in the soup pot.

Imagine a block of wood floating on a tub of honey. If you heated the honey so that it started to convect, the horizontal flow of honey along the surface would drag the block of wood along with it. Some geologists suggest that lithospheric plates are dragged along in a similar manner by a convecting mantle (Fig. 2–13).

GRAVITATIONAL SLIDING AS A CAUSE OF PLATE MOVEMENT

In Section 2.3, we explained why the lithosphere becomes thicker as it moves away from a spreading center. As a result of this thickening, the base of the lithosphere slopes downward from the spreading center with a grade as steep as 8 percent, steeper than most paved roads in North America (Fig. 2–14). Calculations show that if the slope is as slight as 0.3 percent, gravity would cause a plate to slide away from a spreading center at a rate of a few centimeters per year, like a sled gliding slowly down a snowy hill.

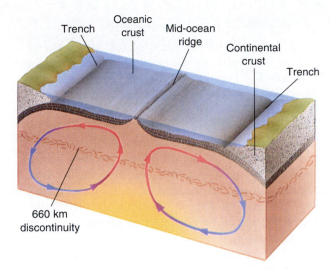

Figure 2–13 According to one explanation, a convecting mantle drags lithospheric plates.

In addition, the lithosphere becomes more dense as it cools and moves away from a spreading center. Eventually, old lithosphere may become denser than the asthenosphere below. Consequently, it can no longer float on the asthenosphere and begins to sink into the mantle, initiating subduction.

As an old, cold lithospheric plate sinks into the mantle, it pulls on the rest of the plate, like a weight pulling on the edge of a tablecloth. Many geologists now think that plates move because they glide downslope from a spreading center and, at the same time, are pulled along by their sinking ends. This combined mechanism is called the push-pull model of plate movement.

Some geologists now feel that this mechanism causes movement of lithospheric plates, and, in turn, the plate movements cause mantle convection. Return to our analogy of the block of wood and the tub of honey. If you dragged the block of wood across the honey, friction between the block and the honey would make the honey flow. Similarly, if the push-pull forces caused the plates to move, their motion would cause the mantle to flow.

(Continued on p. 33)

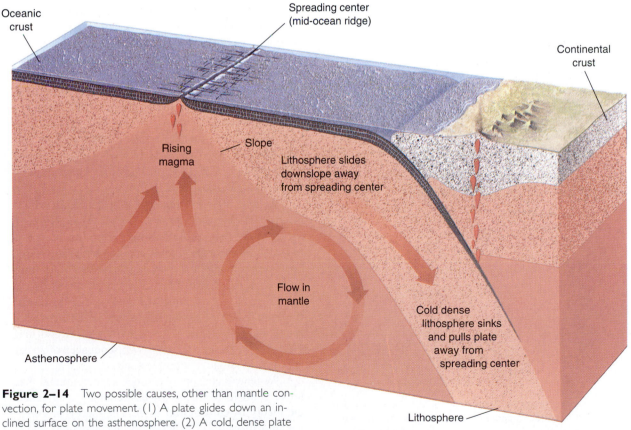

Figure 2–14 Two possible causes, other than mantle convection, for plate movement. (1) A plate glides down an inclined surface on the asthenosphere. (2) A cold, dense plate sinks at a subduction zone, pulling the rest of the plate along with it. In this drawing, both mechanisms are operating simultaneously.

ALFRED WEGENER AND THE ORIGIN OF AN IDEA

In the early twentieth century, a young German scientist named Alfred Wegener noticed that the African and South American coastlines on opposite sides of the Atlantic Ocean seemed to fit as if they were adjacent pieces of a jigsaw puzzle (Fig. 1). He realized that the apparent fit suggested that the continents had once been joined together and had later separated to form the Atlantic Ocean.

Although Wegener was not the first to make this suggestion, he was the first scientist to pursue it with additional research. Studying world maps, Wegener realized that not only did the continents on both sides of the Atlantic fit together, but other continents, when moved properly, also fit like additional pieces of the same jigsaw puzzle (Fig. 2). On his map, all the continents together formed one supercontinent that he called Pangea, from the Greek root words for "all lands." The northern part of Pangea is commonly called Laurasia and the southern part Gondwanaland.

Wegener understood that the fit of the continents alone did not prove that a supercontinent had existed. Therefore, he began seeking additional evidence in 1910 and continued work on the project until his death in 1930.

He mapped the locations of fossils of several species of animals and plants that could neither swim well nor fly. Fossils of the same species are now found in Antarctica, Africa, Australia, South America, and India. Why would the same species be found on continents separated by thousands of kilometers of ocean? When Wegener plotted the same fossil localities on his Pangea map, he found that they all lie in the same region of Pangea (Fig. 2). Wegener then suggested that each species had evolved and spread over that part of Pangea rather than mysteriously migrating across thousands of kilometers of open ocean.

Certain types of sedimentary rocks form in specific climatic zones. Glaciers and gravel deposited by glacial ice, for example, form in cold climates and are therefore found at high latitudes and high altitudes. Sandstones that preserve the structures of desert sand dunes form where deserts are common, near latitudes 30° north and south. Coral reefs and coal swamps thrive in near-equatorial tropical climates. Thus, each of these rocks reflect the latitudes at which they formed.

Wegener plotted 300-million-year-old glacial deposits on a map showing the modern distribution of continents (Fig. 3a). The area inside the line shows how large the ice mass would have been if the continents had been in their present positions. Notice that the glacier would have crossed the equator, and glacial deposits would have formed in tropical and subtropical zones. Figure 3b shows the same glacial deposits, and other climate-indicating rocks, plotted on Wegener's Pangea map. Here the glaciers cluster neatly about the South Pole. The other rocks are also found in logical locations.

Wegener also noticed several instances in which an uncommon rock type or a distinctive sequence of rocks on one side of the Atlantic Ocean was identical to rocks on the other side. When he plotted the rocks on a Pangea map, those on the east side of the Atlantic were continuous with their counterparts on the west side (Fig. 1). For example, the deformed rocks of the Cape Fold belt of South Africa are similar to rocks found in the Buenos Aires province of Argentina. Plotted on a Pangea map, the two sequences of rocks appear as a single, continuous belt.

Wegener's concept of a single supercontinent that broke apart to form the modern continents is called the theory of **continental drift**. The theory of continental drift was so revolutionary that skeptical scientists demanded an explanation of how continents could

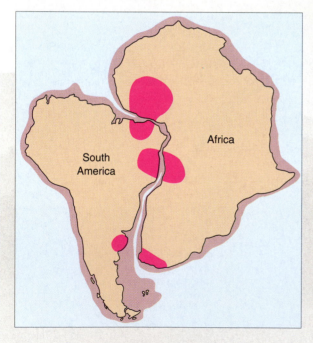

Figure 1 The African and South American coastlines appear to fit together like adjacent pieces of a jigsaw puzzle. The pink areas show locations of distinctive rock types in South America and Africa.

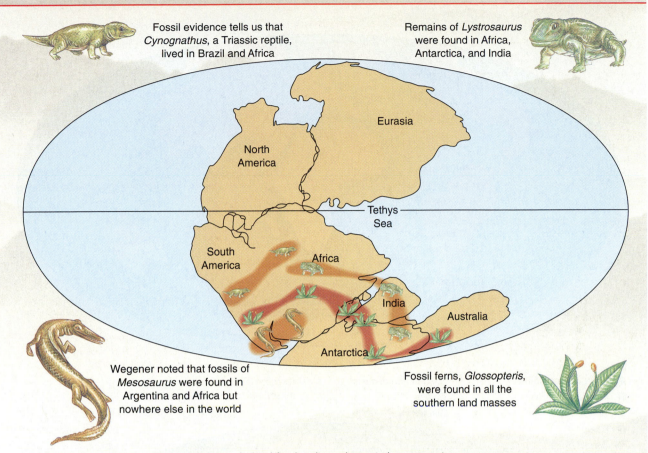

Fossil evidence tells us that *Cynognathus*, a Triassic reptile, lived in Brazil and Africa

Remains of *Lystrosaurus* were found in Africa, Antarctica, and India

Eurasia

North America

Tethys Sea

South America

Africa

India

Australia

Antarctica

Wegener noted that fossils of *Mesosaurus* were found in Argentina and Africa but nowhere else in the world

Fossil ferns, *Glossopteris*, were found in all the southern land masses

Figure 2 Geographic distributions of plant and animal fossils indicate that a single supercontinent, called Pangea, existed about 200 million years ago.

move. They wanted an explanation of the mechanism of continental drift. Wegener had concentrated on developing evidence that continents had drifted, not on how they moved. Finally, perhaps out of exasperation and as an afterthought to what he considered the important part of his theory, Wegener suggested two alternative possibilities: first, that continents plow their way through oceanic crust, shoving it aside as a ship plows through water; or second, that continental crust slides over oceanic crust. These suggestions turned out to be ill considered.

Physicists immediately proved that both of Wegener's mechanisms were impossible. Oceanic crust is too strong for continents to plow through it. The attempt would be like trying to push a matchstick boat through heavy tar. The boat, or the continents, would break apart. Furthermore, frictional resistance is too great for continents to slide over oceanic crust.

These conclusions were quickly adopted by most scientists as proof that Wegener's theory of continen-

tal drift was wrong. Notice, however, that the physicists' calculations proved only that the mechanism proposed by Wegener was incorrect. They did not disprove, or even consider, the huge mass of evidence indicating that the continents were once joined together. During the 30-year period from about 1930 to 1960, a few geologists supported the continental drift theory, but most ignored it.

Much of the theory of continental drift is similar to plate tectonics theory. Modern evidence indicates that the continents *were* together much as Wegener had portrayed them in his map of Pangea. Today, most geologists recognize the importance of Wegener's contributions.

DISCUSSION QUESTION

Compare the manner in which Wegener developed the theory of continental drift with the processes of the scientific method described in the "Focus On" box in Chapter 1. Explain why Wegener's theory was later rejected and, more recently, revived.

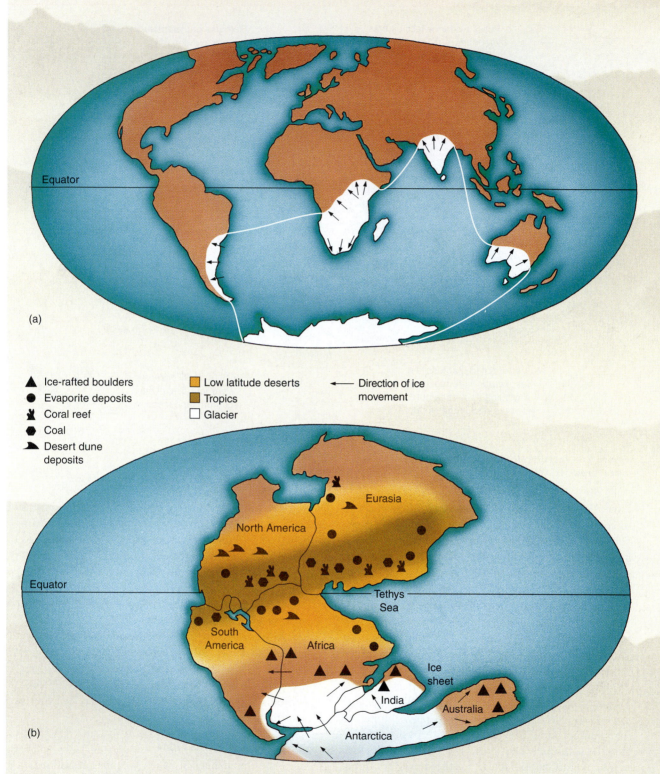

(a)

Ice-rafted boulders
Evaporite deposits
Coral reef
Coal
Desert dune deposits

Low latitude deserts
Tropics
Glacier

Direction of ice movement

(b)

Figure 3 (a) Three-hundred-million-year-old glacial deposits plotted on a map showing the modern distribution of continents. (b) Three-hundred-million-year-old glacial deposits and other climate-sensitive sedimentary rocks plotted on a map of Pangea.

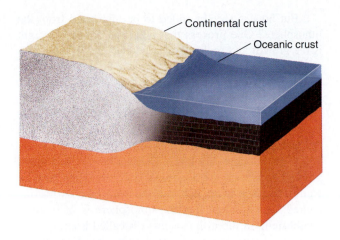

(a)

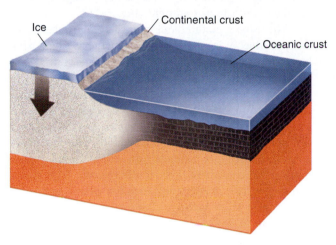

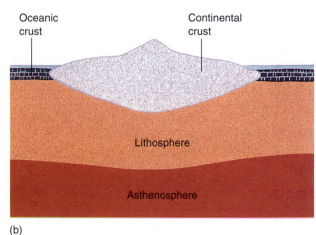

(b)

Figure 2–15 The weight of an ice sheet causes continental crust to sink isostatically.

Figure 2–16 (a) Icebergs illustrate some of the effects of isostasy. The large iceberg has a deep root and also a high peak. (b) Lithosphere covered by continental crust extends more deeply into the asthenosphere than lithosphere covered by oceanic crust.

MANTLE PLUMES

The push-pull model provides a mechanism to maintain plate movement once a lithospheric plate has begun to move, but does not explain why a plate should start moving. A **mantle plume** is a rising column of hot, plastic mantle rock that originates deep within the mantle. It rises because rock in a certain part of the mantle becomes hotter and more buoyant than surrounding regions of the mantle. The Earth's core may provide the heat source to create a mantle plume, or the heat may come from radioactive decay within the mantle. Large quantities of magma form in mantle plumes, and rise to erupt from volcanoes at locations called **hot spots** at the Earth's surface. The island of Hawaii is an example of a volcanic center at such a hot spot. Because mantle plumes form deep within the mantle, hot spot volcanoes commonly erupt within tectonic plates, and well away from plate boundaries.

Some geologists have suggested that a mantle plume might cause a new spreading center to develop within the lithosphere. Once spreading started, the push-pull mechanism would then keep the plates moving, even if the original mantle plume died out. Mantle plumes and their consequences are discussed further in Chapter 4.

▶ 2.7 SUPERCONTINENTS

Many geologists now suggest that movements of tectonic plates have periodically swept the world's continents together to form a single supercontinent. Each supercontinent lasted for a few hundred million years and then broke into fragments, each riding away from the others on its own tectonic plate.

Prior to 2 billion years ago, large continents as we know them today may not have existed. Instead, many—

perhaps hundreds—of small masses of continental crust and island arcs similar to Japan, New Zealand, and the modern island arcs of the southwestern Pacific Ocean dotted a global ocean basin. Then, between 2 billion and 1.8 billion years ago, tectonic plate movements swept these microcontinents together, forming the first **supercontinent**, which we call **Pangea I** after Alfred Wegener's Pangea, a word meaning "all lands" (see Focus On: Alfred Wegener and the Origin of an Idea).

After Pangea I split up about 1.3 billion years ago, the fragments of continental crust reassembled, forming a second supercontinent called **Pangea II**, about 1 billion years ago. In turn, this continent fractured and the continental fragments reassembled into **Pangea III** about 300 million years ago, 70 million years before the appearance of dinosaurs. Pangea III is Alfred Wegener's Pangea, described in the "Focus On" box.

▶ **2.8 ISOSTASY: VERTICAL MOVEMENT OF THE LITHOSPHERE**

If you have ever used a small boat, you may have noticed that the boat settles in the water as you get into it and rises as you step out. The lithosphere behaves in a similar manner. If a large mass is added to the lithosphere, it sinks and the underlying asthenosphere flows laterally away from that region to make space for the settling lithosphere.

But how is weight added to or subtracted from the lithosphere? One process that adds and removes weight is the growth and melting of large glaciers. When a glacier grows, the weight of ice forces the lithosphere downward. For example, in the central portion of Greenland, a 3000-meter-thick ice sheet has depressed the continental crust below sea level. Conversely, when a glacier melts, the continent rises—it rebounds. Geologists have discovered Ice Age beaches in Scandinavia tens of meters above modern sea level. The beaches formed when glaciers depressed the Scandinavian crust. They now lie well above sea level because the land rose as the ice melted. The concept that the lithosphere is in floating equilibrium on the asthenosphere is called **isostasy**, and the vertical movement in response to a changing burden is called **isostatic adjustment** (Fig. 2–15).

The iceberg pictured in Figure 2–16 illustrates an additional effect of isostasy. A large iceberg has a high peak, but its base extends deep below the surface of the water.

The lithosphere behaves in a similar manner. Continents rise high above sea level, and the lithosphere beneath a continent has a "root" that extends 125 kilometers into the asthenosphere. In contrast, most ocean crust lies approximately 5 kilometers below sea level, and oceanic lithosphere extends only about 75 kilometers into the asthenosphere. For similar reasons, high mountain ranges have deeper roots than low plains, just as the bottom of a large iceberg is deeper than the base of a small one.

SUMMARY

The **plate tectonics theory** provides a unifying framework for much of modern geology. It is the concept that the **lithosphere**, the outer, 75 to 125-kilometer-thick layer of the Earth, floats on the **asthenosphere**. The lithosphere is segmented into seven major **plates**, which move relative to one another by gliding over the asthenosphere. Most of the Earth's major geological activity occurs at **plate boundaries**. Three types of plate boundaries exist: (1) New lithosphere forms and spreads outward at a **divergent boundary**, or **spreading center**; (2) two lithospheric plates move toward each other at a **convergent boundary**, which develops into a **subduction zone** if at least one plate carries oceanic crust; and (3) two plates slide horizontally past each other at a **transform plate boundary**. Volcanoes, earthquakes, mountain building, and **oceanic trenches** occur near plate boundaries. Interior parts of lithospheric plates are tectonically stable. Tectonic plates move horizontally at rates that vary from 1 to 16 centimeters per year. Plate movements carry continents across the globe and cause ocean basins to open and close. The Earth is a layered planet. The **crust** is its outermost layer and varies from 5 to 70 kilometers thick. The **mantle** extends from the base of the crust to a depth of 2900 kilometers, where the core begins. The **lithosphere** is the cool outer 75 to 125 kilometers of the Earth; it includes all of the crust and the uppermost mantle. The lithosphere floats on the hot, plastic **asthenosphere**, which extends to 350 kilometers in depth. The **core** is mostly iron and nickel and consists of a liquid outer layer and a solid inner sphere.

Mantle convection may cause plate movement. Alternatively, a plate may move because it slides downhill from a spreading center, as its cold leading edge sinks into the mantle and drags the rest of the plate along. **Supercontinents** may assemble, split apart, and reassemble every 500 million to 700 million years.

The concept that the lithosphere floats on the asthenosphere is called **isostasy**. When weight such as a glacier is added to or removed from the Earth's surface, the lithosphere sinks or rises. This vertical movement in response to changing burdens is called **isostatic adjustment**.

KEY WORDS

plate tectonics theory *16*	transform plate	rift zone *22*	mantle convection *27*
lithosphere *18*	boundary *18*	magma *22*	mantle plume *33*
tectonic plate *18*	crust *18*	mid-oceanic ridge *23*	hot spot *33*
asthenosphere *18*	basalt *18*	continental rifting *23*	supercontinent *34*
plate boundary *18*	granite *21*	rift valley *23*	Pangea *34*
divergent boundary *18*	mantle *21*	subduction *24*	isostasy *34*
convergent boundary	core *21*	subduction zone *25*	isostatic adjustment *34*
18	spreading center *22*	oceanic trench *27*	

REVIEW QUESTIONS

1. Draw a cross-sectional view of the Earth. List all the major layers and the thickness of each.

2. Describe the physical properties of each of the Earth's layers.

3. Describe and explain the important differences between the lithosphere and the asthenosphere.

4. What properties of the asthenosphere allow the lithospheric plates to glide over it?

5. Describe some important differences between the crust and the lithosphere.

6. Describe some important differences between oceanic crust and continental crust.

7. How is it possible for the solid rock of the mantle to flow and convect?

8. Summarize the important aspects of the plate tectonics theory.

9. How many major tectonic plates exist? List them.

10. Describe the three types of tectonic plate boundaries.

11. Explain why tectonic plate boundaries are geologically active and the interior regions of plates are geologically stable.

12. Describe some differences between the lithosphere beneath a continent and that beneath oceanic crust.

13. Describe a reasonable model for a mechanism that causes movement of tectonic plates.

14. Why would a lithospheric plate floating on the asthenosphere suddenly begin to sink into the mantle to create a new subduction zone?

15. How many supercontinents have formed in Earth's history?

16. Describe the mid-Atlantic ridge and the mid-oceanic ridge.

17. Why are the oldest sea-floor rocks only about 200 million years old, whereas some continental rocks are 3.96 billion years old?

DISCUSSION QUESTIONS

1. Discuss why a unifying theory, such as the plate tectonics theory, is desirable in any field of science.

2. Central Greenland lies below sea level because the crust is depressed by the ice cap. If the glacier were to melt, would Greenland remain beneath the ocean? Why or why not?

3. At a rate of 5 centimeters per year, how long would it take for a continent to drift the width of your classroom? The distance between your apartment or dormitory and your classroom? The distance from New York to London?

4. Why do most major continental mountain chains form at convergent plate boundaries? What topographic and geologic features characterize divergent and transform plate boundaries in continental crust? Where do these types of boundaries exist in continental crust today?

5. If you were studying photographs of another planet, what features would you look for to determine whether or not the planet is or has been tectonically active?

6. The largest mountain in the Solar System is Olympus Mons, a volcano on Mars. It is 25,000 meters high, nearly three times the elevation of Mount Everest. Speculate on the factors that might permit such a large mountain on Mars.

7. The core's radius is 3470 kilometers, and that of the mantle is 2900 kilometers, yet the mantle contains 80 percent of the Earth's volume. Explain this apparent contradiction.

8. If you built a model of the Earth 1 meter in radius, how thick would the crust, lithosphere, asthenosphere, mantle, and core be?

9. Look at the map in Figure 2–1 and name a tectonic plate that is covered mostly by continental crust. Name one that is mostly ocean. Name two plates that are about half ocean and half continent.

Minerals

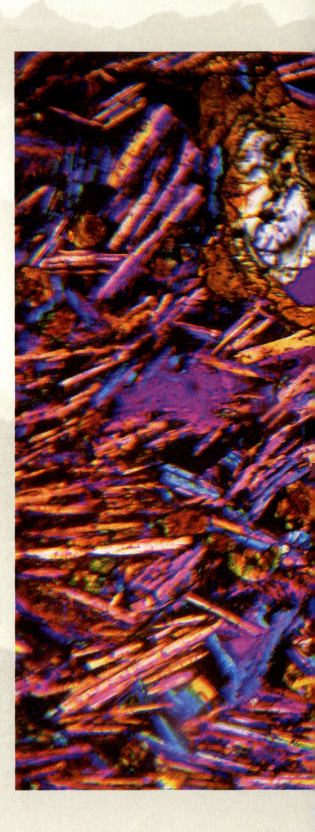

Pick up any rock and look at it carefully. You will probably see small, differently colored specks like those in granite (Fig. 3–1). Each speck is a mineral. A **rock** is an aggregate of **minerals**. Some rocks are made of only one mineral, but most contain two to five abundant minerals plus minor amounts of several others.

Figure 3–1   Each of the differently colored grains in this granite is a different mineral. The pink grains are feldspar, the black ones are biotite, and the glassy-white ones are quartz.

A sample of basalt, one of the most abundant rocks in the Earth's crust, viewed through a microscope. The intense colors are produced by polarized light. (© 1993 Kent Wood)

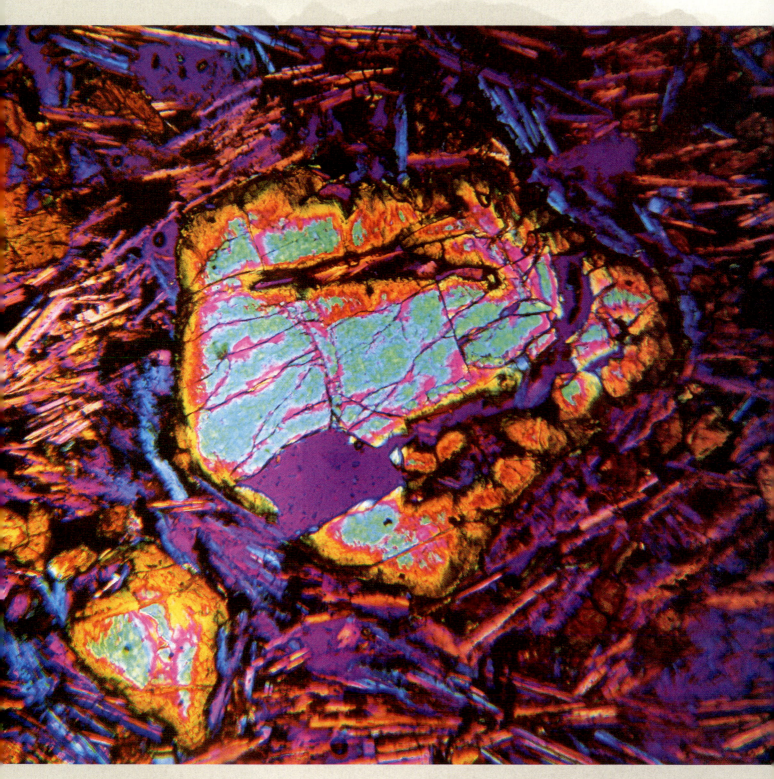

▶ 3.1 WHAT IS A MINERAL?

A mineral is a naturally occurring inorganic solid with a characteristic chemical composition and a crystalline structure. Chemical composition and crystalline structure are the two most important properties of a mineral: They distinguish any mineral from all others. Before discussing them, however, let us briefly consider the other properties of minerals described by this definition.

NATURAL OCCURRENCE

A synthetic diamond can be identical to a natural one, but it is not a true mineral because a mineral must form by natural processes. Like diamond, most gems that occur naturally can also be manufactured by industrial processes. Natural gems are valued more highly than manufactured ones. For this reason, jewelers should always tell their customers whether a gem is natural or artificial, and they usually preface the name of a manufactured gem with the term *synthetic*.

INORGANIC SOLID

Organic substances are made up mostly of carbon that is chemically bonded to hydrogen or other elements. Although organic compounds can be produced in laboratories and by industrial processes, plants and animals create most of the Earth's organic material. In contrast, inorganic compounds do not contain carbon-hydrogen bonds and generally are not produced by living organisms. All minerals are inorganic and most form independently of life. An exception is the calcite that forms limestone. Limestone is commonly composed of the shells of dead corals, clams, and similar marine organisms. Shells, in turn, are made of the mineral calcite or a similar mineral called aragonite. Although produced by organisms and containing carbon, the calcite and aragonite are true minerals.

▶ 3.2 ELEMENTS, ATOMS, AND THE CHEMICAL COMPOSITION OF MINERALS

To consider the chemical composition and crystalline structure of minerals, we must understand the nature of chemical **elements**—the fundamental components of matter. An element cannot be broken into simpler particles by ordinary chemical processes. Most common minerals consist of a small number—usually two to five—of different chemical elements.

A total of 88 elements occur naturally in the Earth's crust. However, eight elements—oxygen, silicon, aluminum, iron, calcium, magnesium, potassium, and sodium—make up more than 98 percent of the crust (Table 3–1).

A complete list of all elements is given in Table 3–2. Each element is represented by a one- or two-letter symbol, such as O for oxygen and Si for silicon. The table shows a total of 108 elements, not 88, because 20 elements are produced in nuclear reactors but do not occur naturally.

An **atom** is the basic unit of an element. An atom is tiny; the diameter of the average atom is about 10^{-10} meters (1/10,000,000,000). A single copper penny contains about 1.56×10^{22} (1.56 followed by 22 zeros) copper atoms. An atom consists of a small, dense, positively charged center called a **nucleus** surrounded by negatively charged **electrons** (Fig. 3–2).

An electron is a fundamental particle; it is not made up of smaller components. An electron orbits the nucleus, but not in a clearly defined path like that of the Earth around the Sun. Rather, an electron travels in a rapidly undulating path and is usually portrayed as a cloud of negative charge surrounding the nucleus. Electrons concentrate in spherical layers, or *shells*, around the nucleus. Each shell can hold a certain number of electrons.

The nucleus is made up of several kinds of particles; the two largest are positively charged **protons** and uncharged **neutrons**. A neutral atom contains equal numbers of protons and electrons. Thus, the positive and negative charges balance each other so that a neutral atom has no overall electrical charge.

An atom is most stable when its outermost shell is completely filled with electrons. But in their neutral

Table 3–1 • THE EIGHT MOST ABUNDANT CHEMICAL ELEMENTS IN THE EARTH'S CRUST

	WEIGHT PERCENT	ATOM PERCENT	VOLUME PERCENT*
O	46.60	62.55	93.8
Si	27.72	21.22	0.9
Al	8.13	6.47	0.5
Fe	5.00	1.92	0.4
Ca	3.63	1.94	1.0
Na	2.83	2.64	1.3
K	2.59	1.42	1.8
Mg	2.09	1.84	0.3
Total	98.59	100.00	100.00

From *Principles of Geochemistry* by Brian Mason and Carleton B. Moore. Copyright © 1982 by John Wiley & Sons, Inc.

*These numbers will vary somewhat as a function of the ionic radii chosen for the calculations.

Table 3–2 • THE PERIODIC TABLE

Groups of Main-Group Elements

Transition Elements

Periods	1	2												3	4	5	6	7	8
1	H 1																		He 2
2	Li 3	Be 4												B 5	C 6	N 7	O 8	F 9	Ne 10
3	Na 11	Mg 12												Al 13	Si 14	P 15	S 16	Cl 17	Ar 18
4	K 19	Ca 20		Sc 21	Ti 22	V 23	Cr 24	Mn 25	Fe 26	Co 27	Ni 28	Cu 29	Zn 30	Ga 31	Ge 32	As 33	Se 34	Br 35	Kr 36
5	Rb 37	Sr 38		Y 39	Zr 40	Nb 41	Mo 42	Tc 43	Ru 44	Rh 45	Pd 46	Ag 47	Cd 48	In 49	Sn 50	Sb 51	Te 52	I 53	Xe 54
6	Cs 55	Ba 56	*	La 57	Hf 72	Ta 73	W 74	Re 75	Os 76	Ir 77	Pt 78	Au 79	Hg 80	Tl 81	Pb 82	Bi 83	Po 84	At 85	Rn 86
7	Fr 87	Ra 88	†	Ac 89	Unq 104	Unp 105	Unh 106	Uns 107	108	Une 109									

*Lanthanoids	Ce 58	Pr 59	Nd 60	Pm 61	Sm 62	Eu 63	Gd 64	Tb 65	Dy 66	Ho 67	Er 68	Tm 69	Yb 70	Lu 71
†Actinoids	Th 90	Pa 91	U 92	Np 93	Pu 94	Am 95	Cm 96	Bk 97	Cf 98	Es 99	Fm 100	Md 101	No 102	Lr 103

Of the 108 elements that appear here, only 88 occur naturally in the Earth's crust. The other 20 are synthetic. The eight most abundant elements are shaded in orange. Elements with filled outer electron shells are shaded in violet.

states, most atoms do not have a filled outer shell. Such an atom may fill its outer shell by acquiring extra electrons until the shell becomes full. Alternatively, an atom may give up electrons until the outermost shell becomes empty. In this case, the next shell in, which is full, then becomes the outermost shell. When an atom loses one or more electrons, its protons outnumber its electrons and it develops a positive charge. If an atom gains one or more extra electrons, it becomes negatively charged. A charged atom is called an **ion**.

A positively charged ion is a **cation**. All of the abundant crustal elements except oxygen release electrons to become cations, as shown in Table 3–3. For example, each potassium atom (K) loses one electron to form a cation with a charge of 1+. Each silicon atom loses four electrons, forming a cation with a 4+ charge. In contrast,

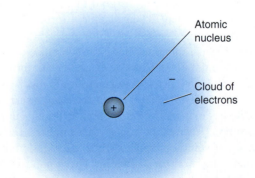

Figure 3–2 An atom consists of a small, dense, positive nucleus surrounded by a much larger cloud of negative electrons.

Table 3–3 • THE MOST COMMON IONS OF THE EIGHT MOST ABUNDANT CHEMICAL ELEMENTS IN THE EARTH'S CRUST

ELEMENT	CHEMICAL SYMBOL	COMMON ION(S)
Oxygen	O	O^{2-}
Silicon	Si	Si^{4+}
Aluminum	Al	Al^{3+}
Iron	Fe	Fe^{2+} and Fe^{3+}
Calcium	Ca	Ca^{2+}
Magnesium	Mg	Mg^{2+}
Potassium	K	K^{1+}
Sodium	Na	Na^{1+}

oxygen *gains* two extra electrons to acquire a 2− charge. Atoms with negative charges are called **anions**.

Atoms and ions rarely exist independently. Instead, they unite to form **compounds**. The forces that hold atoms and ions together to form compounds are called **chemical bonds** (see the "Focus On" box entitled "Chemical Bonds").

Most minerals are compounds. When ions bond together to form a mineral, they do so in proportions so that the total number of negative charges exactly balances the total number of positive charges. Thus, minerals are always electrically neutral. For example, the mineral quartz proportionally consists of one (4+) silicon cation and two (2−) oxygen anions.

Recall that a mineral has a definite chemical composition. A substance with a definite chemical composition is made up of chemical elements that are bonded together in definite proportions. Therefore, the composition can be expressed as a chemical formula, which is written by combining the symbols of the individual elements.

A few minerals, such as gold and silver, consist of only a single element. Their chemical formulas, respectively, are Au (the symbol for gold) and Ag (the symbol for silver). Most minerals, however, are made up of two to five essential elements. For example, the formula of quartz is SiO_2: It consists of one atom of silicon (Si) for every two of oxygen (O). Quartz from anywhere in the Universe has that exact composition. If it had a different composition, it would be some other mineral. The compositions of some minerals, such as quartz, do not vary by even a fraction of a percent. The compositions of other minerals vary slightly, but the variations are limited, as explained in Section 3.7.

The 88 elements that occur naturally in the Earth's crust can combine in many ways to form many different minerals. In fact, about 3500 minerals are known. However, the eight abundant elements commonly combine in only a few ways. As a result, only nine **rock-forming minerals** (or mineral "groups") make up most rocks of the Earth's crust. They are olivine, pyroxene, amphibole, mica, the clay minerals, quartz, feldspar, calcite, and dolomite.

▶ 3.3 CRYSTALS: THE CRYSTALLINE NATURE OF MINERALS

A **crystal** is any substance whose atoms are arranged in a regular, periodically repeated pattern. All minerals are crystalline. The mineral halite (common table salt) has the composition NaCl: one sodium ion (Na^+) for every chlorine ion (Cl^-). Figure 3–3a is an "exploded" view of the ions in halite. Figure 3–3b is more realistic, showing the ions in contact. In both sketches the sodium and chlorine ions alternate in orderly rows and columns intersecting at right angles. This arrangement is the **crystalline structure** of halite.

Think of a familiar object with an orderly, repetitive pattern, such as a brick wall. The rectangular bricks repeat themselves over and over throughout the wall. As a result, the whole wall also has the shape of a rectangle or some modification of a rectangle. In every crystal, a small group of atoms, like a single brick in a wall, repeats itself over and over. This small group of atoms is called a **unit cell**. The unit cell for halite is shown in Figure 3–3a. If you compare Figures 3–3a and 3–3b, you will notice that the simple halite unit cell repeats throughout the halite crystal.

Most minerals initially form as tiny crystals that grow as layer after layer of atoms is added to their surfaces. A halite crystal might grow, for example, as salty seawater evaporates from a tidal pool. At first, a tiny grain might form, similar to the sketch of halite in Figure 3–3b. This model shows a halite crystal containing 125 atoms; it would be only about one millionth of a millimeter long on each side. As evaporation continued, more and more sodium and chlorine ions would precipitate onto the faces of the growing crystal. Minerals crystallize from cooling magma in a similar manner.

The shape of a large, well-formed crystal like that of halite in Figure 3–3c is determined by the shape of the unit cell and the manner in which the crystal grows. For example, it is obvious from Figure 3–4a that the stacking of small cubic unit cells can produce a large cubic crystal. Figure 3–4b shows that a different kind of stacking of the same cubes can also produce an eight-sided crystal, called an octahedron. Halite can crystallize as a cube or as an octahedron. All minerals consist of unit

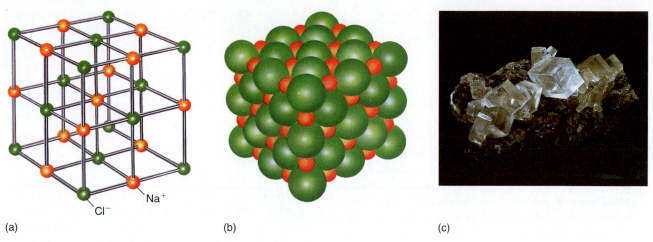

(a) (b) (c)

Figure 3–3 (a and b) The orderly arrangement of sodium and chlorine ions in halite. (c) Halite crystals. The crystal model in (a) is exploded so that you can see into it; the ions are actually closely packed as in (b). Note that ions in (a) and (b) form a cube, and the crystals in (c) are also cubes. (c, American Museum of Natural History)

cells stacked face to face as in halite, but not all unit cells are cubic.

A **crystal face** is a planar surface that develops if a crystal grows freely in an uncrowded environment. The sample of halite in Figure 3–3c has well-developed crystal faces. In nature, the growth of crystals is often impeded by adjacent minerals that are growing simultaneously or that have formed previously. For this reason, minerals rarely show perfect development of crystal faces.

▶ 3.4 PHYSICAL PROPERTIES OF MINERALS

How does a geologist identify a mineral in the field? Chemical composition and crystal structure distinguish each mineral from all others. For example, halite always consists of sodium and chlorine in a one-to-one ratio, with the atoms arranged in a cubic fashion. But if you pick up a crystal of halite, you cannot see the ions. You

(continued on p. 44)

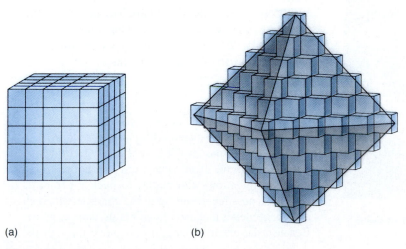

(a) (b)

Figure 3–4 Both a cubic crystal (a) and an "octahedron" (b) can form by different kinds of stacking of identical cubes.

CHEMICAL BONDS

Electrons concentrate in **shells** around the nucleus of an atom (Fig. 1). The laws of quantum physics allow only certain shells, or energy levels, to exist around a nucleus. Electrons can only occupy those shells and cannot orbit in areas between shells.

Look at the periodic table (Table 3–2). The elements in the right-hand column, colored blue, have a filled outer shell in their normal, neutral state. Those elements are stable and chemically nonreactive because their outer electron shells are full. Other elements gain or lose electrons so that they acquire a filled outer shell like those shaded in blue.

For example, sodium has one electron *more* than a filled outer shell. Therefore, it tends to give up one electron. When it loses that electron, its outer shell is perfectly filled, but it becomes a positively charged cation because it then has one more proton than it has electrons. Oxygen acquires a filled shell by gaining two electrons and forming an anion. The common anions and cations of the eight abundant elements are shown in Table 3–3.

Four types of chemical bonds are found in minerals: ionic, covalent, metallic, and van der Waals forces.

Ionic Bonds

Cations and anions are attracted by their opposite electronic charges and thus bond together. This union is called an **ionic bond**. An ionic compound (made up of two or more ions) is neutral because the positive and negative charges balance each other. For example, when sodium and chlorine form an ionic bond, the sodium atom loses one electron to become a cation and chlorine gains one to become an anion. When they combine, the +1 charge balances the −1 charge (Fig. 2).

Covalent Bonds

A **covalent bond** develops when two or more atoms share their electrons to produce the effect of filled outer electron shells. For example, carbon needs four electrons to fill its outermost shell. It can achieve this by forming four covalent bonds with four adjacent carbon atoms. It "gains" four electrons by sharing one with another carbon atom at each of the four bonds. Diamond consists of a three-dimensional network of carbon atoms bonded into a network of tetrahedra, similar to the framework structure of quartz (Fig. 3). The strength and homogeneity of the bonds throughout the crystal make diamond the hardest of all minerals.

In most minerals, the bonds between atoms are partly covalent and partly ionic. The combined characteristics of the different bond types determine the physical properties of those minerals.

Metallic Bonds

In a **metallic bond**, the outer electrons are loose; that is, they are not associated with particular atoms. The metal atoms sit in a "sea" of outer-level electrons that are free to move from one atom to another. That arrangement allows the nuclei to pack together as closely as possible, resulting in the characteristic high density of metals and metallic minerals, such as pyrite. Because the electrons are free to move through the

Figure I Electrons concentrate in spherical layers, or shells, around the nucleus of an atom.

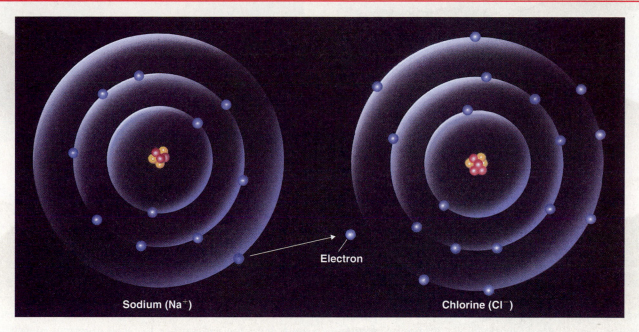

Figure 2 When sodium and chlorine atoms combine, sodium loses one electron, becoming a cation, Na^+. Chlorine acquires the electron to become an anion, Cl^-.

entire crystal, metallic minerals are excellent conductors of electricity and heat.

Van der Waals Forces

Weak electrical forces called **van der Waals forces** also bond molecules together. These weak bonds result from an uneven distribution of electrons around individual molecules, so that one portion of a molecule may have a greater density of negative charge while another portion has a partial positive charge. Because van der Waals forces are weak, minerals in which these bonds are important, such as talc and graphite, tend to be soft and cleave easily along planes of van der Waals bonds.

DISCUSSION QUESTION

Why do some minerals, such as native gold, silver, and graphite, conduct electricity, whereas others, such as quartz and feldspar, do not? Discuss relationships among other physical properties of minerals and the types of chemical bonds found in those minerals.

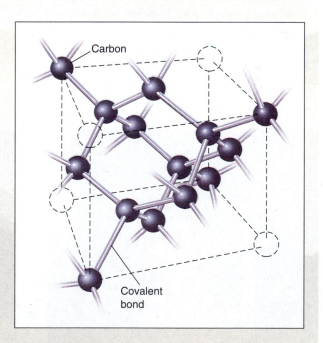

Figure 3 Carbon atoms in diamond form a tetrahedral network similar to that of quartz.

(a)

(c)

(b)

Figure 3–5 (a) *Equant* garnet crystals have about the same dimensions in all directions. (b) Asbestos is *fibrous*. (c) Kyanite forms *bladed* crystals. (Geoffrey Sutton)

could identify a sample of halite by measuring its chemical composition and crystal structure using laboratory procedures, but such analyses are expensive and time-consuming. Instead, geologists commonly identify minerals by visual recognition, and they confirm the identification with simple tests.

Most minerals have distinctive appearances. Once you become familiar with common minerals, you will recognize them just as you recognize any familiar object. For example, an apple just *looks* like an apple, even though apples come in many colors and shapes. In the same way, quartz looks like quartz to a geologist. The color and shape of quartz may vary from sample to sample, but it still looks like quartz. Some minerals, however, look enough alike that their physical properties must be examined further to make a correct identification. Geologists commonly use physical properties such as crystal habit, cleavage, and hardness to identify minerals.

CRYSTAL HABIT

Crystal habit is the characteristic shape of a mineral and the manner in which aggregates of crystals grow. If a

(a)

(b)

Figure 3–6 (a) *Prismatic* quartz grows as elongated crystals. (b) *Massive* quartz shows no characteristic shape. (Arkansas Geological Commission, J. M. Howard, Photographer)

Figure 3–7 A photomicrograph of a thin slice of granite. When crystals grow simultaneously, they commonly interlock and show no characteristic habit. To make this photo, a thin slice of granite was cut with a diamond saw, glued to a microscope slide, and ground to a thickness of 0.02 millimeters. Most minerals are transparent when such thin slices are viewed through a microscope.

crystal grows freely, it develops a characteristic shape controlled by the arrangement of its atoms, as in the cubes of halite shown in Figure 3–3c. Figure 3–5 shows three common minerals with different crystal habits. Some minerals occur in more than one habit. For example, Figure 3–6a shows quartz with a prismatic (pencil-shaped) habit, and Figure 3–6b shows massive quartz.

When crystal growth is obstructed by other crystals, a mineral cannot develop its characteristic habit. Figure 3–7 is a photomicrograph (a photo taken through a microscope) of a thin slice of granite in which the crystals fit like pieces of a jigsaw puzzle. This interlocking texture developed because some crystals grew around others as the magma solidified.

CLEAVAGE

Cleavage is the tendency of some minerals to break along flat surfaces. The surfaces are planes of weak bonds in the crystal. Some minerals, such as mica and graphite, have one set of parallel cleavage planes (Fig. 3–8). Others have two, three, or even four different sets, as shown in Figure 3–9. Some minerals, like the micas, have excellent cleavage. You can peel sheet after sheet from a mica crystal as if you were peeling layers from an onion. Others have poor cleavage. Many minerals have no cleavage at all because they have no planes of

weak bonds. The number of cleavage planes, the quality of cleavage, and the angles between cleavage planes all help in mineral identification.

A flat surface created by cleavage and a crystal face can appear identical because both are flat, smooth surfaces. However, a cleavage surface is duplicated when a crystal is broken, whereas a crystal face is not. So if you are in doubt, break the sample with a hammer—unless, of course, you want to save it.

FRACTURE

Fracture is the pattern in which a mineral breaks other than along planes of cleavage. Many minerals fracture into characteristic shapes. Conchoidal fracture creates smooth, curved surfaces (Fig. 3–10). It is characteristic of quartz and olivine. Glass, although not a mineral because it has no crystalline structure, also typically fractures in a conchoidal pattern. Some minerals break into splintery or fibrous fragments. Most fracture into irregular shapes.

HARDNESS

Hardness is the resistance of a mineral to scratching. It is easily measured and is a fundamental property of each mineral because it is controlled by bond strength between the atoms in the mineral. Geologists commonly

Figure 3–8 Cleavage in mica. This large crystal is the variety of mica called muscovite. (Geoffrey Sutton)

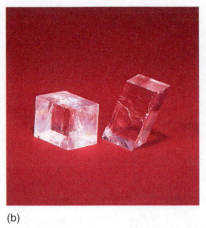

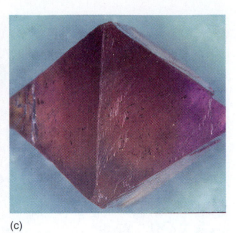

(a)   (b)   (c)

Figure 3–9 Some minerals have more than one cleavage plane. (a) Feldspar has two cleavages intersecting at right angles. (b) Calcite has three cleavage planes. (c) Fluorite has four cleavage planes. (Arthur R. Hill, Visuals Unlimited)

gauge hardness by attempting to scratch a mineral with a knife or other object of known hardness. If the blade scratches the mineral, the mineral is softer than the knife. If the knife cannot scratch the mineral, the mineral is harder.

To measure hardness more accurately, geologists use a scale based on ten minerals, numbered 1 through 10. Each mineral is harder than those with lower numbers on the scale, so 10 (diamond) is the hardest and 1 (talc) is the softest. The scale is known as the **Mohs hardness scale** after F. Mohs, the Austrian mineralogist who developed it in the early nineteenth century.

The Mohs hardness scale shows, for example, that a mineral scratched by quartz but not by orthoclase has a hardness between 6 and 7 (Table 3–4). Because the min-

erals of the Mohs scale are not always handy, it is useful to know the hardness values of common materials. A fingernail has a hardness of slightly more than 2, a copper penny about 3, a pocketknife blade slightly more than 5, window glass about 5.5, and a steel file about 6.5. If you practice with a knife and the minerals of the Mohs scale, you can develop a "feel" for minerals with hardnesses of 5 and under by how easily the blade scratches them.

When testing hardness, it is important to determine whether the mineral has actually been scratched by the object, or whether the object has simply left a trail of its own powder on the surface of the mineral. To check, simply rub away the powder trail and feel the surface of the mineral with your fingernail for the groove of the

Figure 3–10 Quartz typically fractures along smoothly curved surfaces, called conchoidal fractures. This sample is smoky quartz. (Breck P. Kent)

Table 3–4 • THE MINERALS OF THE MOHS HARDNESS SCALE

MINERALS OF MOHS SCALE	COMMON OBJECTS
1. Talc	
2. Gypsum	Fingernail
3. Calcite	Copper penny
4. Fluorite	
5. Apatite	Knife blade
	Window glass
6. Orthoclase	Steel file
7. Quartz	
8. Topaz	
9. Corundum	
10. Diamond	

scratch. Fresh, unweathered mineral surfaces must be used in hardness measurements because weathering often produces a soft rind on minerals.

SPECIFIC GRAVITY

Specific gravity is the weight of a substance relative to that of an equal volume of water. If a mineral weighs 2.5 times as much as an equal volume of water, its specific gravity is 2.5. You can estimate a mineral's specific gravity simply by hefting a sample in your hand. If you practice with known minerals, you can develop a feel for specific gravity. Most common minerals have specific gravities of about 2.7. Metals have much greater specific gravities; for example, gold has the highest specific gravity of all minerals, 19. Lead is 11.3, silver is 10.5, and copper is 8.9.

COLOR

Color is the most obvious property of a mineral, but it is commonly unreliable for identification. Color would be a reliable identification tool if all minerals were pure and had perfect crystal structures. However, both small amounts of chemical impurities and imperfections in crystal structure can dramatically alter color. For example, corundum (Al_2O_3) is normally a cloudy, translucent, brown or blue mineral. Addition of a small amount of chromium can convert corundum to the beautiful, clear, red gem known as ruby. A small quantity of iron or titanium turns corundum into the striking blue gem called sapphire.

STREAK

Streak is the color of a fine powder of a mineral. It is observed by rubbing the mineral across a piece of unglazed porcelain known as a streak plate. Many minerals leave a streak of powder with a diagnostic color on the plate. Streak is commonly more reliable than the color of the mineral itself for identification.

LUSTER

Luster is the manner in which a mineral reflects light. A mineral with a metallic look, irrespective of color, has a metallic luster. The luster of nonmetallic minerals is usually described by self-explanatory words such as glassy, pearly, earthy, and resinous.

OTHER PROPERTIES

Properties such as reaction to acid, magnetism, radioactivity, fluorescence, and phosphorescence can be characteristic of specific minerals. Calcite and some other carbonate minerals dissolve rapidly in acid, releasing visible bubbles of carbon dioxide gas. Minerals containing radioactive elements such as uranium emit radioactivity that can be detected with a scintillometer. Fluorescent materials emit visible light when they are exposed to ultraviolet light. Phosphorescent minerals continue to emit light after the external stimulus ceases.

▶ 3.5 ROCK-FORMING MINERALS, ACCESSORY MINERALS, GEMS, ORE MINERALS, AND INDUSTRIAL MINERALS

Although about 3500 minerals are known to exist in the Earth's crust, only a small number—between 50 and 100—are important because they are common or valuable.

ROCK-FORMING MINERALS

The rock-forming minerals make up the bulk of most rocks in the Earth's crust. They are important to geologists simply because they are the most common minerals. They are olivine, pyroxene, amphibole, mica, the clay minerals, feldspar, quartz, calcite, and dolomite. The first six minerals in this list are actually mineral "groups," in which each group contains several varieties

Figure 3–11 Pyrite is a common accessory mineral. (American Museum of Natural History)

Figure 3–12 Sapphire is one of the most costly precious gems. (Smithsonian Institution)

with very similar chemical compositions, crystalline structures, and appearances. The rock-forming minerals are described in Section 3.6.

ACCESSORY MINERALS

Accessory minerals are minerals that are common but usually are found only in small amounts. Chlorite, garnet, hematite, limonite, magnetite, and pyrite are common accessory minerals (Fig. 3–11).

GEMS

A **gem** is a mineral that is prized primarily for its beauty, although some gems, like diamonds, are also used industrially. Depending on its value, a gem can be either precious or semiprecious. Precious gems include diamond, emerald, ruby, and sapphire (Fig. 3–12). Several varieties of quartz, including amethyst, agate, jasper, and tiger's eye, are semiprecious gems. Garnet, olivine, topaz, turquoise, and many other minerals sometimes occur as aesthetically pleasing semiprecious gems (Fig. 3–13).

ORE MINERALS

Ore minerals are minerals from which metals or other elements can be profitably recovered. A few, such as native gold and native silver, are composed of a single element. However, most metals are chemically bonded to anions. Copper, lead, and zinc are commonly bonded to sulfur to form the important ore minerals chalcopyrite, galena (Fig. 3–14), and sphalerite.

INDUSTRIAL MINERALS

Several minerals are industrially important, although they are not considered ore because they are mined for purposes other than the extraction of metals. Halite is mined for table salt, and gypsum is mined as the raw material for plaster and sheetrock. Apatite and other phosphorus minerals are sources of the phosphate fertilizers crucial to modern agriculture. Many limestones are made up of nearly pure calcite and are mined as the raw material of cement.

Figure 3–13 Topaz is a popular semiprecious gem. (American Museum of Natural History)

Figure 3–14 Galena is the most important ore of lead and commonly contains silver. (Ward's Natural Science Establishment, Inc.)

► 3.6 MINERAL CLASSIFICATION

Geologists classify minerals according to their anions (negatively charged ions). Anions can be either simple or complex. A simple anion is a single negatively charged ion, such as O^{2-}. Alternatively, two or more atoms can bond firmly together and acquire a negative charge to form a complex anion. Two common examples are the silicate, $(SiO_4)^{4-}$, and carbonate, $(CO_3)^{2-}$, complex anions.

Each mineral group (except the native elements) is named for its anion. For example, the oxides all contain O^{2-}, the silicates contain $(SiO_4)^{4-}$, and the carbonates contain $(CO_3)^{2-}$.

NATIVE ELEMENTS

About 20 elements occur naturally in their native states as minerals. Fewer than ten, however, are common enough to be of economic importance. Gold, silver, platinum, and copper are all mined in their pure forms. Iron is rarely found in its native state in the Earth's crust, but metallic iron is common in certain types of meteorites. Native iron and nickel are thought to comprise most of the Earth's core. Native sulfur, used to manufacture sulfuric acid, insecticides, fertilizer, and rubber, is mined from volcanic craters, where it is deposited from gases emanating from the vents (Fig. 3–15).

Pure carbon occurs as both graphite and diamond. The minerals have identical compositions but different crystalline structures and are called **polymorphs**, after the ancient Greek for "several forms." Graphite is one of the softest minerals and is opaque and an electrical conductor. Diamond, the hardest mineral known, is transparent and an electrical insulator. The contrasting characteristics of graphite and diamond emphasize the importance of crystalline structure in determining the physical properties of minerals.

OXIDES

The oxides are a large group of minerals in which oxygen is combined with one or more metals. Oxide minerals are the most important ores of iron, manganese, tin, chromium, uranium, titanium, and several other industrial metals. Hematite (iron oxide, Fe_2O_3) occurs widely in many types of rocks and is the most abundant ore of iron. Although typically red in color, it occasionally occurs as black crystals used as semiprecious gems. Magnetite (Fe_3O_4), a naturally magnetic iron oxide, is another ore of iron. Spinel ($MgAl_2O_4$) often occurs as attractive red or blue crystals that are used as inexpensive, semiprecious gems. Synthetic spinels are also commonly used in jewelry. Ice, the oxide of hydrogen (H_2O), is a common mineral at the Earth's surface.

SULFIDES

Sulfide minerals consist of sulfur combined with one or more metals. Many sulfides are extremely important ore minerals. They are the world's major sources of copper, lead, zinc, molybdenum, silver, cobalt, mercury, nickel, and several other metals. The most common sulfides are pyrite (FeS_2), chalcopyrite ($CuFeS_2$), galena (PbS), and sphalerite (ZnS).

SULFATES

The sulfate minerals contain the sulfate complex anion $(SO_4)^{2-}$. Gypsum ($CaSO_4 \cdot 2H_2O$) and anhydrite ($CaSO_4$) are two important industrial sulfates used to manufacture plaster and sheetrock. Both form by evaporation of seawater or salty lake water.

PHOSPHATES

Phosphate minerals contain the complex anion $(PO_4)^{3-}$. Apatite, $Ca_5(F,Cl,OH)(PO_4)_3$, is the substance that makes up both teeth and bones. Phosphate is an essential fertilizer in modern agriculture. It is mined from fossil bone beds near Tampa, Florida, and from great sedimentary apatite deposits in the northern Rocky Mountains.

CARBONATES

The complex carbonate anion $(CO_3)^{2-}$ is the basis of two common rock-forming minerals, calcite ($CaCO_3$)

Figure 3–15 Native sulfur is forming today in the vent of Ollagüe Volcano on the Chile–Bolivia border.

(a)

(b)

Figure 3–16 Calcite (a) and dolomite (b) are two rock-forming carbonate minerals. (Ward's Natural Science Establishment, Inc.)

and dolomite [CaMg(CO$_3$)$_2$] (Figs. 3–16a and 3–16b). Most limestone is composed of calcite, and dolomite makes up the similar rock that is also called dolomite or sometimes dolostone. Limestone is mined as a raw in-

gredient of cement. Aragonite is a polymorph of calcite that makes up the shells of many marine animals.

SILICATES

The silicate minerals contain the (SiO$_4$)$^{4-}$ complex anion. Silicates make up about 95 percent of the Earth's crust. They are so abundant for two reasons. First, silicon and oxygen are the two most plentiful elements in the crust. Second, silicon and oxygen combine readily. To understand the silicate minerals, remember four principles:

1. Every silicon atom surrounds itself with four oxygens. The bonds between each silicon and its four oxygens are very strong.

2. The silicon atom and its four oxygens form a pyramid-shaped structure called the **silicate tetrahedron** with silicon in the center and oxygens at the four corners (Fig. 3–17). The silicate tetrahedron has a 4− charge and forms the (SiO$_4$)$^{4-}$ complex anion. The silicate tetrahedron is the fundamental building block of all silicate minerals.

(a)

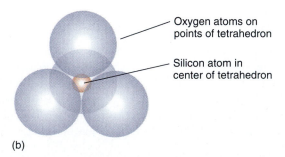

— Oxygen atoms on points of tetrahedron

— Silicon atom in center of tetrahedron

(b)

Figure 3–17 The silicate tetrahedron consists of one silicon atom surrounded by four oxygens. It is the fundamental building block of all silicate minerals. (a) A ball-and-stick representation. (b) A proportionally accurate model.

Figure 3–18 The five silicate structures are based on sharing of oxygens among silicate tetrahedra. (A) Independent tetrahedra share no oxygens. (B) In single chains, each tetrahedron shares two oxygens with adjacent tetrahedra, forming a chain. (C) A double chain is a pair of single chains that are crosslinked by additional oxygen sharing. (D) In the sheet silicates, each tetrahedron shares three oxygens with adjacent tetrahedra. (E) A three-dimensional silicate framework shares all four oxygens of each tetrahedron.

Class	Arrangement of SiO_4 tetrahedron	Unit composition	Mineral examples
A Independent tetrahedra		$(SiO_4)^{4-}$	Olivine: The composition varies between Mg_2SiO_4 and Fe_2SiO_4.
B Single chains		$(SiO_3)^{2-}$	Pyroxene: The most common pyroxene is augite, $Ca(Mg, Fe, Al) (Al, Si)_2O_6$.
C Double chains		$(Si_4O_{11})^{6-}$	Amphibole: The most common amphibole is hornblende, $NaCa_2(Mg, Fe, Al)_5(Si, Al)_8O_{22}(OH)_2$.
D Sheet silicates		$(Si_2O_5)^{2-}$	Mica, clay minerals, chlorite, e.g.: muscovite $KAl_2(Si_3Al)O_{10}(OH)_2$
E Framework silicates		SiO_2	Quartz: SiO_2 Feldspar: As an example, potassium feldspar is $KAlSi_3O_8$.

Figure 3–19 The seven rock-forming silicate mineral groups. (a) Olivine. (Geoffrey Sutton) (b) Pyroxene. (Jeffrey Scovill) (c) Amphibole. (Jeffrey Scovill) (d) Black biotite is one common type of mica. White muscovite (Fig. 3–8) is the other. (e) Clay. (f) Feldspar, represented here by orthoclase feldspar. (Breck P. Kent) (g) Quartz. (Jeffrey Scovill)

3. To make silicate minerals electrically neutral, other cations must combine with the silicate tetrahedra to balance their negative charges. (The lone exception is quartz, in which the positive charges on the silicons exactly balance the negative ones on the oxygens. How this occurs is described later.)

4. Silicate tetrahedra commonly link together by sharing oxygens. Thus, two tetrahedra may share a single oxygen, bonding the tetrahedra together.

Rock-Forming Silicate Minerals

The rock-forming silicates (and most other silicate minerals) fall into five classes, based on five ways in which tetrahedra share oxygens (Fig. 3–18). Each class contains at least one of the rock-forming mineral groups.

1. In independent tetrahedra silicates, adjacent tetrahedra do not share oxygens (Fig. 3–18A). **Olivine** is an independent tetrahedra mineral that occurs in small quantities in basalt of both continental and

oceanic crust (Fig. 3–19a). However, rocks composed mostly of olivine and pyroxene are thought to make up most of the mantle.

2. In the single-chain silicates, each tetrahedron links to two others by sharing oxygens, forming a continuous chain of tetrahedra (Fig. 3–18B). The **pyroxenes** are a group of similar minerals with single chain structures (Fig. 3–19b). Pyroxenes are a major component of both oceanic crust and the mantle and are also abundant in some continental rocks.

3. The double-chain silicates consist of two single chains crosslinked by the sharing of additional oxygens between them (Fig. 3–18C). The **amphiboles** (Fig. 3–19c) are a group of double-chain silicates with similar properties. They occur commonly in many continental rocks. One variety of amphibole grows as sharply pointed needles and is a type of asbestos.

Pyroxene and amphibole can resemble each other so closely that they are difficult to tell apart. Both groups have similar chain structures and similar chemical compositions. In addition, both commonly grow as pencil-shaped crystals.

4. In the sheet silicates, each tetrahedron shares oxygens with three others in the same plane, forming a continuous sheet (Fig. 3–18D). All of the atoms within each sheet are strongly bonded, but each sheet is only weakly bonded to those above and below. Therefore, sheet silicates have excellent cleavage. The **micas** are sheet silicates and typically grow as plate-shaped crystals, with flat surfaces (Fig. 3–19d). Mica is common in continental rocks. The **clay minerals** (Fig. 3–19e) are similar to mica in structure, composition, and platy habit. Individual clay crystals are so small that they can barely be seen with a good optical microscope. Most clay forms when other minerals weather at the Earth's surface. Thus, clay minerals are abundant near the Earth's surface and are an important component of soil and of sedimentary rocks.

5. In the framework silicates, each tetrahedron shares all four of its oxygens with adjacent tetrahedra (Fig. 3–18E). Because tetrahedra share oxygens in all directions, minerals with the framework structure tend to grow blocky crystals that have similar dimensions in all directions. **Feldspar** and **quartz** have framework structures.

The feldspars (Fig. 3–19f) make up more than 50 percent of the Earth's crust. The different varieties of feldspar are named according to whether potassium or a mixture of sodium and calcium is present in the mineral. **Orthoclase** is a common feldspar containing potassium. Feldspar containing calcium and/or sodium is called **plagioclase**. Plagioclase and orthoclase often look alike and can be difficult to tell apart.

Quartz is the only common silicate mineral that contains no cations other than silicon; it is pure SiO_2 (Fig. 3–19g). It has a ratio of one $4+$ silicon for every two $2-$ oxygens, so the positive and negative charges neutralize each other perfectly. Quartz is widespread and abundant in continental rocks but rare in oceanic crust and the mantle.

▶ **3.7 IONIC SUBSTITUTION**

Ionic substitution is the replacement of one ion by another in the crystal structure of a mineral. Generally, one ion can substitute for another if the ions are of similar size and if their charges are within $1+$ or $1-$.

Many minerals show variations in composition because of ionic substitution, although the variation is restricted to well-defined limits for each mineral. For example, olivine can be pure Mg_2SiO_4 or pure Fe_2SiO_4, and it can also have any proportion of magnesium (Mg) to iron (Fe) between the two extremes. As we mentioned earlier in this chapter, ionic substitution of small amounts of chromium for aluminum in the mineral corundum is responsible for the characteristic red color of ruby, and substitution of iron or titanium produces the blue of sapphire.

In the tetrahedral framework structure of potassium feldspar, one Al^{3+} ion substitutes for every fourth Si^{4+}. Because aluminum has only a $3+$ charge and silicon is $4+$, a charge of $1-$ develops for every fourth tetrahedron. One $1+$ potassium ion enters the feldspar structure to maintain electrical neutrality. Similar substitution of aluminum for silicon occurs in all other feldspars and in many other silicate minerals.

SUMMARY

Minerals are the substances that make up rocks. A mineral is a naturally occurring inorganic solid with a definite chemical composition and a crystalline structure. Each mineral consists of chemical elements bonded together in definite proportions, so that its chemical composition can be given as a chemical formula. The **crystalline structure** of a mineral is the orderly, periodically repeated arrangement of its atoms. A **unit cell** is a small structural and compositional module that repeats itself throughout a crystal. The shape of a crystal is determined by the shape and arrangement of its unit cells. Every mineral is distinguished from others by its chemical composition and crystal structure.

Most common minerals are easily recognized and identified visually. Identification is aided by observing a few physical properties, including **crystal habit**, **cleavage**, **fracture**, **hardness**, **specific gravity**, **color**, **streak**, and **luster**.

Although about 3500 minerals are known in the Earth's crust, only the nine **rock-forming mineral groups** are abundant in most rocks. They are **feldspar**, **quartz**, **pyroxene**, **amphibole**, **mica**, **the clay minerals**, **olivine**, **calcite**, and **dolomite**. The first seven on this list are **silicates**; their structures and compositions are based on the **silicate tetrahedron**, in which a silicon atom is surrounded by four oxygens to form a pyramid-shaped structure. Silicate tetrahedra link together by sharing oxygens to form the basic structures of the silicate minerals. The silicates are the most abundant minerals because silicon and oxygen are the two most abundant elements in the Earth's crust and bond together readily to form the silicate tetrahedron. Two carbonate minerals, calcite and dolomite, are also sufficiently abundant to be called rock-forming minerals.

Accessory minerals are commonly found, but in small amounts. **Ore minerals**, **industrial minerals**, and **gems** are important for economic reasons. Many minerals show compositional variation because of ionic substitution. In general, one element can substitute for another if the two are similar in charge and size.

Rock-Forming Mineral Groups

Feldspar	**Amphibole**	**Olivine**
Quartz	**Mica**	**Calcite**
Pyroxene	**Clay minerals**	**Dolomite**

KEY WORDS

mineral *36*	cation *39*	crystal face *41*	luster *47*
element *38*	anion *40*	crystal habit *44*	accessory mineral *48*
atom *38*	compound *40*	cleavage *45*	gem *48*
nucleus *38*	chemical bond *40*	fracture *45*	ore mineral *48*
electron *38*	rock-forming mineral *40*	hardness *45*	polymorph *49*
proton *38*	crystal *40*	Mohs hardness scale *46*	silicates *50*
neutron *38*	crystalline structure *40*	specific gravity *47*	silicate tetrahedron *50*
ion *39*	unit cell *40*	streak *47*	ionic substitution *53*

REVIEW QUESTIONS

1. What properties distinguish minerals from other substances?

2. Explain why oil and coal are not minerals.

3. What does the chemical formula for quartz, SiO_2, tell you about its chemical composition? What does $KAlSi_3O_8$ tell you about orthoclase feldspar?

4. What is an atom? An ion? A cation? An anion? What roles do they play in minerals?

5. What is a chemical bond? What role do chemical bonds play in minerals?

6. Every mineral has a crystalline structure. What does this mean?

7. What factors control the shape of a well-formed crystal?

8. What is a crystal face?

9. What conditions allow minerals to grow well-formed crystals? What conditions prevent their growth?

10. List and explain the physical properties of minerals most useful for identification.

11. Why do some minerals have cleavage and others do not? Why do some minerals have more than one set of cleavage planes?

12. Why is color often an unreliable property for mineral identification?

13. List the rock-forming mineral groups. Why are they called "rock-forming"? Which are silicates? Why are so many of them silicates?

14. Draw a three-dimensional view of a single silicate tetrahedron. Draw the five different arrangements of tetrahedra found in the rock-forming silicate minerals. How many oxygen ions are shared between adjacent tetrahedra in each of the five configurations?

15. Make a table with two columns. In the left column list the basic silicate structures. In the right column list one or more rock-forming minerals with each structure.

DISCUSSION QUESTIONS

1. Diamond and graphite are two minerals with identical chemical compositions, pure carbon (C). Diamond is the hardest of all minerals, and graphite is one of the softest. If their compositions are identical, why do they have such profound differences in physical properties?

2. List the eight most abundant chemical elements in the Earth's crust. Are any unfamiliar to you? List familiar elements that are not among the eight. Why are they familiar?

3. Table 3–1 shows that silicon and oxygen together make up nearly 75 percent by weight of the Earth's crust. But silicate minerals make up more than 95 percent of the crust. Explain the apparent discrepancy.

4. Quartz is SiO_2. Why does no mineral exist with the composition SiO_3?

5. If you were given a crystal of diamond and another of quartz, how would you tell which is diamond?

6. Would you expect minerals found on the Moon, Mars, or Venus to be different from those of the Earth's crust? Explain your answer.

Igneous Rocks

The Earth is almost entirely rock to a depth of 2900 kilometers, where the solid mantle gives way to the liquid outer core. Even casual observation reveals that rocks are not all alike. The great peaks of the Sierra Nevada in California are hard, strong granite. The red cliffs of the Utah desert are soft sandstone. The top of Mount Everest is limestone, composed of clamshells and the remains of other small marine animals.

The marine fossils of Mount Everest tell us that the limestone formed in the sea. What forces lifted the rock to the highest point of the Himalayas? Where did the vast amounts of sand in the Utah sandstone come from? How did the granite of the Sierra Nevada form? All of these questions ask about the processes that formed the rocks and about the events that moved and shaped them throughout geologic history. In the following five chapters, we will study rocks: how they form and what they are made of. In later chapters we will use our understanding of rocks to interpret the Earth's geologic history.

The Minaret Peaks in the eastern Sierra are composed of volcanic rocks.

57

▶ 4.1 ROCKS AND THE ROCK CYCLE

Geologists group rocks into three categories on the basis of how they form: **igneous rocks**, **sedimentary rocks**, and **metamorphic rocks**.

Under certain conditions, rocks of the upper mantle and lower crust melt, forming a hot liquid called **magma** (Fig. 4–1). An igneous rock forms when magma solidifies. About 95 percent of the Earth's crust consists of igneous rock and metamorphosed igneous rock. Although much of this igneous foundation is buried by a relatively thin layer of sedimentary rock, igneous rocks are conspicuous because they make up some of the world's most spectacular mountains. **Granite** and **basalt** are two common and familiar igneous rocks (Fig. 4–2).

Rocks of all kinds decompose, or weather, at the Earth's surface. Weathering breaks rocks into smaller fragments such as gravel, sand, and clay. At the same time, rainwater may dissolve some of the rock. Streams, wind, glaciers, and gravity then erode the weathered particles, carry them downhill, and deposit them at lower elevations. All such particles, formed by weathering and then eroded, transported, and deposited in layers, are called sediment. The sand on a beach and mud on a mud flat are examples of sediment that accumulated by these processes.

A sedimentary rock forms when sediment becomes cemented or compacted into solid rock. When the beach sand is cemented, it becomes **sandstone**; the mud becomes **shale**. Sedimentary rocks make up less than 5 percent of the Earth's crust. However, because sediment accumulates on the Earth's surface, sedimentary rocks form a thin layer over about 80 percent of all land. For this reason, sedimentary rocks seem more abundant than they really are (Fig. 4–3).

A metamorphic rock forms when any preexisting rock is altered by heating, increased pressure, or tectonic deformation. Tectonic processes can depress the Earth's surface to form a basin that may be hundreds of kilometers in diameter and thousands of meters deep. Sediment accumulates in the depression, burying the lowermost layers to great depths. When a rock is buried, its temperature and pressure increase, causing changes in both the minerals and the texture of the rock. These changes are called metamorphism, and the rock formed by these processes is a metamorphic rock. Metamorphism also occurs when magma heats nearby rock, or when tectonic forces deform rocks (Fig. 4–4). Schist, gneiss, and marble are common metamorphic rocks.

No rock is permanent over geologic time; instead, all rocks change slowly from one of the three rock types to another. This continuous process is called the **rock**

Figure 4–1 Magma rises from Pu'u O'o vent during an eruption in June 1986. (U.S. Geological Survey, J. D. Griggs)

Figure 4–2 The peaks of Sam Ford Fiord, Baffin Island, are part of a large granite pluton.

Figure 4–3 *Sedimentary layers of sandstone and coal form steep cliffs near Bryce, Utah.*

Figure 4–4 Metamorphic rocks are commonly contorted as a result of tectonic forces that deform them.

which then becomes cemented to form a sedimentary rock. An igneous rock may be metamorphosed. The rock cycle simply expresses the idea that rock is not permanent but changes over geologic time.

▶ 4.2 IGNEOUS ROCKS: THE ORIGINS OF MAGMA

If you drilled a well deep into the crust, you would find that Earth temperature rises about 30°C for every kilometer of depth. Below the crust, temperature continues to rise, but not as rapidly. In the asthenosphere (between depths of about 100 to 350 kilometers), the temperature

cycle (Fig. 4–5). The transformations from one rock type to another can follow many different paths. For example, weathering may reduce a metamorphic rock to sediment,

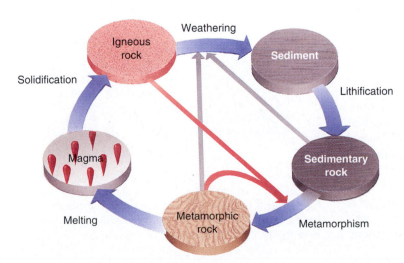

Figure 4–5 The rock cycle shows that rocks change continuously over geologic time. The arrows show paths that rocks can follow as they change.

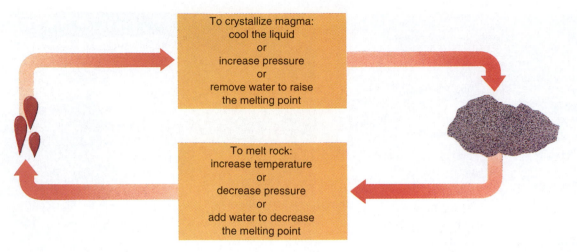

Figure 4–6 The lower box shows that increasing temperature, addition of water, and decreasing pressure all melt rock to form magma. The upper box shows that cooling, increasing pressure, and water loss all solidify magma to form an igneous rock.

is so high that rocks melt in certain environments to form magma.

PROCESSES THAT FORM MAGMA

Three different processes melt the asthenosphere: rising temperature, decreasing pressure, and addition of water (Fig. 4–6). We will consider these processes and then look at the tectonic environments in which they generate magma.

Rising Temperature

Everyone knows that a solid melts when it becomes hot enough. Butter melts in a frying pan, and snow melts under the spring sun. For similar reasons, an increase in temperature will melt a hot rock. Oddly, however, increasing temperature is the least important cause of magma formation in the asthenosphere.

Decreasing Pressure

A mineral is composed of an ordered array of atoms bonded together. When a mineral melts, the atoms become disordered and move freely, taking up more space than when they were in the solid mineral. Consequently, magma occupies about 10 percent more volume than the rock that melted to form it. As an analogy, think of a crowd of people sitting in an auditorium listening to a concert. At first, they sit in closely packed, orderly rows. But if everyone gets up to dance, they need more room

because spaces open up between the dancers as they move.

If a rock is heated to its melting point on the Earth's surface, it melts readily because there is little pressure to keep it from expanding. The temperature in the asthenosphere is more than hot enough to melt rock, but there, the high pressure prevents the rock from expanding, and it cannot melt (Fig. 4–7). However, if the pressure were to decrease, large volumes of the asthenosphere would melt. Melting caused by decreasing pressure is called **pressure-release melting**. In the section entitled "Environments of Magma Formation" we will see how certain tectonic processes decrease pressure on parts

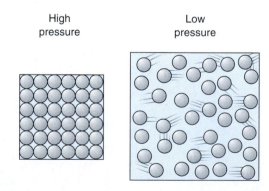

Figure 4–7 When most minerals melt, the volume increases. If a deeply buried mineral is near its melting point, high pressure prevents expansion and it doesn't melt. If the pressure decreases, the mineral can expand more easily and it melts, even though the temperature remains constant.

of the asthenosphere and cause large-scale melting and magma formation.

Addition of Water

A wet rock generally melts at a lower temperature than an otherwise identical dry rock. Thus, addition of water to rock that is near its melting temperature can melt the rock. Certain tectonic processes add water to the hot rock of the asthenosphere to form magma.

ENVIRONMENTS OF MAGMA FORMATION

Magma forms in three tectonic environments: spreading centers, mantle plumes, and subduction zones. Let us consider each environment to see how the three aforementioned processes melt the asthenosphere to create magma.

Magma Production in a Spreading Center

As lithospheric plates separate at a spreading center, hot, plastic asthenosphere oozes upward to fill the gap (Fig. 4–8). As the asthenosphere rises, pressure drops and pressure-release melting forms basaltic magma (the terms *basaltic* and *granitic* refer to magmas with the chemical compositions of basalt and granite, respectively). Because the magma is of lower density than the surrounding rock, it rises toward the surface.

Most of the world's spreading centers are in the ocean basins, where they form the mid-oceanic ridge. The magma created by pressure-release melting forms new oceanic crust at the ridge. The oceanic crust then spreads outward, riding atop the separating tectonic plates. Nearly all of the Earth's oceanic crust is created in this way at the mid-oceanic ridge. Some spreading centers, like the East African rift, occur in continents, and here, too, basaltic magma erupts onto the Earth's surface.

Magma Production at a Hot Spot

Recall from Chapter 2 that a mantle plume is a rising column of hot, plastic mantle rock that originates deep within the mantle. The plume rises because it is hotter than the surrounding mantle and, consequently, is buoy-

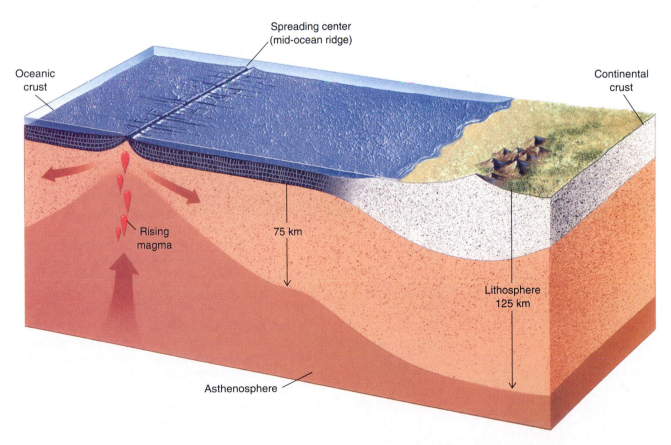

Figure 4–8 Pressure-release melting occurs where hot asthenosphere rises beneath a spreading center.

ant. As a mantle plume rises, pressure-release melting forms magma that erupts onto the Earth's surface (Fig. 4–9). A **hot spot** is a volcanically active place at the

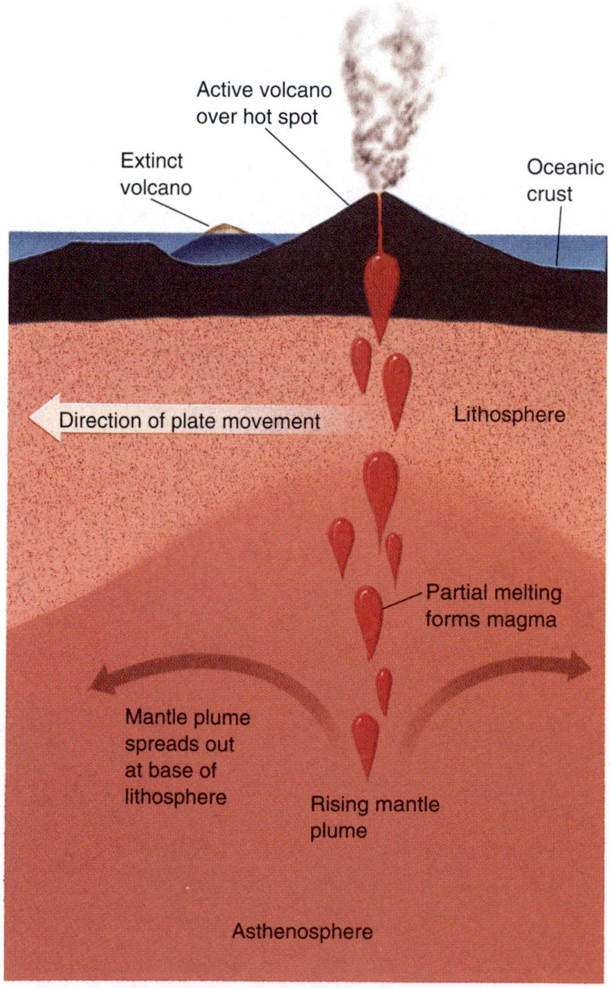

Figure 4–9 Pressure-release melting occurs in a rising mantle plume, and magma rises to form a volcanic hot spot.

Earth's surface directly above a mantle plume. Because mantle plumes form below the asthenosphere, hot spots can occur within a tectonic plate. For example, the Yellowstone hot spot, responsible for the volcanoes and hot springs in Yellowstone National Park, lies far from the nearest plate boundary. If a mantle plume rises beneath the sea, volcanic eruptions build submarine volcanoes and volcanic islands.

Magma Production in a Subduction Zone

At a subduction zone, a lithospheric plate sinks hundreds of kilometers into the mantle (Fig. 4–10). As you learned in Chapter 2, a subducting plate is covered by oceanic crust, which, in turn, is saturated with seawater. As the wet rock dives into the mantle, rising temperature drives off the water, which ascends into the hot asthenosphere directly above the sinking plate.

As the subducting plate descends, it drags plastic asthenosphere rock down with it, as shown by the elliptical arrows in Figure 4–10. Rock from deeper in the asthenosphere then flows upward to replace the sinking rock. Pressure decreases as this hot rock rises.

Finally, friction generates heat in a subduction zone as one plate scrapes past the opposite plate. Figure 4–10 shows that addition of water, pressure release, and frictional heating combine to melt portions of the asthenosphere, at a depth of about 100 kilometers, where the subducting plate passes into the asthenosphere. Addition of water is probably the most important factor in magma production in a subduction zone, and frictional heating is probably the least important.

As a result of these processes, igneous rocks are common features of a subduction zone. The volcanoes of the Pacific Northwest, the granite cliffs of Yosemite, and the Andes Mountains are all examples of igneous rocks formed at subduction zones. The "ring of fire" is a zone of concentrated volcanic activity that traces the subduction zones encircling the Pacific Ocean basin. About 75

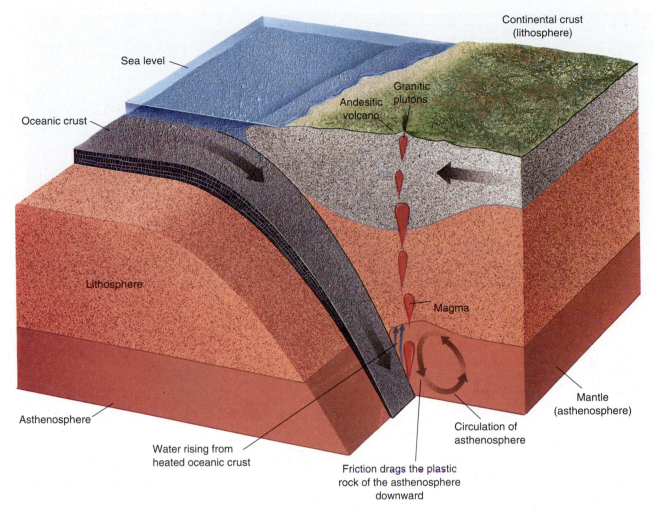

Figure 4–10 Three factors contribute to melting of the asthenosphere and production of magma at a subduction zone: (1) Friction heats rocks in the subduction zone; (2) water rises from oceanic crust on top of the subducting plate; and (3) circulation in the asthenosphere decreases pressure on hot rock.

percent of the Earth's active volcanoes (exclusive of the submarine volcanoes at the mid-oceanic ridge) lie in the ring of fire (Fig. 4–11).

CHARACTERISTICS OF MAGMA

We have just described how, why, and where magma forms. Now we consider its properties and behavior.

Temperature

The temperature of magma varies from about 600° to 1400°C, depending on its chemical composition and the depth at which it forms. Generally, basaltic magma forms at great depth and has a temperature near the high end of this scale. Granitic magmas, which form at shallower depths, tend to lie near the cooler end of the scale. As a comparison, an iron bar turns red hot at about 600°C and melts at slightly over 1500°C.

Chemical Composition

Because oxygen and silicon are the two most abundant elements in the crust and mantle, nearly all magmas are silicate magmas. In addition to oxygen and silicon, they also contain lesser amounts of the six other common elements of the Earth's crust: aluminum, iron, magnesium, calcium, potassium, and sodium. The main variations among different types of magmas are differences in the relative proportions of these eight elements. For example, basaltic magma contains more iron and magnesium than granitic magma, but granitic magma is richer in silicon, potassium, and sodium. A few rare magmas are of carbonate composition. The rocks that form from these

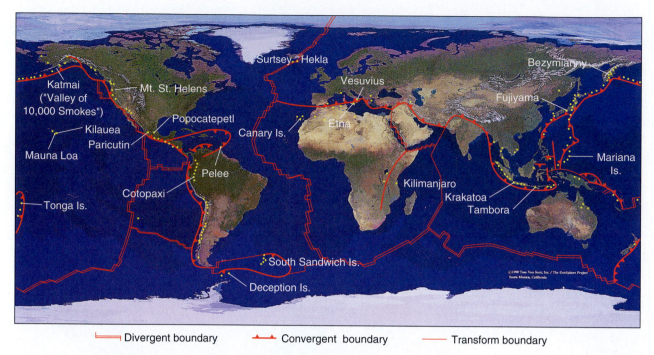

Divergent boundary · Convergent boundary · Transform boundary

Figure 4–11 Seventy-five percent of the Earth's active volcanoes (yellow dots) lie in the "ring of fire," a chain of subduction zones (heavy red lines with teeth) that encircles the Pacific Ocean. (Tom Van Sant, Geosphere Project)

are called carbonatites and contain carbonate minerals such as calcite and dolomite.

Behavior

When a silicate rock melts, the resulting magma expands by about 10 percent. It is then of lower density than the rock around it, so magma rises as it forms—much as a hot air balloon ascends in the atmosphere. When magma rises, it enters the cooler, lower-pressure environment near the Earth's surface. When temperature and pressure drop sufficiently, it solidifies to form solid igneous rock.

▶ 4.3 CLASSIFICATION OF IGNEOUS ROCKS

Magma can either rise all the way through the crust to erupt onto the Earth's surface, or it can solidify within the crust. An **extrusive igneous rock** forms when magma erupts and solidifies on the Earth's surface. Because extrusive rocks are so commonly associated with volcanoes, they are also called **volcanic rocks** after Vulcan, the Greek god of fire.

An **intrusive igneous rock** forms when magma solidifies *within* the crust. Intrusive rocks are sometimes called **plutonic rocks** after Pluto, the Greek god of the underworld.

TEXTURES OF IGNEOUS ROCKS

The **texture** of a rock refers to the size, shape, and arrangement of its mineral grains, or crystals (Table 4–1). Some igneous rocks consist of mineral grains that are too small to be seen with the naked eye; others are made up of thumb-size or even larger crystals. Volcanic rocks are usually fine grained, whereas plutonic rocks are medium or coarse grained.

Table 4–1 · IGNEOUS ROCK TEXTURES BASED ON GRAIN SIZE

GRAIN SIZE	NAME OF TEXTURE
No mineral grains (obsidian)	Glassy
Too fine to see with naked eye	Very fine grained
Up to 1 millimeter	Fine grained
1–5 millimeters	Medium grained
More than 5 millimeters	Coarse grained
Relatively large grains in a finer-grained matrix	Porphyritic

Figure 4–12 Basalt is a fine-grained volcanic rock. The holes are gas bubbles that were preserved as the magma solidified in southeastern Idaho.

Extrusive (Volcanic) Rocks

After magma erupts onto the relatively cool Earth surface, it solidifies rapidly—perhaps over a few days or years. Crystals form but do not have much time to grow. The result is a very fine-grained rock with crystals too small to be seen with the naked eye. Basalt is a common very fine-grained volcanic rock (Fig. 4–12).

If magma rises slowly through the crust before erupting, some crystals may grow while most of the magma remains molten. If this mixture of magma and crystals then erupts onto the surface, it solidifies quickly, forming **porphyry**, a rock with the large crystals, called **phenocrysts**, embedded in a fine-grained matrix (Fig. 4–13).

In unusual circumstances, volcanic magma may solidify within a few hours of erupting. Because the magma hardens so quickly, the atoms have no time to align themselves to form crystals. The result is the volcanic glass called **obsidian** (Fig. 4–14).

Figure 4–13 Porphyry is an igneous rock containing large crystals embedded in a fine-grained matrix. This rock is rhyolite porphyry with large pink feldspar phenocrysts.

Figure 4–14 Obsidian is natural volcanic glass. It contains no crystals. (Geoffrey Sutton)

(a) Granite

(b) Rhyolite

Figure 4–15 Although granite (a) and rhyolite (b) contain the same minerals, they have very different textures because granite cools slowly and rhyolite cools rapidly.

Intrusive (Plutonic) Rocks

When magma solidifies within the crust, the overlying rock insulates the magma like a thick blanket. The magma then crystallizes slowly, and the crystals may have hundreds of thousands or even millions of years in which to grow. As a result, most plutonic rocks are medium to coarse grained. Granite, the most abundant rock in continental crust, is a medium- or coarse-grained plutonic rock.

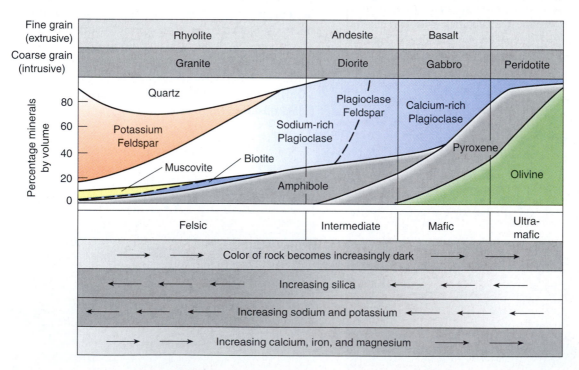

Figure 4–16 The names of common igneous rocks are based on the minerals and texture of a rock. In this figure, a mineral's abundance in a rock is proportional to the thickness of its colored band beneath the rock name. If a rock has a fine grain texture, its name is found in the top row of rock names; if it has a coarse grain texture, its name is in the second row.

NAMING IGNEOUS ROCKS

Geologists use both the minerals and texture to classify and name igneous rocks. For example, any medium- or coarse-grained igneous rock consisting mostly of feldspar and quartz is called granite. **Rhyolite** also consists mostly of feldspar and quartz but is very fine grained (Fig. 4–15). The same magma that erupts onto the Earth's surface to form rhyolite can also solidify slowly within the crust to form granite.

Like granite and rhyolite, most common igneous rocks are classified in pairs, each member of a pair containing the same minerals but having a different texture. The texture depends mainly on whether the rock is volcanic or plutonic. Figure 4–16 shows the minerals and textures of common igneous rocks.

The chemical compositions of common igneous rocks are summarized in Figure 4–17. Granite and rhyolite contain large amounts of feldspar and silica, and so are called **felsic rocks**. Basalt and gabbro are called **mafic rocks** because of their high magnesium and iron contents (*ferrum* is the Latin word for iron). Rocks with especially high magnesium and iron concentrations are called **ultramafic**. Rocks with compositions between those of granite and basalt are called **intermediate rocks**.

Once you learn to identify the rock-forming minerals, it is easy to name a plutonic rock using Figure 4–17 because the minerals are large enough to be seen. It is more difficult to name many volcanic rocks because the minerals are too small to identify. A field geologist often uses color to name a volcanic rock. Figure 4–17 shows that rhyolite is usually light in color: White, tan, red, and pink are common. Many andesites are gray or green, and basalt is commonly black. The minerals in many volcanic rocks cannot be identified even with a microscope because of their tiny crystal sizes. In this case, definitive identification is based on chemical and X-ray analyses carried out in the laboratory.

▶ 4.4 COMMON IGNEOUS ROCKS

GRANITE AND RHYOLITE

Granite is a felsic rock that contains mostly feldspar and quartz. Small amounts of dark biotite or hornblende of-

Descriptive Terms	Felsic (granitic)	Intermediate (andesitic)	Mafic (basaltic)	Ultramafic
Intrusive	Granite	Diorite	Gabbro	Peridotite
Extrusive	Rhyolite	Andesite	Basalt	
Composition	Aluminum oxide 14%, Iron oxides 3%, Magnesium oxide 1%, Other 10%, Silica 72%	Iron oxides 8%, Magnesium oxide 3%, Other 13%, Aluminum oxide 17%, Silica 59%	Magnesium oxide 7%, Other 16%, Silica 50%, Iron oxides 11%, Aluminum oxide 16%	Other 8%, Magnesium oxide 31%, Silica 45%, Iron oxides 12%, Aluminum oxide 4%
Major minerals	Quartz, Potassium feldspar, Sodium feldspar (plagioclase)	Amphibole, Intermediate plagioclase feldspar	Calcium feldspar (plagioclase), Pyroxene	Olivine, Pyroxene
Minor minerals	Muscovite, Biotite, Amphibole	Pyroxene	Olivine, Amphibole	Calcium feldspar (plagioclase)
Most common color	Light colored	Medium gray or medium green	Dark gray to black	Very dark green to black

Figure 4–17 *Chemical compositions, minerals, and typical colors of common igneous rocks.*

Figure 4–18 The authors on Inugsuin Point Buttress, a granite wall on Baffin Island. (Steve Sheriff)

ten give it a black and white speckled appearance. Granite (and metamorphosed granitic rocks) are the most common rocks in continental crust. They are found nearly everywhere beneath the relatively thin veneer of sedimentary rocks and soil that cover most of the continents. Geologists often call these rocks **basement rocks** because they make up the foundation of a continent. Granite is hard and resistant to weathering; it forms steep, sheer cliffs in many of the world's great mountain ranges. Mountaineers prize granite cliffs for the steepness and strength of the rock (Fig. 4–18).

As granitic magma rises through the Earth's crust, some of it may erupt from a volcano to form rhyolite, while the remainder solidifies beneath the volcano, forming granite. Most obsidian forms from magma with a granitic (rhyolitic) composition.

BASALT AND GABBRO

Basalt is a mafic rock that consists of approximately equal amounts of plagioclase feldspar and pyroxene. It makes up most of the oceanic crust as well as huge **basalt plateaus** on continents (Fig. 4–19). **Gabbro** is the plutonic counterpart of basalt; it is mineralogically identical but consists of larger crystals. Gabbro is uncommon at the Earth's surface, although it is abundant in deeper parts of oceanic crust, where basaltic magma crystallizes slowly.

ANDESITE AND DIORITE

Andesite is a volcanic rock intermediate in composition between basalt and granite. It is commonly gray or green and consists of plagioclase and dark minerals (usually biotite, amphibole, or pyroxene). It is named for the Andes mountains, the volcanic chain on the western edge of South America, where it is abundant. Because it is volcanic, andesite is typically very fine grained.

Diorite is the plutonic equivalent of andesite. It forms from the same magma as andesite and, consequently, often underlies andesitic mountain chains such as the Andes.

PERIDOTITE

Peridotite is an ultramafic igneous rock that makes up most of the upper mantle but is rare in the Earth's crust. It is coarse grained and composed of olivine, and it usually contains pyroxene, amphibole, or mica but no feldspar.

▶ 4.5 PARTIAL MELTING AND THE ORIGINS OF COMMON IGNEOUS ROCKS

BASALT AND BASALTIC MAGMA

Recall that oceanic crust is mostly basalt and that basaltic magma forms by melting of the asthenosphere. However, the asthenosphere is made of peridotite. Figure 4–18 shows that basalt and peridotite are quite different in composition: Peridotite contains about 40 percent silica, but basalt contains about 50 percent. Peridotite contains considerably more iron and magnesium than basalt. How does peridotite melt to create basaltic magma? Why does the magma have a composition different from that of the rock that melted to produce it?

Any pure substance, such as ice, has a definite melting point. Ice melts at exactly 0°C. In addition, ice melts to form water, which has exactly the same composition as the ice, pure H_2O. A rock does not behave in this way because it is a *mixture* of several minerals, each of which melts at a different temperature. If you heat peridotite slowly, the minerals with the lowest melting point begin to melt first, while the other minerals remain solid. This phenomenon is called **partial melting**. (Of course, if the temperature is high enough, the whole rock will melt.)

In general, minerals with the highest silica contents melt at the lowest temperatures. Silica-poor minerals melt only at higher temperatures. In parts of the asthenosphere where magma forms, the temperature is only hot enough to melt the minerals with the lowest melting points. As a result, magma is always richer in silica than the rock that melted to produce it. In this way, basaltic magma forms from peridotite rock at a temperature of about 1100°C. When the basaltic magma rises toward the Earth's surface, it leaves silica-depleted peridotite in the asthenosphere.

GRANITE AND GRANITIC MAGMA

Granite contains more silica than basalt and therefore melts at a lower temperature—typically between 700° and 900°C. Thus, basaltic magma is hot enough to melt granitic continental crust. In certain tectonic environments, the asthenosphere melts beneath a continent, forming basaltic magma that rises into continental crust. These environments include a subduction zone, a continental rift zone, and a mantle plume rising beneath a continent.

Because the lower continental crust is hot, even a small amount of basaltic magma melts large quantities of the continent to form granitic magma. Typically, the granitic magma then rises a short distance and then solidifies within the crust to form plutonic rocks. Most granitic plutons solidify at depths between a few kilometers and about 20 kilometers. Some of the magma may rise to the Earth's surface to erupt rhyolite and similar volcanic rocks. Small amounts of the original basaltic magma may erupt with the rhyolite or solidify at depth with the granite.

ANDESITE AND ANDESITIC MAGMA

Igneous rocks of intermediate composition, such as andesite and diorite, form by processes similar to those that generate granitic magma. Their magmas contain less silica than granite, either because they form by melting of continental crust that is lower in silica or because the basaltic magma from the mantle has contaminated the granitic magma.

Figure 4–19 Lava flows of the Columbia River basalt plateau are well exposed along the Columbia River.

SUMMARY

Geologists separate rocks into three classes based on how they form: **igneous rocks**, **sedimentary rocks**, and **metamorphic rocks**. Igneous rocks form when a hot, molten liquid called **magma** solidifies. Sedimentary rocks form when loose **sediment**, such as sand and clay, becomes cemented to form a solid rock. Metamorphic rocks form when older igneous, sedimentary, or other metamorphic rocks change because of high temperature and/or pressure or are deformed during mountain building. The **rock cycle** shows that all rocks change slowly over geologic time from one of the three rock types to another.

Three different processes—rising temperature, lowering of pressure, and addition of water—melt portions of the Earth's asthenosphere. These processes form great quantities of magma in three geologic environments: spreading centers, mantle plumes, and subduction zones. The temperature of magma varies from about 600° to 1400°C. Nearly all magmas are silicate magmas. Magma usually rises toward the Earth's surface because it is of lower density than rocks that surround it.

An **extrusive**, or **volcanic**, igneous rock forms when magma erupts and solidifies on the Earth's surface. An **intrusive**, or **plutonic**, rock forms when magma cools and solidifies below the surface. Plutonic rocks typically have medium- to coarse-grained textures, whereas volcanic rocks commonly have very fine- to fine-grained textures. A **porphyry** consists of larger crystals imbedded in a fine-grained matrix.

The two most common types of igneous rocks in the Earth's crust are **granite**, which comprises most of the continental crust, and **basalt**, which makes up oceanic crust. The upper mantle is composed of **peridotite**.

An igneous rock is classified and named according to its texture and mineral composition. The textures, mineral contents, and names of the common igneous rocks are summarized in Figures 4–16 and 4–17.

A **mafic** rock is low in silica, high in iron and magnesium, and dark in color. Basalt is a common mafic rock. A **felsic** rock is rich in feldspar and silicon, low in iron and magnesium, and light in color. Granite is a common felsic rock. An **intermediate** rock has a composition and color that lie between those of mafic and felsic rocks. The most common intermediate rock is **andesite**. **Ultramafic** rocks have the lowest silicon and aluminum content and the highest amounts of magnesium and iron. Peridotite, an ultramafic rock, is rare in the crust but abundant in the mantle.

Magmas invariably have a higher silica content than the rocks that melt to produce them, due to the phenomenon of **partial melting**.

Important Igneous Rocks

| Extrusive | Rhyolite | Andesite | Basalt | |
| Intrusive | Granite | Diorite | Gabbro | Peridotite |

KEY WORDS

magma *58*

igneous rock *58*

sedimentary rock *58*

metamorphic rock *58*

rock cycle *58*

pressure-release melting *60*

hot spot *62*

extrusive igneous rock *64*

volcanic rock *64*

intrusive igneous rock *64*

plutonic rocks *64*

texture *64*

porphyry *65*

phenocryst *65*

obsidian *65*

felsic rock *67*

mafic rock *67*

ultramafic rock *67*

intermediate rock *67*

basalt plateau *68*

partial melting *68*

REVIEW QUESTIONS

1. Describe the three main classes of rocks.

2. What criteria are used to categorize rocks into the three classes that you described in question 1?

3. List two common rock types in each of the three main classes of rocks. Were these rock names familiar to you before you read this chapter?

4. What is the most important concept described by the rock cycle?

5. What is magma?

6. Describe and explain each of the three processes that melt rock to form magma.

7. Describe each of the three main geologic environments in which magma forms in large quantities.

8. Describe the processes that melt rock to generate magma in each of the three environments that you discussed in the previous question.

9. Explain how oceanic crust forms only at the mid-oceanic ridge but makes up the entire sea floor.

10. Describe the locations of some volcanically active regions associated with subduction zones.

11. What is the temperature of magma?

12. What is the general chemical composition of most magmas?

13. Why do magmas begin to rise through the Earth's outer layers as soon as they form?

14. How would you distinguish a plutonic rock from a volcanic rock in the field?

15. What factor distinguishes obsidian from all other types of igneous rocks?

16. What are the most common minerals in igneous rocks? Why?

17. What do the terms *mafic*, *ultramafic*, *felsic*, and *intermediate* mean?

18. Describe the mineralogy, texture, and common geologic occurrence of the following types of igneous rocks: granite, rhyolite, basalt, gabbro, andesite, and peridotite.

19. What type of igneous rock is the most abundant constituent of continental crust? What type makes up most oceanic crust?

20. Why is it sometimes difficult to identify a volcanic rock accurately in the field?

21. Why does magma normally have a higher silica content than the rock from which it formed?

DISCUSSION QUESTIONS

1. The temperature of most magma is thought to be approximately equal to the initial melting temperature of the rock from which the magma formed, rarely much hotter. Why do magmas not get much hotter than their initial melting points?

2. Why is oceanic crust predominantly basalt, whereas continental crust is mainly of granitic composition?

3. Devise a scheme for naming and classifying igneous rocks that is different from the one based on mineral content presented in Figure 4–16.

4. Explain why feldspar is the most abundant mineral in the Earth's crust and yet is nearly completely absent from the peridotite of the upper mantle.

5. What could you infer about the history of another planet if you discovered extrusive igneous rocks but no intrusive igneous rocks on its surface?

6. Draw a graph with silica content on the Y-axis and felsic, intermediate, mafic, and ultramafic rocks on the X-axis. Draw similar graphs for aluminum, magnesium, iron, and potassium contents.

Plutons and Volcanoes

In Chapter 4 you learned that magma forms deep within the Earth. In some instances, it solidifies within the crust to form plutonic rocks. In others, it erupts onto the Earth's surface to form volcanic rocks.

Because plutonic rocks crystallize within the crust, we cannot see them form. However, tectonic forces commonly raise them, and erosion exposes these intrusive rocks in many of the world's greatest mountain ranges. California's Sierra Nevada, portions of the European Alps, and parts of the Himalayas are made up of plutonic rocks.

In contrast, a volcanic eruption can be one of the most conspicuous and violent of all geologic events. During the past 100 years, eruptions have killed approximately 100,000 people and caused about $10 billion in damage. Some eruptions have buried towns and cities in hot lava or volcanic ash. For example, the 1902 eruption of Mount Pelée on the Caribbean island of Martinique buried the city of Saint Pierre in glowing volcanic ash that killed 29,000 people. Other volcanoes erupt gently. Tourists flock to Hawaii to photograph flowing lava and fire fountains erupting into the sky (Fig. 5–1).

Volcanic eruptions can trigger other deadly events. The 1883 eruption of Krakatoa in the southwest Pacific Ocean generated tsunamis (large sea waves commonly but incorrectly called tidal waves) that killed 36,000 people. In 1985, a small eruption of Nevado del Ruiz in Colombia triggered a mudflow that buried the town of Armero, killing more than 22,000 people.

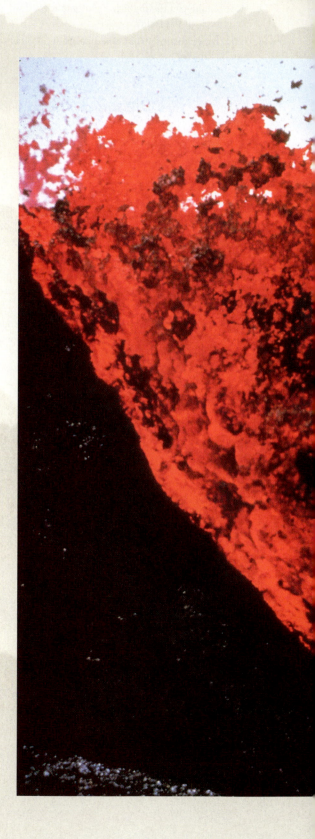

A cascade of molten lava pours from Mauna Loa volcano during the July 1993 eruption.   (J. D. Griggs/USGS)

Figure 5–1   Two geologists retreat from a slowly advancing lava flow on the island of Hawaii.   (U. S. Geological Survey)

▶ 5.1   THE BEHAVIOR OF MAGMA

Why do some volcanoes explode violently but others erupt gently? Why do some magmas crystallize within the Earth to form plutonic rocks and others rise all the way to the surface to erupt from volcanoes? To answer these questions, we must consider the properties and behavior of magma.

Once magma forms, it rises toward the Earth's surface because it is less dense than surrounding rock. As it rises, two changes occur. First, it cools as it enters shallower and cooler levels of the Earth. Second, pressure drops because the weight of overlying rock decreases. As you learned in Chapter 4, cooling and decreasing pressure have opposite effects on magma: Cooling tends to solidify it, but decreasing pressure tends to keep it liquid.

So does magma solidify or remain liquid as it rises toward the Earth's surface? The answer depends on the type of magma. Basaltic magma commonly rises to the surface to erupt from a volcano. In contrast, granitic magma usually solidifies within the crust.

The contrasting behavior of granitic and basaltic magmas is a result of their different compositions. Granitic magma contains about 70 percent silica, whereas the silica content of basaltic magma is only about 50 percent. In addition, granitic magma generally contains up to 10 percent water, whereas basaltic magma contains only 1 to 2 percent water (see the following chart).

TYPICAL GRANITIC MAGMA	**TYPICAL BASALTIC MAGMA**
70% silica	50% silica
Up to 10% water	1 to 2% water

EFFECTS OF SILICA ON MAGMA BEHAVIOR

In the silicate minerals, silicate tetrahedra link together to form the chains, sheets, and framework structures described in Chapter 3. Silicate tetrahedra link together in a similar manner in magma. They form long chains and similar structures if silica is abundant in the magma, but shorter chains if less silica is present. Because of its higher silica content, granitic magma contains longer chains than does basaltic magma. **Viscosity** is resistance to flow. In granitic magma, the long chains become tangled, making the magma stiff, or viscous. It rises slowly because of its viscosity and has ample time to solidify within the crust before reaching the surface. In contrast, basaltic magma, with its shorter silicate chains, is less viscous and flows easily. Because of its fluidity, it rises rapidly to erupt at the Earth's surface.

EFFECTS OF WATER ON MAGMA BEHAVIOR

A second, and more important, difference is that granitic magma contains more water than basaltic magma. Water lowers the temperature at which magma solidifies. Thus, if dry granitic magma solidifies at 700°C, the same magma with 10 percent water may remain liquid at 600°C.

Water tends to escape as steam from hot magma. But deep in the crust where granitic magma forms, high pressure prevents the water from escaping. As magma rises, pressure decreases and water escapes. Because the magma loses water, its solidification temperature *rises*, causing it to crystallize. Thus, water loss causes rising granitic magma to solidify within the crust. For this reason, most granitic magmas solidify at depths of 5 to 20 kilometers beneath the Earth's surface.

Because basaltic magmas have only 1 to 2 percent water to begin with, water loss is relatively unimportant. As a result, rising basaltic magma remains liquid all the way to the Earth's surface, and basalt volcanoes are common.

▶ 5.2 PLUTONS

In most cases, granitic magma solidifies within the Earth's crust to form a **pluton** (Fig. 5–2). Many granite plutons are large, measuring tens of kilometers in diameter. To form a large pluton, a huge volume of granitic magma must rise through continental crust. How can such a large mass of magma rise through solid rock?

If you place oil and water in a jar, screw the lid on, and shake the jar, oil droplets disperse throughout the water. When you set the jar down, the droplets coalesce to form larger bubbles, which rise toward the surface, easily displacing the water as they ascend. Granitic magma rises in a similar way. It forms near the base of continental crust, where surrounding rock behaves plastically because it is hot. As the magma rises, it shoulders aside the hot, plastic rock, which then slowly flows back to fill in behind the rising bubble.

After a pluton forms, tectonic forces may push it upward, and erosion may expose parts of it at the Earth's surface. A **batholith** is a pluton exposed over more than

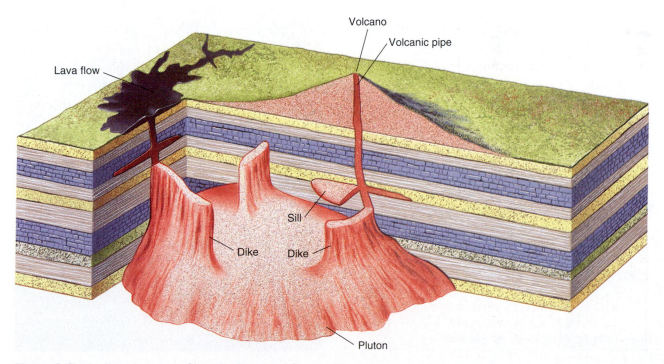

Figure 5–2 A pluton is a mass of intrusive igneous rock.

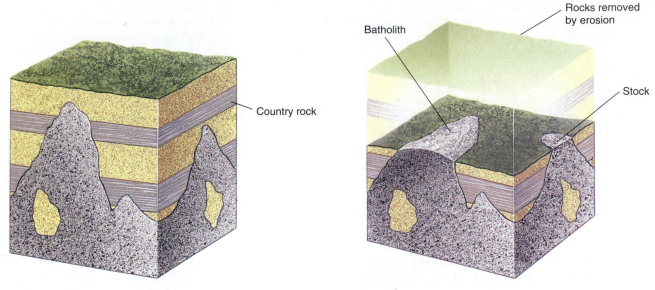

Figure 5–3 A batholith is a pluton with more than 100 square kilometers exposed at the Earth's surface. A stock is similar to a batholith but has a smaller surface area.

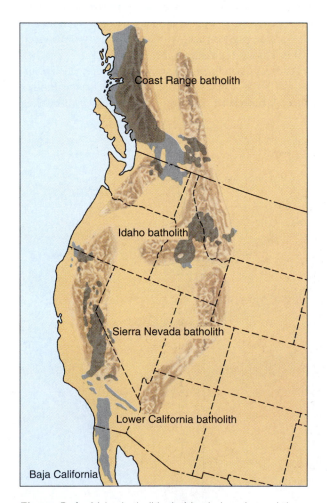

Figure 5–4 Major batholiths in North America and the mountain ranges associated with them.

Figure 5–5 An uplifted and exposed portion of the Sierra Nevada batholith, Yosemite National Park.

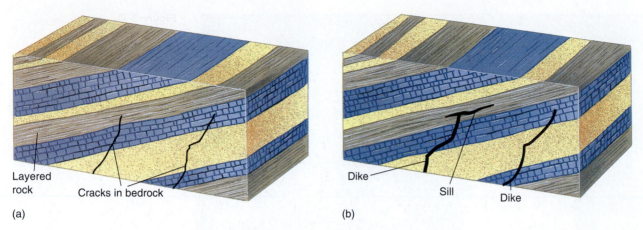

Figure 5–6 A dike cuts across the grain of country rock. A sill is parallel to the grain, or layering, of country rock.

100 square kilometers of the Earth's surface (Fig. 5–3). A large batholith may be as much as 20 kilometers thick, but an average one is about 10 kilometers thick. A **stock** is similar to a batholith but is exposed over less than 100 square kilometers.

Figure 5–4 shows the major batholiths of western North America. Many mountain ranges, such as California's Sierra Nevada, contain large granite batholiths. A batholith is commonly composed of numerous smaller plutons intruded sequentially over millions of years. For example, the Sierra Nevada batholith contains about 100 smaller plutons most of which were intruded over a period of 50 million years. The formation of this complex batholith ended about 80 million years ago (Fig. 5–5).

A large body of magma pushes country rock aside as it rises. In contrast, a smaller mass of magma may flow into a fracture or between layers in country rock. A **dike** is a tabular, or sheetlike, intrusive rock that forms when magma oozes into a fracture (Fig. 5–6). Dikes cut *across* sedimentary layers or other features in country rock and range from less than a centimeter to more than a kilometer thick (Fig. 5–7). Dikes commonly occur in parallel or radiating sets called a dike swarm, where magma has intruded a set of fractures. A dike is commonly more resistant to weathering than surrounding

Figure 5–7 A basalt dike cross-cutting sedimentary rock in Grand Canyon.

Figure 5–8 A large dike in central Colorado has been left standing after softer sandstone country rock eroded away. (Copyright Science Graphics, Inc./Ward's Natural Science Establishment, Inc.)

rock. As the country rock erodes away, the dike is left standing on the surface (Fig. 5–8).

Magma that oozes between layers of country rock forms a sheetlike rock *parallel* to the layering, called a **sill** (Fig. 5–9). Like dikes, sills vary in thickness from less than a centimeter to more than a kilometer and may extend for tens of kilometers in length and width.

▶ 5.3 VOLCANIC ROCKS AND VOLCANOES

The material erupted from volcanoes creates a wide variety of rocks and landforms, including lava plateaus and several types of volcanoes. Many islands, including the Hawaiian Islands, Iceland, and most islands of the south-

Figure 5–9 A basaltic sill has intruded between sedimentary rock layers on Mt. Gould in Glacier National Park, Montana. The white rock above and below the sill was metamorphosed by heat from the magma. (Breck P. Kent)

Figure 5–10 A car buried in pahoehoe lava, Hawaii. (Kenneth Neuhauser)

western Pacific Ocean, were built entirely by volcanic eruptions.

LAVA AND PYROCLASTIC ROCKS

Lava is fluid magma that flows onto the Earth's surface; the word also describes the rock that forms when the magma solidifies. Lava with low viscosity may continue to flow as it cools and stiffens, forming smooth, glassy-surfaced, wrinkled or "ropy" ridges. This type of lava is called **pahoehoe** (Fig. 5–10). If the viscosity of lava is higher, its surface may partially solidify as it flows. The solid crust breaks up as the deeper, molten lava continues to move, forming **aa** lava, with a jagged, rubbly, broken surface. When lava cools, escaping gases such as water and carbon dioxide form bubbles in the lava. If the lava solidifies before the gas escapes, the bubbles are preserved as holes called **vesicles** (Fig. 5–11).

Hot lava shrinks as it cools and solidifies. The shrinkage pulls the rock apart, forming cracks that grow as the rock continues to cool. In Hawaii geologists have watched fresh lava cool. When a solid crust only 0.5 centimeter thick had formed on the surface of the glowing liquid, five- or six-sided cracks developed. As the lava continued to cool and solidify, the cracks grew downward through the flow. Such cracks, called **columnar joints**, are regularly spaced and intersect to form five- or six-sided columns (Fig. 5–12).

When basaltic magma erupts under water, the rapid cooling causes it to contract into pillow-shaped struc-

tures called **pillow lava** (Fig. 5–13). Pillow lava is abundant in oceanic crust, where it forms as basaltic magma oozes onto the sea floor at the mid-oceanic ridge.

Figure 5–11 Aa lava showing vesicles, gas bubbles frozen into the flow, in Shoshone, Idaho.

(a)

(b)

Figure 5–12 Columnar joints at Devil's Postpile National Monument. (a) A view from the top, where the columns have been polished by glaciers. (b) Side view.

Figure 5–13 Pillow lava, formed on the sea floor, was thrust onto land during a tectonic collision in western Oregon.

If a volcano erupts explosively, it may eject both liquid magma and solid rock fragments. A rock formed from particles of magma that were hurled into the air from a volcano is called a **pyroclastic rock**. The smallest particles, called **volcanic ash**, consist of tiny fragments of glass that formed when liquid magma exploded into the air. **Cinders** vary in size from 2 to 64 millimeters. Still larger fragments called **volcanic bombs** form

Figure 5–14 The streaky surface and spindle shape of this volcanic bomb formed as a blob of magma whirled through the air.

as blobs of molten lava spin through the air, cooling and solidifying as they fall back to Earth (Fig. 5–14).

FISSURE ERUPTIONS AND LAVA PLATEAUS

The gentlest type of volcanic eruption occurs when magma is so fluid that it oozes from cracks in the land surface called **fissures** and flows over the land like water. Basaltic magma commonly erupts in this manner because of its low viscosity. Fissures and fissure eruptions vary greatly in scale. In some cases, lava pours from small cracks on the flank of a volcano. Fissure flows of this type are common on Hawaiian and Icelandic volcanoes.

In other cases, however, fissures extend for tens or hundreds of kilometers and pour thousands of cubic kilometers of lava onto the Earth's surface. A fissure eruption of this type creates a **flood basalt**, which covers the landscape like a flood. It is common for many such fissure eruptions to occur in rapid succession and to create a **lava plateau** (or "basalt plateau") covering thousands of square kilometers.

The Columbia River plateau in eastern Washington, northern Oregon, and western Idaho is a lava plateau containing 350,000 cubic kilometers of basalt. The lava is up to 3000 meters thick and covers 200,000 square kilometers (Fig. 5–15). It formed about 15 million years ago as basaltic magma oozed from long fissures in the Earth's surface. The individual flows are between 15 and 100 meters thick. Similar large lava plateaus occur in western India, northern Australia, Iceland, Brazil, Argentina, and Antarctica.

VOLCANOES

If lava is too viscous to spread out as a flood, it builds a hill or mountain called a **volcano**. Volcanoes differ widely in shape, structure, and size (Table 5–1). Lava and rock fragments commonly erupt from an opening called a **vent** located in a **crater**, a bowl-like depression at the summit of the volcano (Fig. 5–16). As mentioned previously, lava or pyroclastic material may also erupt from a fissure on the flanks of the volcano.

An **active volcano** is one that is erupting or is expected to erupt. A **dormant volcano** is one that is not now erupting but has erupted in the past and will probably do so again. Thus, no clear distinction exists between active and dormant volcanoes. In contrast, an **extinct volcano** is one that is expected never to erupt again.

Shield Volcanoes

Fluid basaltic magma often builds a gently sloping mountain called a **shield volcano** (Fig. 5–17). The sides of a shield volcano generally slope away from the vent at angles between 6° and 12° from horizontal. Although their slopes are gentle, shield volcanoes can be enormous. The height of Mauna Kea volcano in Hawaii, measured from its true base on the sea floor to its top, rivals the height of Mount Everest.

(a)

(b)

Figure 5–15 (a) The Columbia River basalt plateau covers much of Washington, Oregon, and Idaho. (b) Columbia River basalt in eastern Washington. Each layer is a separate lava flow. (Larry Davis)

Table 5–1 • **CHARACTERISTICS OF DIFFERENT TYPES OF VOLCANOES**

TYPE OF VOLCANO	FORM OF VOLCANO	SIZE	TYPE OF MAGMA	STYLE OF ACTIVITY	EXAMPLES
Basalt plateau	Flat to gentle slope	100,000 to 1,000,000 km^2 in area; 1 to 3 km thick	Basalt	Gentle eruption from long fissures	Columbia River Plateau
Shield volcano	Slightly sloped, 6° to 12°	Up to 9000 m high	Basalt	Gentle, some fire fountains	Hawaii
Cinder cone	Moderate slope	100 to 400 m high	Basalt or andesite	Ejections of pyroclastic material	Paricutín, Mexico
Composite volcano	Alternate layers of flows and pyroclastics	100 to 3500 m high	Variety of types of magmas and ash	Often violent	Vesuvius, Mount St. Helens, Aconcagua
Caldera	Cataclysmic explosion leaving a circular depression called a caldera	Less than 40 km in diameter	Granite	Very violent	Yellowstone, San Juan Mountains

Although shield volcanoes, such as those of Hawaii and Iceland, erupt regularly, the eruptions are normally gentle and rarely life threatening. Lava flows occasionally overrun homes and villages, but the flows advance slowly enough to give people time to evacuate.

Cinder Cones

A **cinder cone** is a small volcano composed of pyroclastic fragments. A cinder cone forms when large amounts of gas accumulate in rising magma. When the gas pressure builds up sufficiently, the entire mass erupts

Figure 5–16 Two vents in the crater of Marum volcano, Vanuatu.

Figure 5–17 Mount Skjoldbreidier in Iceland shows the typical low-angle slopes of a shield volcano. (Science Graphics, Inc./Ward's Natural Science Establishment, Inc.)

explosively, hurling cinders, ash, and molten magma into the air. The particles then fall back around the vent to accumulate as a small mountain of volcanic debris. A cinder cone is usually active for only a short time because once the gas escapes, the driving force behind the eruption is gone.

As the name implies, a cinder cone is symmetrical. It also can be steep (about 30°), especially near the vent, where ash and cinders pile up (Fig. 5–18). Most are less

than 300 meters high, although a large one can be up to 700 meters high. A cinder cone erodes easily and quickly because the pyroclastic fragments are not cemented together.

About 350 kilometers west of Mexico City, numerous extinct cinder cones are scattered over a broad plain. Prior to 1943, a hole a meter or two in diameter existed in one part of the plain. The hole had been there for as long as anyone could remember, and people grew corn

Figure 5–18 These cinder cones in southern Bolivia are composed of loose pyroclastic fragments.

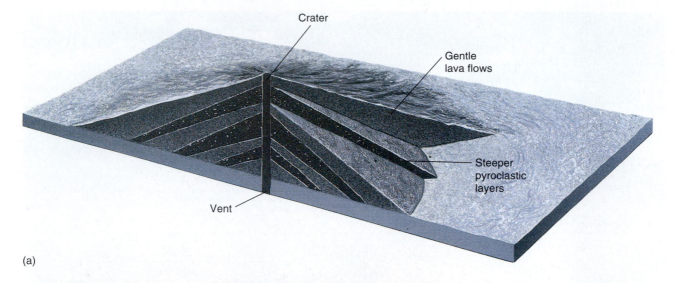

Crater

Gentle
lava flows

Steeper
pyroclastic
layers

Vent

(a)

(b)

Figure 5–19 (a) A schematic
cross section of a composite cone
showing alternating layers of lava
and pyroclastic material. (b) Steam
and ash pouring from Mount
Ngauruhoe, a composite cone in
New Zealand. (Don Hyndman)

just a few meters away. In February 1943, as two farmers were preparing their field for planting, smoke and
sulfurous gases rose from the hole. As night fell, hot,
glowing rocks flew skyward, creating spectacular arcing
flares like a giant fireworks display. By morning, a
40-meter-high cinder cone had grown where the hole
had been. For the next five days, pyroclastic material
erupted 1000 meters into the sky and the cone grew to
100 meters in height. After a few months, a fissure opened
at the base of the cone, extruding lava that buried the
town of San Juan Parangaricutiro. Two years later the
cone had grown to a height of 400 meters. After nine
years, the eruptions ended, and today the volcano, called
El Parícutin, is dormant.

Composite Cones

Composite cones, sometimes called **stratovolcanoes**,
form over a long time by repeated lava flows and pyroclastic eruptions. The hard lava covers the loose pyroclastic material and protects it from erosion (Fig. 5–19).

Many of the highest mountains of the Andes and
some of the most spectacular mountains of western North
America are composite cones. Repeated eruption is a
trademark of a composite volcano. Mount St. Helens
erupted dozens of times in the 4500 years preceding its
most recent eruption in 1980. Mount Rainier, also in
Washington, has been dormant in recent times but could
become active again at any moment.

Figure 5–20   Shiprock, New Mexico, is a volcanic neck. The great rock was once the core of a volcano. The softer flanks of the cone have now eroded away. A dike several kilometers long extends to the left.   (Dougal McCarty)

Volcanic Necks and Pipes

After a volcano's final eruption, magma remaining in the vent may cool and solidify. This **volcanic neck** is commonly harder than surrounding rock. Given enough time, the slopes of the volcano may erode, leaving only the tower-like neck exposed (Fig. 5–20).

In some locations, cylindrical dikes called **pipes** extend from the asthenosphere to the Earth's surface. They are conduits that once carried magma on its way to erupt from a volcano, but they are now filled with solidified magma.

An unusual type of pipe contains a rock called **kimberlite**, which is the only known source of diamonds. Under asthenosphere pressure, small amounts of carbon found in the mantle crystallize as diamond, but at shallower levels of the Earth carbon forms graphite. If the kimberlite's journey to the surface were slow, the pipes would contain graphite rather than diamonds. Thus, the presence of diamonds suggests that the kimberlite magma shot upward through the lithosphere at very high, perhaps even supersonic, speed. It is thought that high mantle pressure drove the kimberlite magma through the lithosphere at such high speed. Most known pipes formed between 70 and 140 million years ago, and most occur in continental crust older than 2.5 billion years. The most famous diamond-rich kimberlite pipes are located in South Africa, although others are known in Canada's Northwest Territories, Arkansas, and Russia.

▶ 5.4   VIOLENT MAGMA: ASH-FLOW TUFFS AND CALDERAS

Although granitic magma usually solidifies within the crust, under certain conditions it rises to the Earth's surface, where it erupts violently. The granitic magmas that rise to the surface probably contain only a few percent water, like basaltic magma. They reach the surface because, like basaltic magma, they have little water to lose. "Dry" granitic magma ascends more slowly than basaltic magma because of its higher viscosity. As it rises, decreasing pressure allows the small amount of dissolved water to separate and form steam bubbles in the magma. The bubbles create a frothy mixture of gas and liquid magma that may be as hot as 900°C (Fig. 5–21a). As the mixture rises to within a few kilometers of the Earth's surface, it fractures overlying rocks and explodes skyward through the fractures (Fig. 5–21b).

As an analogy, think of a bottle of beer or soda pop. When the cap is on and the contents are under pressure, carbon dioxide gas is dissolved in the liquid. When you remove the cap, pressure decreases and bubbles rise to the surface. If conditions are favorable, the frothy mixture erupts through the bottleneck.

A large and violent eruption might blast a column of pyroclastic material 10 or 12 kilometers into the sky, and the column might be several kilometers in diameter. A cloud of fine ash may rise even higher—into the upper atmosphere. The force of material streaming out of the magma chamber can hold the column up for hours or even days. Several recent eruptions of Mount Pinatubo blasted ash columns high into the atmosphere and held them up for hours.

After an eruption, upper layers of the remaining magma are depleted in gas and the explosive potential is low. However, deeper parts of the magma continue to release gas, which rises and builds pressure again to begin another cycle of eruption. Time intervals between successive eruptions vary from a few thousand to about half a million years.

In some cases, the gas-charged magma does not explode, but simply oozes from the fractures. It then flows over the land like root beer foam overflowing the edge of a mug. Some of the frothy magma may solidify to form **pumice**, a rock so full of gas bubbles that it floats on water.

ASH FLOWS

When most of the gas has escaped from the upper layers of magma, the eruption ends. The column of ash, rock, and gas that had been sustained by the force of the eruption then falls back to the Earth's surface (Fig. 5–21c). The falling material spreads over the land and flows down stream valleys. Such a flow is called an **ash flow**, or **nuée ardente**, a French term for "glowing cloud." Small ash flows move at speeds up to 200 kilometers per hour. Large flows have traveled distances exceeding 100 kilometers. The 2000-year-old Taupo flow on New Zealand's South Island leaped over a 700-meter-high ridge as it crossed from one valley into another.

When an ash flow stops, most of the gas escapes into the atmosphere, leaving behind a chaotic mixture of volcanic ash and rock fragments called **ash-flow tuff** (Fig. 5–22). **Tuff** includes all pyroclastic rocks—that is, rocks composed of volcanic ash or other material formed in a volcanic explosion. Some ash flows are hot enough to melt partially after they stop moving. This mixture then cools and solidifies to form a tough, hard rock called **welded tuff**, which often shows structures formed by plastic flow of the melted ash (Fig. 5–23).

The largest known ash-flow tuff from a single eruption is located in the San Juan Mountains of southwestern Colorado and has a volume greater than 3000 cubic

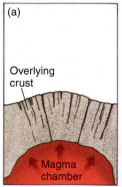

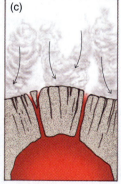
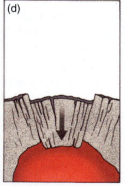

Figure 5–21 (a) When granitic magma rises to within a few kilometers of the Earth's surface, it stretches and fractures overlying rock. Gas separates from the magma and rises to the upper part of the magma body. (b) The gas-rich magma explodes through fractures, rising as a vertical column of hot ash, rock fragments, and gas. (c) When the gas is used up, the column collapses and spreads outward as a high-speed ash flow. (d) Because so much material has erupted from the top of the magma chamber, the roof collapses to form a caldera.

Figure 5–22 Ash-flow tuff forms when an ash flow comes to a stop. The fragments in the tuff are pieces of rock that were carried along with the volcanic ash and gas. (Geoffrey Sutton)

ash-flow tuffs that erupted from them. The oldest eruption took place 1.9 million years ago and ejected 2500 cubic kilometers of pyroclastic material. The next major eruption occurred 1.3 million years ago. The most recent, 0.6 million years ago, ejected 1000 cubic kilometers of ash and other debris and produced the Yellowstone caldera in the center of the park.

Intervals of 0.6 to 0.7 million years separate the three Yellowstone eruptions; 0.6 million years have passed since the most recent one. The park's geysers and hot springs are heated by hot magma beneath Yellowstone, and numerous small earthquakes indicate that the magma is moving. Geologists would not be surprised if another eruption occurred at any time.

A geologic environment similar to that of Yellowstone is found near Yosemite National Park in eastern California. Here the 170-cubic-kilometer Bishop Tuff erupted from the Long Valley caldera 0.7 million years ago. Although only one major eruption has occurred, seismic monitoring indicates that magma lies beneath Mammoth Mountain, a popular California ski area, on the southwest edge of the Long Valley caldera (Fig. 5–25). Unusual amounts of carbon dioxide—a common

kilometers. Another of comparable size lies in southern Nevada.

CALDERAS

After the gas-charged magma erupts, nothing remains to hold up the overlying rock, and the roof of the magma chamber collapses (Fig. 5–21d). Because most magma bodies are circular when viewed from above, the collapsing roof forms a circular depression called a **caldera**. A large caldera may be 40 kilometers in diameter and have walls as much as a kilometer high. Some calderas fill up with volcanic debris as the ash column collapses; others maintain the circular depression and steep walls. We usually think of volcanic landforms as mountain peaks, but the topographic depression of a caldera is an exception.

Figure 5–24 shows that calderas, ash-flow tuffs, and related rocks occur over a large part of western North America. Consider two well-known examples.

Yellowstone National Park and the Long Valley Caldera

Yellowstone National Park in Wyoming and Montana is the oldest national park in the United States. Its geology consists of three large overlapping calderas and the

Figure 5–23 This welded tuff formed when an ash flow became hot enough to melt and flow as a plastic mass. The streaky texture formed when rock fragments similar to those in Figure 5–22 melted and smeared out as the rock flowed. (Geoffrey Sutton)

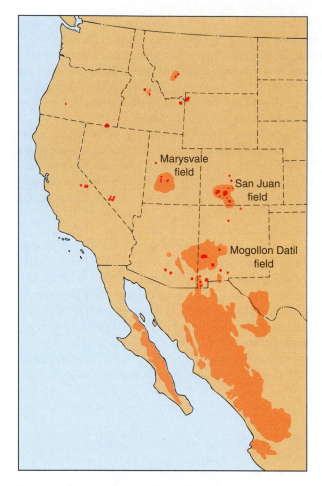

Figure 5--24   Calderas (red dots) and ash-flow tuffs (orange areas) in western North America.

volcanic gas—have escaped from the vicinity of the caldera since 1994. As in the case of Yellowstone, another eruption would not surprise most geologists.

Figure 5–25   California's popular Mammoth Mountain Ski Area lies on the edge of the Long Valley Caldera.

► 5.5   RISK ASSESSMENT: PREDICTING VOLCANIC ERUPTIONS

Approximately 1300 active volcanoes are recognized globally, and 5564 eruptions have occurred in the past 10,000 years. Many volcanoes have erupted recently, and we are certain that others will erupt soon. How can geologists predict an eruption and reduce the risk of a volcanic disaster?

REGIONAL PREDICTION

Volcanoes concentrate near subduction zones, spreading centers, and hot spots and are rare in other places. Thus, the first step in assessing the volcanic hazard of an area is to understand its tectonic environment. Western Washington and Oregon are near a subduction zone and in a region likely to experience future volcanic activity. Kansas and Nebraska are not.

Furthermore, the potential violence of a volcanic eruption is related to the environment of the volcano. If an active volcano lies on continental crust, the eruptions may be violent because granitic magma may form. In contrast, if the region lies on oceanic crust, the eruptions may be gentle because basaltic volcanism is more likely. Violent eruptions are likely in Western Washington and Oregon, but less so on Hawaii or Iceland.

Risk assessment is based both on frequency of past eruptions and on potential violence. However, regional prediction just outlines probabilities and cannot be used to determine when a particular volcano will erupt or the intensity of a particular eruption.

SHORT-TERM PREDICTION

In contrast to regional predictions, short-term predictions attempt to forecast the specific time and place of an impending eruption. They are based on instruments that monitor an active volcano to detect signals that the volcano is about to erupt. The signals include changes in the shape of the mountain and surrounding land, earthquake swarms indicating movement of magma beneath the mountain, increased emissions of ash or gas, increasing temperatures of nearby hot springs, and any other signs that magma is approaching the surface.

In 1978, two United States Geological Survey (USGS) geologists, Dwight Crandall and Don Mullineaux, noted that Mount St. Helens had erupted more frequently and violently during the past 4500 years than any other volcano in the contiguous 48 states. They predicted that the volcano would erupt again before the end of the century.

In March 1980, about two months before the great May eruption, puffs of steam and volcanic ash rose from

Figure 5–26 Eruption of Mount St. Helens, May 18, 1980. (USGS)

the crater of Mount St. Helens, and swarms of earthquakes occurred beneath the mountain. This activity convinced USGS geologists that Crandall and Mullineaux's prediction was correct. In response, they installed networks of seismographs, tiltmeters, and surveying instruments on and around the mountain.

In the spring of 1980, the geologists warned government agencies and the public that Mount St. Helens showed signs of an impending eruption. The U.S. Forest Service and local law enforcement officers quickly evacuated the area surrounding the mountain and averted a much larger tragedy that might have occurred when the mountain exploded (Fig. 5–26). Using similar kinds of information, geologists predicted the 1991 Mount Pinatubo eruption in the Philippines, saving many lives.

SUMMARY

Basaltic magma usually erupts in a relatively gentle manner onto the Earth's surface from a **volcano**. In contrast, granitic magma typically solidifies within the Earth's crust. When granitic magma does erupt onto the surface, it often does so violently. These contrasts in behavior of the two types of magma are caused by differences in silica and water content.

Any intrusive mass of igneous rock is a **pluton**. A **batholith** is a pluton with more than 100 square kilometers of exposure at the Earth's surface. A **dike** and a **sill** are both sheetlike plutons. Dikes cut across layering in country rock, and sills run parallel to layering.

Magma may flow onto the Earth's surface as **lava** or may erupt explosively as **pyroclastic** material. Fluid lava forms **lava plateaus** and **shield volcanoes**. A pyroclas-

tic eruption may form a **cinder cone**. Alternating eruptions of fluid lava and pyroclastic material from the same vent create a **composite cone**. When granitic magma rises to the Earth's surface, it may erupt explosively, forming **ash-flow tuffs** and **calderas**.

Volcanic eruptions are common near a subduction zone, near a spreading center, and at a hot spot over a mantle plume but are rare in other tectonic environments. Eruptions on a continent are often violent, whereas those in oceanic crust are gentle. Such observations form the basis of **regional predictions** of volcanic hazards. **Short-term predictions** are made on the basis of earthquakes caused by magma movements, swelling of a volcano, increased emissions of gas and ash from a vent, and other signs that magma is approaching the surface.

KEY WORDS

viscosity 75	columnar joint 79	vent 81	volcanic neck 85
pluton 75	pillow lava 79	crater 81	pipe 85
batholith 75	pyroclastic rock 80	active volcano 81	kimberlite 85
stock 77	volcanic ash 80	dormant volcano 81	pumice 86
dike 77	cinder 80	extinct volcano 81	ash flow 86
sill 78	volcanic bomb 80	shield volcano 81	nuée ardente 86
lava 79	fissure 81	cinder cone 82	tuff 86
pahoehoe 79	flood basalt 81	composite cone 84	welded tuff 86
aa 79	lava plateau 81	stratovolcano 84	caldera 87
vesicle 79	volcano 81		

REVIEW QUESTIONS

1. Describe several different ways in which volcanoes and volcanic eruptions can threaten human life and destroy property.

2. What has been the death toll from volcanic activity during the past 2000 years? During the past 100 years?

3. How much silica does average granitic magma contain? How much does basaltic magma contain?

4. Why does magma rise soon after it forms?

5. What happens to most basaltic magma after it forms?

6. What happens to most granitic magma after it forms?

7. Explain why basaltic magma and granitic magma behave differently as they rise toward the Earth's surface.

8. Many rocks, and even entire mountain ranges, at the Earth's surface are composed of granite. Does this observation imply that granite forms at the surface?

9. Do batholiths and stocks differ chemically or physically, or both chemically and physically?

10. Explain the difference between a dike and a sill.

11. How do columnar joints form in a basalt flow?

12. How do a shield volcano, a cinder cone, and a composite cone differ from one another? How are they similar?

13. Which type of volcanic mountain has the shortest life span? Why is this structure a transient feature of the landscape?

14. How does a composite cone form?

15. What is a volcanic neck? How is it formed?

16. Explain why and how granitic magma forms ash-flow tuffs and calderas.

17. What is pumice, and how does it form?

18. How does welded tuff form?

19. How does a caldera form?

20. How much pyroclastic material can erupt from a large caldera?

21. Explain why additional eruptions in Yellowstone Park seem likely. Describe what such an eruption might be like.

DISCUSSION QUESTIONS

1. How and why does pressure affect the melting point of rock and, conversely, the solidification temperature of magma? How does the explanation differ for basaltic and granitic magma?

2. Why does water play an important role in magma generation in subduction zones, but not in the other two major environments of magma generation?

3. How could you distinguish between a sill exposed by erosion and a lava flow?

4. Imagine that you detect a volcanic eruption on a distant planet but have no other data. What conclusions could you draw from this single bit of information? What types of information would you search for to expand your knowledge of the geology of the planet?

5. Explain why some volcanoes have steep, precipitous faces, but many do not.

6. Parts of the San Juan Mountains of Colorado are composed of granite plutons, and other parts are volcanic rock. Explain why these two types of rock are likely to occur in proximity.

7. Compare and contrast the danger of living 5 kilometers from Yellowstone National Park with the danger of living an equal distance from Mount St. Helens. Would your answer differ for people who live 50 kilometers or those who live 500 kilometers from the two regions?

8. Use long-term prediction methods to evaluate the volcanic hazards in the vicinity of your college or university.

9. Discuss some possible consequences of a large caldera eruption in modern times. What is the probability that such an event will occur?

Weathering and Soil

Early in Earth history, between 4.5 and 3.5 billion years ago, swarms of meteorites crashed into all of the planets and their moons. Today, the craters created by these impacts are abundant on the Moon but are completely gone from the Earth's surface. Why has the Moon retained its craters, and why have the craters vanished from the Earth?

Tectonic activity such as mountain building and volcanic eruptions has continually renewed the Earth's surface over geologic time. In addition, Earth has an atmosphere and water, which decompose and erode bedrock. The combination of tectonic activity, weathering, and erosion has eliminated all traces of early meteorite impacts from the Earth's surface. In contrast, the smaller Moon has lost most of its heat, so tectonic activity is nonexistent. In addition, the Moon has no atmosphere or water to weather and erode its surface. As a result, the lunar surface is covered with meteorite craters, many of which are billions of years old.

Delicate Arch, in Utah, formed as sandstone weathered and eroded.

Figure 6–1   This boulder weathered in place.

wind or water slows down and loses energy or, in the case of glaciers, when the ice melts, transport stops and sediment is deposited. These four processes—weathering, erosion, transportation, and deposition—work together to modify the Earth's surface (Fig. 6–2).

MECHANICAL AND CHEMICAL WEATHERING

The environment at the Earth's surface is corrosive to most materials. An iron tool left outside will rust. Even stone is vulnerable to corrosion. As a result, ancient stone cities have fallen to ruin. Over longer periods of time, rock outcrops and entire mountain ranges wear away. Weathering occurs by both mechanical and chemical processes. **Mechanical weathering** reduces solid rock to rubble but does not alter the chemical composition of rocks and minerals. In contrast, **chemical weathering** occurs when air and water chemically react with rock to alter its composition and mineral content. These chemical changes are analogous to rusting in that the final products differ both physically and chemically from the starting material.

► 6.1   WEATHERING

Weathering is the decomposition and disintegration of rocks and minerals at the Earth's surface. Weathering itself involves little or no movement of the decomposed rocks and minerals. This material accumulates where it forms and overlies unweathered bedrock (Fig. 6–1).

   Erosion is the removal of weathered rocks and minerals by moving water, wind, glaciers, and gravity. After a rock fragment has been eroded from its place of origin, it may be transported large distances by those same agents: flowing water, wind, ice, and gravity. When the

► 6.2   MECHANICAL WEATHERING

Mechanical weathering breaks large rocks into smaller ones but does not alter the rock's chemical nature or its minerals. Think of grinding a rock in a crusher; the fragments are no different from the parent rock, except that they are smaller.

   Five major processes cause mechanical weathering: pressure-release fracturing, frost wedging, abrasion, organic activity, and thermal expansion and contraction. Two additional processes—salt cracking and hydrolysis expansion—result from combinations of mechanical and chemical processes.

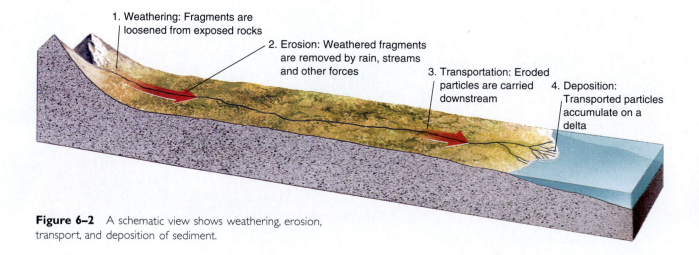

Figure 6–2   A schematic view shows weathering, erosion, transport, and deposition of sediment.

PRESSURE-RELEASE FRACTURING

Many igneous and metamorphic rocks form deep below the Earth's surface. Imagine, for example, that a granitic pluton solidifies from magma at a depth of 15 kilometers. At that depth, the pressure from the weight of overlying rock is about 5000 times that at the Earth's surface. Over millennia, tectonic forces may raise the pluton to form a mountain range. The overlying rock erodes away as the pluton rises and the pressure on the buried rock decreases. As the pressure diminishes, the rock expands, but because the rock is now cool and brittle, it fractures as it expands. This process is called **pressure-release fracturing**. Many igneous and metamorphic rocks that formed at depth, but now lie at the Earth's surface, have been fractured in this manner (Fig. 6–3).

FROST WEDGING

Water expands when it freezes. If water accumulates in a crack and then freezes, its expansion pushes the rock apart in a process called **frost wedging**. In a temperate climate, water may freeze at night and thaw during the day. Ice cements the rock together temporarily, but when it melts, the rock fragments may tumble from a steep cliff. If you hike or climb in mountains when the daily freeze–thaw cycle occurs, be careful; rockfall due to frost wedging is common. Experienced climbers travel in the early morning when the water is still frozen and ice holds the rock together.

Large piles of loose angular rocks, called **talus slopes**, lie beneath many cliffs (Fig. 6–4). These rocks fell from the cliffs mainly as a result of frost wedging.

ABRASION

Many rocks along a stream or beach are rounded and smooth. They have been shaped by collisions with other rocks as they tumbled downstream and with silt and sand carried by moving water. As particles collide, their sharp edges and corners wear away. The mechanical wearing and grinding of rock surfaces by friction and impact is called **abrasion** (Fig. 6–5). Note that pure water itself is not abrasive; the collisions among rock, sand, and silt cause the weathering.

Wind also hurls sand and other small particles against rocks, often sandblasting unusual and beautiful landforms (Fig. 6–6). Glaciers (discussed in Chapter 17) also cause much abrasion as they drag particles ranging in size from clay to boulders across bedrock. In this case, both the rock fragments embedded in the ice and the bedrock beneath are abraded.

ORGANIC ACTIVITY

If soil collects in a crack in solid rock, a seed may fall there and sprout. The roots work their way down into the crack, expand, and may eventually push the rock apart (Fig. 6–7). City dwellers often see the results of organic activity in sidewalks, where tree roots push from underneath, raising the concrete and frequently cracking it.

THERMAL EXPANSION AND CONTRACTION

Rocks at the Earth's surface are exposed to daily and yearly cycles of heating and cooling. They expand when they are heated and contract when they cool. When tem-

Figure 6–3 Pressure-release fracturing contributed to the formation of these cracks in a granite cliff in Tuolumne Meadows, California.

perature changes rapidly, the surface of a rock heats or cools faster than its interior and, as a result, the surface expands or contracts faster than the interior. The resulting forces may fracture the rock.

In mountains or deserts at mid-latitudes, temperature may fluctuate from −5°C to +25°C during a spring day. Is this 30° difference sufficient to fracture rocks?

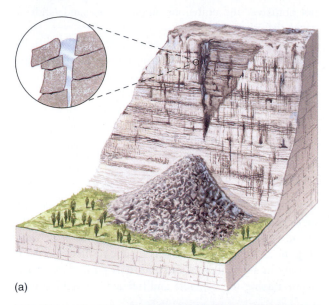

(a)

(b)

Figure 6–4   (a) Frost wedging dislodges rocks from cliffs and creates talus slopes. (b) Frost wedging has produced this talus cone in Valley of the Ten Peaks, Canadian Rockies.

Figure 6–5   Abrasion rounded these rocks in a streambed in Yellowstone National Park, Wyoming.

The answer is uncertain. In one laboratory experiment, scientists heated and cooled granite repeatedly by more than 100°C and they did not observe any fracturing. These results imply that normal temperature changes might not be an important cause of mechanical weathering. However, the rocks used in the experiment were small and the experiment was carried out over a brief period of time. Perhaps thermal expansion and contraction are more significant in large outcrops. Or perhaps daily heating–cooling cycles repeated over hundreds of thousands of years may promote fracturing.

In contrast to a small atmospheric temperature fluctuation, fire heats rock by hundreds of degrees. If you line a campfire with granite stones, the rocks commonly break as you cook your dinner. In a similar manner,

Figure 6–6   Wind abrasion selectively eroded the base of this rock in Lago Poopo, Bolivia, because windblown sand moves mostly near the ground surface.

Figure 6–7 As this tree grew from a crack in bedrock, its roots forced the crack to widen.

forest fires or brush fires occur commonly in many ecosystems and are an important agent of mechanical weathering.

▶ 6.3   CHEMICAL WEATHERING

Rock is durable over a single human lifetime. Return to your childhood haunts and you will see that the rock outcrops in woodlands or parks have not changed. Over longer expanses of geologic time, however, rocks decompose chemically at the Earth's surface.

The most important processes of chemical weathering are dissolution, hydrolysis, and oxidation. Water, carbon dioxide, acids and bases, and oxygen are common substances that cause these processes to decompose rocks.

DISSOLUTION

If you put a crystal of halite (rock salt) in water, it dissolves and the ions disperse to form a solution. Halite dissolves so rapidly and completely that this mineral is rare in moist environments.

A small proportion of water molecules spontaneously dissociate (break apart) to form an equal number of hydrogen ions (H^+) and hydroxyl ions (OH^-).[1] Many common chemicals dissociate in water to increase either the hydrogen or the hydroxyl ion concentration. For example, HCl (hydrochloric acid) dissociates to release H^+ and Cl^- ions. The H^+ ions increase the hydrogen ion concentration and the solution becomes acid. In a similar manner, NaOH dissociates to increase the hydroxyl ion concentration and the solution becomes a base. Hydrogen and hydroxyl ions are chemically reactive and therefore acids and bases are much more corrosive than pure water.

To understand how acids and bases dissolve minerals, think of an atom on the surface of a crystal. It is held in place because it is attracted to the other atoms in the

[1]Hydrogen ions react instantaneously and completely with water, H_2O, to form the hydronium ion, H_3O^+, but for the sake of simplicity, we will consider the hydrogen ion, H^+, as an independent entity.

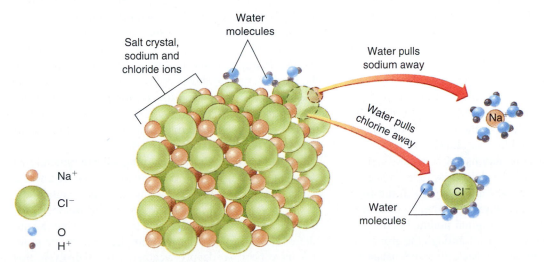

Figure 6–8 Halite dissolves in water because the attractions between the water molecules and the sodium and chloride ions are greater than the strength of the chemical bonds in the crystal.

F O C U S   O N

REPRESENTATIVE REACTIONS IN CHEMICAL WEATHERING

Dissolution of Calcite

Calcite, the mineral that comprises limestone and marble, weathers in natural environments in a three-step process. In the first two steps, water reacts with carbon dioxide in the air to produce carbonic acid, which dissociates to release hydrogen ions:

$$CO_2 + H_2O \longrightarrow H_2CO_3 \longrightarrow H^+ + HCO_3^-$$

Carbon dioxide   Water   Carbonic acid   Hydrogen ion   Bicarbonate ion

In the third step, calcite dissolves in the carbonic acid solution.

$$CaCO_3 + H^+ \longrightarrow Ca^{2+} + HCO_3^-$$

Calcite   Hydrogen ion   Calcium ion   Bicarbonate ion

Hydrolysis

An example of a reaction in which feldspar hydrolyzes to clay is as follows:

$$2\ KAlSi_3O_8 + 2\ H^+ + H_2O \longrightarrow$$

Orthoclase feldspar   Hydrogen ion   Water

$$Al_2Si_2O_5(OH)_4 + 2\ K^+ + 4\ SiO_2$$

Clay mineral   Potassium ion   Silica

---

crystal by electrical forces. At the same time, electrical attractions to the outside environment pull the atom away from the crystal. The result is like a tug-of-war. If the bonds between the atom and the crystal are stronger than the attraction of the atom to its outside environment, then the crystal remains intact. If outside attractions are stronger, they pull the atom away from the crystal and the mineral dissolves (Fig. 6–8). Acids and bases are generally more effective at dissolving minerals than pure water because they provide more electrically charged hydrogen and hydroxyl ions to pull atoms out of crystals. For example, limestone is made of the mineral calcite ($CaCO_3$). Calcite barely dissolves in pure water but is quite soluble in acid. If you place a drop of strong acid on limestone, bubbles of carbon dioxide gas rise from the surface as the calcite dissolves.

Water found in nature is never pure. Atmospheric carbon dioxide dissolves in raindrops and reacts to form a weak acid called carbonic acid. As a result, even the purest rainwater, which falls in the Arctic or on remote mountains, is slightly acidic. As shown in the "Focus On" box "Representative Reactions in Chemical Weathering," this acidic rainwater dissolves limestone. Industrial pollution can make rain even more acidic. Limestone outcrops commonly show signs of intense chemical weathering as a result of natural and polluted rain.

In addition, when water flows through the ground, it dissolves ions from soil and bedrock. In some instances, these ions render the water acidic; in other cases the water becomes basic. Flowing water carries the dissolved ions away from the site of weathering. Weathering by so-

lution produces spectacular caverns in limestone (Fig. 6–9). This topic is discussed further in Chapter 15.

Most solution reactions are reversible. A reversible reaction can proceed in either direction if conditions change. For example, calcite dissolves readily in acid to form a solution. If a base is added to the solution, solid calcite will precipitate again.

HYDROLYSIS

During dissolution, a mineral dissolves but does not otherwise react chemically with the solution. However, during **hydrolysis**, water reacts with a mineral to form a new mineral with the water incorporated into its crystal structure. Many common minerals weather by hydrolysis. For example, feldspar, the most abundant mineral in the Earth's crust, weathers by hydrolysis to form clay. As feldspar converts to clay, flowing water carries off soluble cations such as potassium. The water combines with the less soluble ions to form clay minerals (see the "Focus On" box "Representative Reactions").

Quartz is the only rock-forming silicate mineral that does not weather to form clay. Quartz resists weathering because it is pure silica, SiO_2, and does not contain any of the more soluble cations. When granite weathers, the feldspar and other minerals react to form clay but the unaltered quartz grains fall free from the rock. Some granites have been so deeply weathered by hydrolysis that mineral grains can be pried out with a fingernail to depths of several meters (Fig. 6–10). The rock looks like granite but has the consistency of sand.

Figure 6–9 Stalactites and stalagmites in a limestone cavern. (Hubbard Scientific Co.)

Because quartz is so tough and resistant to weathering, it is the primary component of sand. Much of it is transported to the sea coast, where it concentrates on beaches and eventually forms sandstone.

OXIDATION

Many elements react with atmospheric oxygen, O_2. Iron rusts when it reacts with water and oxygen. Rusting is one example of a more general process called **oxidation**.[2] Oxidation reactions are so common in nature that pure metals are rare in the Earth's crust, and most metallic elements exist in nature as compounds. Only a few metals, such as gold, silver, copper, and platinum commonly occur in their pure states.

Recall from Chapter 3 that iron is abundant in many minerals, including olivine, pyroxene, and amphibole. If the iron in such a mineral oxidizes, the mineral decomposes. Many metallic elements, such as iron, copper, lead, and zinc, occur as sulfide minerals in ore deposits. When metallic sulfides oxidize, the sulfur reacts to form sulfuric acid, a strong acid. For example, pyrite (FeS_2) oxidizes to form sulfuric acid and iron oxide. The sulfuric acid washes into streams and ground water, where it may harm aquatic organisms. Thus, many natural ore deposits generate sulfuric acid when they weather. The

[2]Oxidation is properly defined as the loss of electrons from a compound or element during a chemical reaction. In the weathering of common minerals, this usually occurs when the mineral reacts with molecular oxygen.

same reaction may be accelerated when ore is dug up and exposed at a mine site. This problem is discussed further in Chapter 19.

▶ 6.4 CHEMICAL AND MECHANICAL WEATHERING OPERATING TOGETHER

Chemical and mechanical weathering work together, often on the same rock at the same time. Chemical processes generally act only on the surface of a solid object, so the reaction speeds up if the surface area increases. Think of a burning log; the fire starts on the outside and works its way toward the interior. A split log burns faster because the surface area is greater. Mechanical processes crack rocks, thereby exposing more surface area for chemical agents to work on (Fig. 6–11).

After mechanical processes fracture a rock, water and air seep into the fractures and initiate chemical weathering. Figure 6–12a shows that chemical weathering attacks a rock face from only one direction but attacks an edge from two sides and a corner from three sides. As a result of the multidirectional attack, the corners and edges weather most rapidly; the faces, attacked from only one direction, weather more slowly. Over time, the corners and edges become rounded in a process called **spheroidal weathering** (Fig. 6–12b). It is common to see rounded boulders still lying where they formed by this process.

SALT CRACKING

In environments where ground water is salty, salt water seeps into cracks in bedrock. When the water evaporates,

Figure 6–10 Coarse grains of quartz and feldspar accumulate directly over weathered granite. The lens cap in the middle illustrates scale.

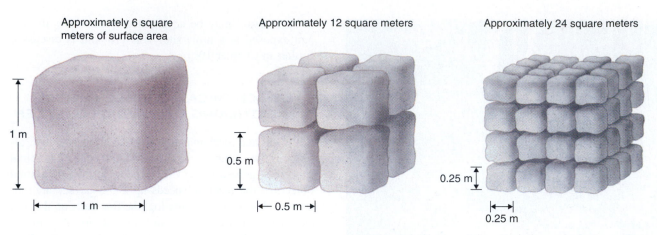

Approximately 6 square meters of surface area

1 m

1 m

Approximately 12 square meters

0.5 m

0.5 m

Approximately 24 square meters

0.25 m

0.25 m

Figure 6–11 When rocks are broken apart by mechanical weathering, more surface is available for chemical weathering.

the dissolved salts crystallize. The growing crystals exert tremendous forces, enough to widen a crack and fracture a rock, a process called **salt cracking**. Thus, a mechanical process such as thermal expansion or pressure release may initially fracture bedrock. Then salt water migrates into the crack, and salt precipitates (a chemical process). Finally, the expanding salt crystals mechanically push the rock apart.

Many sea cliffs show pits and depressions caused by salt cracking because spray from the breaking waves brings the salt to the rock. Salt cracking is also common in deserts, where surface and underground water often contain dissolved salts (Fig. 6–13).

EXFOLIATION

Granite commonly fractures by **exfoliation**, a process in which large plates or shells split away like the layers of an onion (Fig. 6–14). The plates may be only 10 or 20 centimeters thick near the surface, but they thicken with depth. Because exfoliation fractures are usually absent below a depth of 50 to 100 meters, they seem to be a result of exposure of the granite at the Earth's surface.

Exfoliation is frequently explained as a form of pressure-release fracturing. However, many geologists suggest that hydrolysis may contribute to exfoliation. During hydrolysis, feldspars and other silicate minerals react to form clay. As a result of the addition of water, clays have

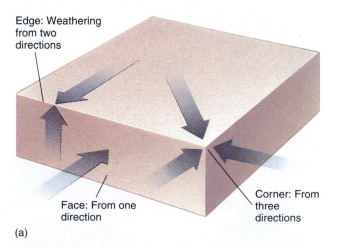

Edge: Weathering from two directions

Face: From one direction

Corner: From three directions

(a)

Figure 6–12 (a) More surface area is available for chemical attack on the corners and edges of a cube than on a face. Therefore, corners and edges are rounded during weathering. (b) Both mechanical and chemical processes have weathered this boulder, along old fractures.

(b)

Figure 6–13 Salt cracking formed this depression in sandstone in Cedar Mesa, Utah. The white patches are salt crystals.

a greater volume than that of the original minerals. Thus, a chemical reaction (hydrolysis) forms clay, and the mechanical expansion of the clay contributes to the exfoliation fractures.

▶ 6.5 SOIL

Mechanical weathering produces both large rock fragments and small particles such as sand and silt. Chemical weathering forms clay and dissolved ions. Some of these weathering products accumulate on the Earth over bedrock. This material is called **regolith**. Soil scientists define **soil** as upper layers of regolith that support plant growth.

Soil commonly consists of sand, silt, clay, and organic material. Clay particles are so small and pack so tightly that water does not flow through them readily. In fact, even gases have trouble passing through clay-rich soils, so plants growing in clay soils suffer from lack of oxygen. In contrast, water and oxygen travel easily through loosely packed sandy soils. The most fertile soils contain a mixture of sand, clay, and silt as well as generous amounts of organic matter. Such a mixture is called **loam**.

If you walk through a forest or prairie, you can find bits of leaves, stems, and flowers on the soil surface. This material is called **litter** (Fig. 6–15). When litter decomposes sufficiently that you can no longer determine the origin of individual pieces, it becomes **humus**.

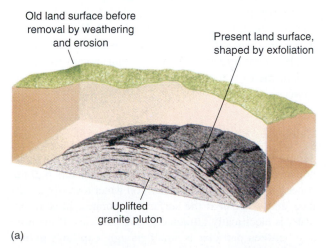

(a)

(b)

Figure 6–14 (a) Formation of an exfoliation dome. The exfoliation slabs are only a few centimeters to a few meters thick. (b) Exfoliation has fractured this granite in Pinkham Notch, New Hampshire.

Humus is an essential component of most fertile soils. If you pour a small amount of water into soil rich in humus, the soil absorbs most of the water. Humus retains so much moisture that humus-rich soil swells after a rain and shrinks during dry spells. This alternate shrinking and swelling loosens the soil, allowing roots to grow into it easily. A rich layer of humus also insulates the soil from excessive heat and cold and reduces water loss by evaporation. Humus also retains nutrients in soil and makes them available to plants.

In intensive agriculture, farmers commonly plow the soil and leave it exposed for weeks or months. Humus oxidizes in air and degrades. Running water frequently dissolves soil nutrients and carries them away. Farmers replace the lost nutrients with chemical fertilizers but frequently do not replenish humus. As a result, much of the natural ability of soil to conserve and regulate water and nutrients is lost. When rainwater flows over the surface, it carries soil particles, excess fertilizer, and pesticide residues, polluting streams and ground water.

SOIL PROFILES

A typical mature soil consists of several layers called **soil horizons**. The uppermost layer is called the **O horizon**, named for its **O**rganic component. This layer consists mostly of litter and humus with a small proportion of minerals (Fig. 6–16). The next layer down, called the **A horizon**, is a mixture of humus, sand, silt, and clay. The combined O and A horizons are called **topsoil**. A kilogram of average fertile topsoil contains about 30 percent by weight organic matter, including approximately 2 trillion bacteria, 400 million fungi, 50 million algae, 30 million protozoa, and thousands of larger organisms such as insects, worms, nematodes, and mites.

Figure 6–15 Litter is organic matter that has fallen to the ground and started to decompose but still retains its original form and shape.

The third layer, the **B horizon** or subsoil, is a transitional zone between topsoil and weathered parent rock below. Roots and other organic material occur in the B horizon, but the total amount of organic matter is low. The lowest layer, called the **C horizon**, consists of partially weathered rock that grades into unweathered parent rock. This zone contains little organic matter.

When rainwater falls on soil, it sinks into the O and A horizons, weathering minerals and carrying dissolved ions to lower levels. This downward movement of dissolved ions is called **leaching**. The A horizon is sandy because water also carries clay downward but leaves the sand behind. Because materials are removed from the A horizon, it is called the **zone of leaching**.

Dissolved ions and clay carried downward from the A horizon accumulate in the B horizon, which is called the **zone of accumulation**. This layer retains moisture because of its high clay content. Although moisture retention may be beneficial, if too much clay accumulates, the B horizon creates a dense, waterlogged soil.

▶ 6.6 SOIL-FORMING FACTORS

Why are some soils rich and others poor, some sandy and others loamy? Six factors control soil characteristics: parent rock, climate, rates of plant growth and decay, slope angle and aspect, time, and transport.

PARENT ROCK

The texture and composition of a soil depends partly on its parent rock. For example, when granite decomposes, the feldspar converts to clay and the rock releases quartz as sand grains. If the clay leaches into the B horizon, a sandy soil forms. In contrast, because basalt contains no quartz, soil formed from basalt is likely to be rich in clay and contain only small amounts of sand. Nutrient abundance also depends in part on the parent rock. For example, a pure quartz sandstone contains no nutrients, and soil formed on it must get its nutrients from outside sources.

CLIMATE

Temperature and rainfall affect soil formation. Rain seeps downward through soil, but several other factors pull the water back upward. Roots suck soil water toward the surface, and water near the surface evaporates. In addition, water is electrically attracted to soil particles. If the pore size between particles is small enough, **capillary action** draws water upward.

During a rainstorm, water seeps through the A horizon, dissolving soluble ions such as calcium, magnesium, potassium, and sodium. In arid and semiarid

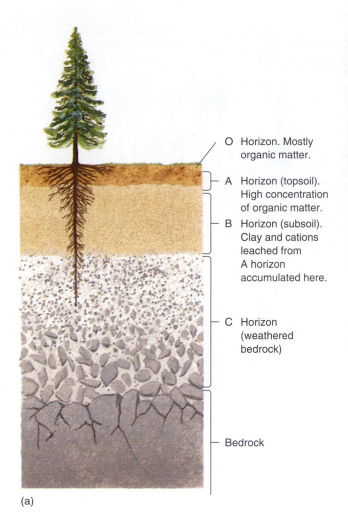

O Horizon. Mostly organic matter.

A Horizon (topsoil). High concentration of organic matter.

B Horizon (subsoil). Clay and cations leached from A horizon accumulated here.

C Horizon (weathered bedrock)

Bedrock

(a)

(b)

Figure 6–16   (a) Schematic soil profile showing typical soil horizons. (b) Soil horizons are often easily distinguished by color and texture. The dark upper layer is the A horizon; the whiter lower layer is the B horizon.   (Soil Conservation Service)

regions, when the water reaches the B horizon, capillary action and plant roots then draw it back up toward the surface, where it evaporates or is incorporated into plant tissue. After the water escapes, many of its dissolved ions precipitate in the B horizon, encrusting the soil with salts. A soil of this type is a **pedocal** (Fig. 6–17a). This process often deposits enough calcium carbonate to form a hard cement called **caliche** in the soil. In the Imperial Valley in California, for example, irrigation water contains high concentrations of calcium carbonate. A thick continuous layer of caliche forms in the soil as the water evaporates. To continue growing crops, farmers must then rip this layer apart with heavy machinery.

Because nutrients concentrate when water evaporates, many pedocals are fertile if irrigation water is available. However, salts often concentrate so much that they become toxic to plants (Fig. 6–18). As mentioned previously, all streams contain small concentrations of dissolved salts. If arid or semiarid soils are intensively irrigated, salts can accumulate until plants cannot grow. This process is called **salinization**. Some historians argue that salinization destroyed croplands and thereby

contributed to the decline of many ancient civilizations, such as the Babylonian Empire.

In a wet climate, water seeping down through the soil leaches soluble ions from both the A and B horizons. The less soluble elements, such as aluminum, iron, and some silicon, remain behind, accumulating in the B horizon to form a soil type called a **pedalfer** (Fig. 6–17b). The subsoil in a pedalfer is commonly rich in clay, which is mostly aluminum and silicon, and has the reddish color of iron oxide.

In regions of very high rainfall, such as a tropical rainforest, so much water seeps through the soil that it leaches away nearly all the soluble cations. Only very insoluble aluminum and iron minerals remain (Fig. 6–17c). Soil of this type is called a **laterite**. Laterites are often colored rust-red by iron oxide (Fig. 6–19). A highly aluminous laterite, called **bauxite**, is the world's main type of aluminum ore.

The second important component of climate, average annual temperature, affects soil formation in two ways. First, chemical reactions proceed more rapidly in warm temperatures than in cooler conditions, so chemi-

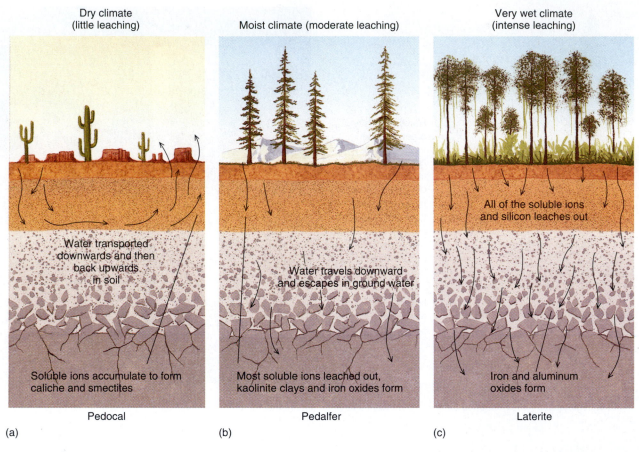

Dry climate (little leaching)

Moist climate (moderate leaching)

Very wet climate (intense leaching)

Water transported downwards and then back upwards in soil

Water travels downward and escapes in ground water

All of the soluble ions and silicon leaches out

Soluble ions accumulate to form caliche and smectites

Most soluble ions leached out, kaolinite clays and iron oxides form

Iron and aluminum oxides form

Pedocal

Pedalfer

Laterite

(a)   (b)   (c)

Figure 6–17 The formation of pedocals, pedalfers, and laterites.

cal weathering is faster in warmer climates than in cold ones. Second, plant growth and decay are temperature dependent as discussed below.

RATES OF GROWTH AND DECAY OF ORGANIC MATERIAL

In the tropics, plants grow and decay rapidly all year long. When leaves and stems decay, the nutrients are quickly absorbed by the growing plants. As a result, little humus accumulates and few nutrients are stored in the soil (Fig. 6–20a). The Arctic, on the other hand, is so cold that plant growth and decay are slow. Therefore, litter and humus form slowly and Arctic soils contain little organic matter (Fig. 6–20b).

The most fertile soils are those of prairies and forests in temperate latitudes. There, large amounts of plant litter drop to the ground in the autumn, but decay is slow during the cold winter months. During the spring and summer, litter decomposes and releases nutrients into the soil. However, in a temperate region, plant growth is not fast enough to remove all the nutrients during the grow-

ing season. As a result, thick layers of humus accumulate and soil contains abundant nutrients.

SLOPE ASPECT AND STEEPNESS

Aspect is the orientation of a slope with respect to the Sun. In the semiarid American West, thick soils and dense forests cover the cool, shady north slopes of hills, but thin soils and grass dominate hot, dry southern exposures. The reason for this difference is that in the Northern Hemisphere more water evaporates from the hot, sunny southern slopes. Therefore, fewer plants grow, weathering occurs slowly, and soil development is retarded. The moister northern slopes weather more deeply to form thicker soils.

In general, hillsides have thin soils and valleys are covered by thicker soil, because soil erodes from hills and accumulates in valleys. When hilly regions were first settled and farmed, people naturally planted their crops in the valley bottoms, where the soil was rich and water was abundant. Recently, as population has expanded, farmers have moved to the thinner, less stable hillside soils.

Figure 6–18   Saline seep on a ranch in Wyoming. Saline water seeps into this depression and then evaporates to deposit white salt crystals on the ground and on the fence posts.

TIME

Most chemical weathering occurs at the relatively low temperatures of the Earth's surface. Consequently, chemical weathering goes on slowly in most places, and time becomes an important factor in determining the extent of weathering.

Recall that feldspars weather to form clay, whereas quartz does not decompose easily. In geologically young soils, the decomposition of feldspars may be incomplete and the soils are likely to be sandy. As soils mature, more feldspars decompose, and the clay content increases.

SOIL TRANSPORT

By studying recent lava flows, scientists have determined how quickly plants return to an area after it has been covered by hard, solid rock. In many cases, plants appear when a lava flow is only a few years old, even before weathering has formed soil. Closer scrutiny shows that the plants have rooted in tiny amounts of soil that were transported from nearby areas by wind or water.

Figure 6–19   Iron oxide colors this Georgia laterite.   (USGS)

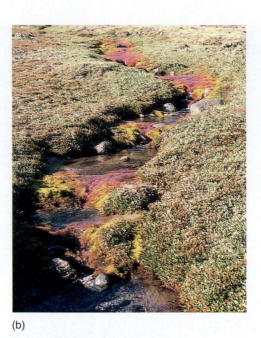

(a)   (b)

Figure 6–20 (a) Tropical soil of Costa Rica supports lush growth, but organic material decays so rapidly that little humus accumulates. (b) Arctic soil of Baffin Island, Canada, supports sparse vegetation and contains little organic matter.

In many of the world's richest agricultural areas, most of the soil was transported from elsewhere. Streams may deposit sediment, wind deposits dust, and soil slides downslope from mountainsides into valleys. These foreign materials mix with residual soil, changing its composition and texture. River deltas and the rich windblown loess soils of China and the North American Great Plains are examples of transported soils.

SUMMARY

Weathering is the decomposition and disintegration of rocks and minerals at the Earth's surface. **Erosion** is the removal of weathered rock or soil by moving water, wind, glaciers, or gravity. After rock or soil has been eroded from the immediate environment, it may be transported large distances and eventually deposited.

Mechanical weathering can occur by **frost wedging**, **abrasion**, **organic activity**, **thermal expansion and contraction**, and **pressure-release fracturing**.

Chemical weathering occurs when chemical reactions decompose minerals. A few minerals dissolve readily in water. Acids and bases often markedly enhance the solubility of minerals. Rainwater is slightly acidic due to reactions between water and atmospheric carbon dioxide. A serious environmental problem is caused by **acid rain**. The **hydrolysis** of feldspar and other common minerals, except quartz, is a form of chemical weathering. **Oxidation** is the reaction with oxygen to decompose minerals.

Chemical and mechanical weathering often operate together. For example, solution seeping into cracks may cause rocks to expand by growth of salts or hydrolysis. Hydrolysis combines with pressure-release fracturing to form **exfoliated** granite.

Soil is the layer of weathered material overlying bedrock. **Sand**, **silt**, **clay**, and **humus** are commonly found in soil. Water **leaches** soluble ions downward through the soil. Clays are also transported downward by water. The uppermost layer of soil, called the **O horizon**, consists mainly of litter and humus. The amount of organic matter decreases downward. The **A horizon** is the **zone of leaching**, and the **B horizon** is the **zone of accumulation**.

Six factors control soil characteristics: parent rock, climate, rates of growth and decay, slope aspect and steepness, time, and transport. In dry climates, **pedocals** form. In pedocals, leached ions precipitate in the B horizon, where they accumulate and may form **caliche**. In moist climates, **pedalfer** soils develop. In these regions, soluble ions are removed from the soil, leaving high concentrations of less soluble aluminum and iron. **Laterite** soils form in very moist climates, where all of the more soluble ions are removed.

KEY WORDS

weathering   *94*
erosion   *94*
mechanical weathering   *94*
chemical weathering   *94*
pressure-release fracturing   *95*
frost wedging   *95*
talus slope   *95*
abrasion   *95*

hydrolysis   *98*
oxidation   *99*
spheroidal weathering   *99*
salt cracking   *100*
exfoliation   *100*
regolith   *101*
soil   *101*
loam   *101*
litter   *101*
humus   *101*

soil horizons   *102*
O horizon   *102*
A horizon   *102*
topsoil   *102*
B horizon   *102*
C horizon   *102*
leaching   *102*
zone of leaching   *102*
zone of accumulation   *102*

capillary action   *102*
pedocal   *103*
caliche   *103*
salinization   *103*
pedalfer   *103*
laterite   *103*
bauxite   *103*
aspect   *104*

REVIEW QUESTIONS

1. Explain the differences among the terms weathering, erosion, transport, and deposition.

2. Explain the differences between mechanical weathering and chemical weathering.

3. List five processes that cause mechanical weathering.

4. Explain how thermal expansion can establish forces that could fracture a rock.

5. What is a talus slope? What conditions favor the formation of talus slopes?

6. What is oxidation? Give an example.

7. Explain why limestone dissolves very slowly in pure water. Why does it dissolve more rapidly in strong acids? Why does it dissolve in rainwater?

8. What is hydrolysis? What happens when granitic rocks undergo hydrolysis? What minerals react? What are the reaction products?

9. What is pressure-release fracturing? Why is pressure-release fracturing an example of chemical and mechanical processes operating together?

10. List the products of weathering in order of decreasing size.

11. What are the components of healthy soil? What is the function of each component?

12. Characterize the four major horizons of a mature soil.

13. List the six soil-forming factors and briefly discuss each one.

14. Imagine that soil forms on granite in two regions, one wet and the other dry. Will the soil in the two regions be the same or different? Explain.

15. Explain how soils formed from granite will change with time.

16. What are laterite soils? How are they formed? Why are they unsuitable for agriculture?

DISCUSSION QUESTIONS

1. What process is responsible for each of the following observations or phenomena? Is the process a mechanical or chemical change?
   a. A board is sawn in half.   b. A board is burned.
   c. A cave is formed when water seeps through a limestone formation.   d. Calcite is formed when mineral-rich water is released from a hot underground spring.   e. Meter-thick sheets of granite peel off a newly exposed pluton.
   f. Rockfall is more common in mountains of the temperate region in the spring than in mid-summer.

2. Most substances contract when they freeze, but water expands. How would weathering be affected if water contracted instead of expanded when it froze?

3. Discuss the similarities and differences between salt cracking and frost wedging.

4. What types of weathering would predominate on the following fictitious planets? Defend your conclusions.

   a. Planet X has a dense atmosphere composed of nitrogen, oxygen, and water vapor with no carbon dioxide. Temperatures range from a low of 10°C in the winter to 75°C in the summer. Windstorms are common. No living organisms have evolved.   b. The atmosphere of Planet Y consists mainly of nitrogen and oxygen with smaller concentrations of carbon dioxide and water vapor. Temperatures range from a low of −60°C in the polar regions in the winter to +35°C in the tropics. Windstorms are common. A lush blanket of vegetation covers most of the land surfaces.

5. The Arctic regions are cold most of the year, and summers are short there. Thus decomposition of organic matter is slow. In contrast, decay is much more rapid in the temperate regions. How does this difference affect the fertility of the soils?

Sedimentary Rocks

Weathering decomposes bedrock. Flowing water, wind, gravity, and glaciers then erode the decomposed rock, transport it downslope, and finally deposit it on the sea coast or in lakes and river valleys. Finally, the loose sediment is cemented to form hard sedimentary rock.

Sedimentary rocks make up only about 5 percent of the Earth's crust. However, because they form on the Earth's surface, they are widely spread in a thin veneer over underlying igneous and metamorphic rocks. As a result, sedimentary rocks cover about 75 percent of continents.

Many sedimentary rocks have high economic value. Oil and gas form in certain sedimentary rocks. Coal, a major energy resource, is a sedimentary rock. Limestone is an important building material, both as stone and as the primary ingredient in cement. Gypsum is the raw material for plaster. Ores of copper, lead, zinc, iron, gold, and silver concentrate in certain types of sedimentary rocks.

Horizontally layered sandstone in eastern Utah has been eroded to produce spectacular towers.

▶ 7.1 TYPES OF SEDIMENTARY ROCKS

Sedimentary rocks are broadly divided into four categories:

1. *Clastic sedimentary rocks* are composed of fragments of weathered rocks, called **clasts**, that have been transported, deposited, and cemented together. Clastic rocks make up more than 85 percent of all sedimentary rocks (Fig. 7–1). This category includes sandstone, siltstone, and shale.

2. *Organic sedimentary rocks* consist of the remains of plants or animals. Coal is an organic sedimentary rock made up of decomposed and compacted plant remains.

3. *Chemical sedimentary rocks* form by direct precipitation of minerals from solution. Rock salt, for example, forms when salt precipitates from evaporating seawater or saline lake water.

4. *Bioclastic sedimentary rocks*. Most limestone is composed of broken shell fragments. The fragments are clastic, but they form from organic material. As a result, limestone formed in this way is called a **bioclastic** rock.

▶ 7.2 CLASTIC SEDIMENTARY ROCKS

Clastic sediment consists of grains and particles that were eroded from weathered rocks and then were transported and deposited in loose, unconsolidated layers at the Earth's surface. The sand on a beach, boulders in a river bed, and mud in a puddle are all clastic sediments.

Clastic sediment is named according to particle size (Table 7–1). **Gravel** includes all rounded particles larger than 2 millimeters in diameter. Angular particles in the

Table 7–1 • SIZES AND NAMES OF SEDIMENTARY PARTICLES AND CLASTIC ROCKS

DIAMETER (mm)	SEDIMENT		CLASTIC SEDIMENTARY ROCK
256–64–	Boulders Cobbles Pebbles	Gravel (rubble)	Conglomerate (rounded particles) or breccia (angular particles)
2–	Sand		Sandstone
1/16–	Silt	Mud	Siltstone
1/256–	Clay		Claystone or shale } Mudstone

same size range are called **rubble**. **Sand** ranges from 1/16 to 2 millimeters in diameter. Sand feels gritty when rubbed between your fingers, and you can see the grains with your naked eye. **Silt** varies from 1/256 to 1/16 millimeter. Individual silt grains feel smooth when rubbed between the fingers but gritty when rubbed between your teeth. **Clay** is less than 1/256 millimeter in diameter. It is so fine that it feels smooth even when rubbed between your teeth. Geologists often rub a small amount of sediment or rock between their front teeth to distinguish between silt and clay. **Mud** is wet silt and clay.

TRANSPORT OF CLASTIC SEDIMENT

After weathering creates clastic sediment, flowing water, wind, glaciers, and gravity erode it and carry it downslope. Streams carry the greatest proportion of clastic

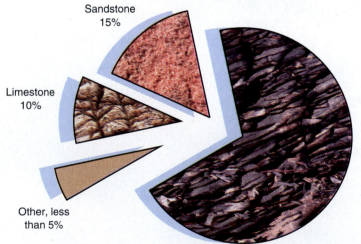

Sandstone
15%

Limestone
10%

Other, less
than 5%

Shale and siltstone
70%

Figure 7–1 Relative abundances of sedimentary rock types.

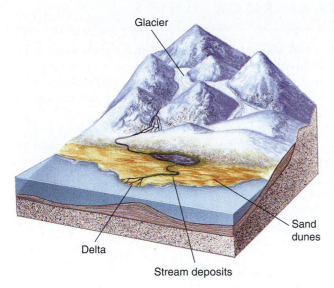

Figure 7–2 Sediment and dissolved ions are transported by water, gravity, wind, and glaciers. They may be deposited temporarily in many different environments along the way, but eventually most sediment reaches the ocean.

Figure 7–3 Rounded cobbles in the West Fork of the Bitterroot River just below Trapper Peak.

sediment. Because most streams empty into the oceans, most sediment accumulates near continental coastlines (Fig. 7–2). However, some streams deposit their sediment in lakes or in inland basins.

Streams and wind modify sediment as they carry it downslope. The **rounded** cobbles shown in Figure 7–3 originally formed as angular rubble in the Bitterroot Range of western Montana. The rubble became rounded as the stream carried it only a few kilometers. Water and wind round clastic particles as fine as silt by tumbling them against each other during transport. Finer particles do not round as effectively because they are so small and light that water and even wind, to some extent, cushion them as they bounce along, minimizing abrasion. Glaciers do not round clastic particles because the ice prevents the particles from abrading each other.

Weathering breaks bedrock into particles of all sizes, ranging from clay to boulders. Yet most clastic sediment and sedimentary rocks are well **sorted**—that is, the grains are of uniform size. Some sandstone formations extend for hundreds of square kilometers and are more than a kilometer thick, but they consist completely of uniformly sized sand grains.

Sorting depends on three factors: the viscosity and velocity of the transporting medium and the durability of the particles. **Viscosity** is resistance to flow; ice has high viscosity, air has low viscosity, and water is intermediate. Ice does not sort effectively because it transports particles of all sizes, from house-sized boulders to clay.

In contrast, wind transports only sand, silt, and clay and leaves the larger particles behind. Thus, wind sorts particles according to size.

A stream transports only small particles when it flows slowly, but larger particles when it picks up speed. For example, a stream transports large and small particles when it is flooding, but only small particles during normal flows. As a flood recedes and the water gradually slows down, the stream deposits the largest particles first and the smallest ones last, creating layers of different-sized particles.

Finally, durability of the particles affects sorting. Sediment becomes abraded as it travels downstream. Thus a stream may transport cobbles from the mountains toward a delta, but the cobbles may never complete the journey because they wear down to smaller grains along the way. This is one reason why mountain streams are frequently boulder choked but deltas are composed of mud and sand.

NAMING SEDIMENTARY ROCK UNITS

A body of rock is commonly given a formal name and referred to as a **formation**. A formation can consist of a single rock type or several different rock types. To qualify as a formation, a body of rock should be easily recognizable in the field and be thick and laterally extensive enough to show up well on a geologic map. Although sedimentary rocks are most commonly designated as formations, bodies of igneous and metamorphic rock that meet these qualifications also are named and are called formations.

Formations are often named for the geographic locality where they are well exposed and were first defined. Names also include the dominant rock type—for example, the Navajo Sandstone, the Mission Canyon Limestone, and the Chattanooga Shale. If the formation contains more than one abundant rock type, the word *formation* is used in the name instead of a rock type, as in the Green River Formation.

A **contact** is the surface between two rocks of different types or ages. Contacts separate formations and separate different rock types or layers within a single formation. In sedimentary rocks, contacts are usually bedding planes.

For convenience, geologists sometimes lump two or more formations together into a **group** or subdivide a formation into **members**. For example, the Madison Group in central Montana consists of three formations deposited about 350 million years ago: the Paine Limestone, the Woodhurst Limestone, and the Mission Canyon Limestone.

DISCUSSION QUESTION

Discuss how you would recognize a contact in the field. Are contacts always horizontal? If not, discuss how a vertical or tilted contact may have formed.

LITHIFICATION

Lithification refers to processes that convert loose sediment to hard rock. Two of the most important processes are **compaction** and **cementation**.

If you fill a container with sand, the sand grains do not fill the entire space. Small voids, called **pores**, exist between the grains (Fig. 7–4a). When sediment is deposited in water, the pores are usually filled with water. The proportion of space occupied by pores depends on particle size, shape, and sorting. Commonly, freshly deposited clastic sediment has about 20 to 40 percent pore space, although a well-sorted and well-rounded sand may have up to 50 percent pore space. Clay-rich mud may have as much as 90 percent pore space occupied by water.

As more sediment accumulates, its weight compacts the buried sediment, decreasing pore space and forcing out some of the water (Fig. 7–4b). This process is called **compaction**. Compaction alone may lithify clay because the platy grains interlock like pieces of a puzzle.

Water normally circulates through the pore space in buried and compacted sediment. This water commonly contains dissolved calcium carbonate, silica, and iron, which precipitate in the pore spaces and **cement** the clastic grains together to form a hard rock (Fig. 7–4c). The red sandstone in Figure 7–5 gets its color from red iron oxide cement.

In some environments, sediment lithifies quickly, whereas the process is slow in others. In the Rocky Mountains, calcite has cemented glacial deposits less

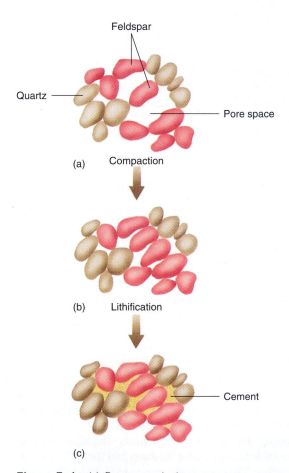

Figure 7–4 (a) Pore space is the open space among grains of sediment. (b) Compaction decreases pore space and lithifies sediment by interlocking the grains. (c) Cement fills pores and lithifies sediment by binding grains together.

Figure 7–5 Red iron oxide cement colors the red sandstone of Indian Creek, Utah.

Figure 7–6 Conglomerate is lithified gravel.

than 20,000 years old. In contrast, some sand and gravel deposited between 30 and 40 million years ago in southwestern Montana can be dug with a hand shovel. The speed of lithification depends mainly on the availability of cementing material and water to carry the dissolved cement through the sediment.

TYPES OF CLASTIC ROCKS

Conglomerate and Breccia

Conglomerate (Fig. 7–6) and **breccia** are coarse-grained clastic rocks. They are the lithified equivalents of gravel and rubble, respectively. In a conglomerate the particles are rounded, and in a sedimentary breccia they are angular. Because large particles become rounded rapidly over short distances of transport, sedimentary breccias are usually found close to the weathering site where the angular rock fragments formed.

Each clast in a conglomerate or breccia is usually much larger than the individual mineral grains in the rock. Therefore, the clasts retain most of the characteristics of the parent rock. If enough is known about the geology of an area where conglomerate or breccia is found, it may be possible to identify exactly where the clasts originated. A granite clast in a sedimentary breccia probably came from nearby granite bedrock.

Gravel typically has large pores between the clasts because the individual particles are large. These pores usually fill with finer sediment such as sand or silt. The next time you walk along a cobbly stream, look carefully between the cobbles. You will probably see sand or silt trapped among the larger clasts. As a result, most conglomerates have fine sediment among the large clasts.

Sandstone

Sandstone consists of lithified sand grains (Fig. 7–7). When granitic bedrock weathers, feldspar commonly converts to clay, but quartz crystals resist weathering. As streams carry the clay and quartz grains toward the sea, the quartz grains become rounded. The flowing water deposits the sand in one environment and the clay in another. Consequently, most sandstones consist predominantly of rounded quartz grains.

The word *sandstone* refers to any clastic sedimentary rock comprising primarily sand-sized grains. Most sandstones are quartz sandstone and contain more than 90 percent quartz. **Arkose** is a sandstone comprising 25 percent or more feldspar grains, with most of the remaining grains being quartz. The sand grains in arkose

(a)

(b)

Figure 7–7 Sandstone is lithified sand. (a) A sandstone cliff above the Colorado River, Canyonlands, Utah. (b) A close-up of sandstone. Notice the well-rounded sand grains.

are commonly coarse and angular. The high feldspar content and the coarse, angular nature of the grains indicate that the rock forms only a short distance from its source area, perhaps adjacent to granite cliffs (Fig. 7–8). **Graywacke** is a poorly sorted sandstone with considerable quantities of silt and clay in its pores. Graywacke is commonly dark in color because of fine clay that coats the sand grains. The grains are usually quartz, feldspar, and fragments of volcanic, metamorphic, and sedimentary rock.

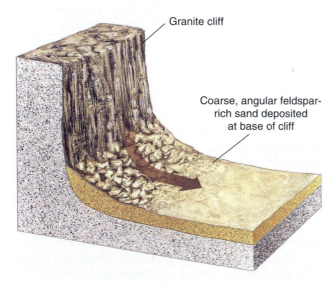

Granite cliff

Coarse, angular feldspar-rich sand deposited at base of cliff

Figure 7–8 Arkose commonly accumulates close to the source of the sediment.

Claystone, Shale, Mudstone, and Siltstone

Claystone, shale, mudstone, and siltstone are all fine-grained clastic rocks. **Claystone** is composed predominantly of clay minerals and small amounts of quartz and other minerals of clay size. **Shale** (Fig. 7–9a) consists of the same material but has a finely layered structure called **fissility**, along which the rock splits easily (Fig. 7–9b). Clay minerals have platy shapes, like mica. When clays are deposited in water, the sediment commonly contains 50 to 60 percent water, and the platelike clay minerals are randomly oriented, as shown in Figure 7–10a. As more sediment accumulates, compaction drives out most of the water and the clay plates rotate so that their flat surfaces lie perpendicular to the pull of gravity (Fig. 7–10b). Thus, they stack like sheets of paper on a shelf. The fissility of shale results from the parallel orientation of the platy clay minerals.

Mudstone is a nonfissile rock composed of clay and silt. In some mudstone and claystone, layering is absent because burrowing animals such as worms, clams, and crabs disrupted it by churning the sediment.

Siltstone is lithified silt. The main component of most siltstones is quartz, although clays are also commonly present. Siltstones often show layering but lack the fine fissility of shales because of their lower clay content.

Shale, mudstone, and siltstone make up 70 percent of all clastic sedimentary rocks (Fig. 7–1). Their abundance reflects the vast quantity of clay produced by weathering. Shale is usually gray to black due to the

(a)

(b)

Figure 7–9   Shale is made up mostly of platy clays. Therefore, it shows very thin layering called fissility. (a) An outcrop of shale near Drummond, Montana. (b) A close-up of shale.

presence of partially decayed remains of plants and animals commonly deposited with clay-rich sediment. This organic material in shales is the source of most oil and natural gas. (The formation of oil and gas from this organic material is discussed in Chapter 19.)

► 7.3   ORGANIC SEDIMENTARY ROCKS

Organic sedimentary rocks, such as chert and coal, form by lithification of the remains of plants and animals.

CHERT

Chert is a rock composed of pure silica. It occurs as sedimentary beds interlayered with other sedimentary rocks and as irregularly shaped lumps called **nodules** in other sedimentary rocks (Fig. 7–11). Microscopic examination of bedded chert often shows that it is made up of the remains of tiny marine organisms that make their skeletons of silica rather than calcium carbonate. In contrast, some nodular chert appears to form by precipitation from silica-rich ground water, most often in limestone. Chert was

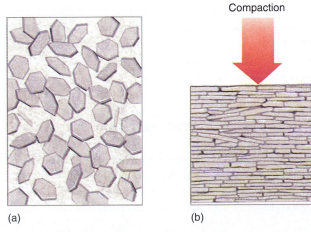

Compaction

(a)   (b)

Figure 7–10   (a) Randomly oriented clay particles in freshly deposited mud. (b) Parallel-oriented clay particles after compaction and dewatering by weight of overlying sediments.

Figure 7–11   Red nodules of chert in light-colored limestone.

one of the earliest geologic resources. Flint, a dark gray to black variety, was frequently used for arrowheads, spear points, scrapers, and other tools chipped to hold a fine edge.

COAL

When plants die, their remains usually decompose by reaction with oxygen. However, in warm swamps and in other environments where plant growth is rapid, dead plants accumulate so rapidly that the oxygen is used up long before the decay process is complete. The undecayed or partially decayed plant remains form **peat**. As peat is buried and compacted by overlying sediments, it converts to **coal**, a hard, black, combustible rock. (Coal formation is discussed in more detail in Chapter 19.)

▶ 7.4 CHEMICAL SEDIMENTARY ROCKS

Some common elements in rocks and minerals, such as calcium, sodium, potassium, and magnesium, dissolve during chemical weathering and are carried by ground water and streams to the oceans or to lakes. Most lakes are drained by streams that carry the salts to the ocean. Some lakes, such as the Great Salt Lake in Utah, are landlocked. Streams flow into the lake, but no streams exit. As a result, water escapes only by evaporation. When water evaporates, salts remain behind and the lake

Figure 7–12 An evaporating lake precipitated thick salt deposits on the Salar de Uyuni, Bolivia.

Figure 7–13 Rocks that precipitate from solution have interlocking grains.

water becomes steadily more salty. The same process can occur if ocean water is trapped in coastal or inland basins, where it can no longer mix with the open sea.

Evaporites form when evaporation concentrates dissolved ions to the point at which they precipitate from solution (Fig. 7–12). As the individual crystals precipitate, they interlock with each other to produce grain boundaries like those of an igneous rock (Fig. 7–13). The interlocking texture forms a solid rock, even though the rock may never have been compacted or cemented.

The most common minerals found in evaporite deposits are gypsum ($CaSO_4 \cdot 2H_2O$)[1] and halite ($NaCl$). Gypsum is used in plaster and wallboard, and halite is common salt. Evaporites form economic deposits in many basins and coastal areas. However, they compose only a small proportion of all sedimentary rocks.

Seawater is so nearly saturated in calcium carbonate that calcium carbonate minerals can precipitate under the proper conditions. This process occurs today on the shallow Bahama Banks, south of Bimini in the Caribbean Sea. As waves and currents roll tiny shell fragments back and forth on the sea bottom, calcium carbonate precipitates in concentric layers on the fragments. This process produces nearly perfect spheres called oöliths. In turn, oöliths may become cemented together to form **oölitic limestone**. Limestone of this type is a chemical sedimentary rock. However, most limestone is bioclastic, as discussed next.

▶ 7.5 BIOCLASTIC ROCKS

Carbonate rocks are made up primarily of carbonate minerals, which contain the carbonate ion $(CO_3)^{2-}$. The most common carbonate minerals are calcite (calcium carbonate, $CaCO_3$) and dolomite (calcium magnesium

---

[1]The $2H_2O$ in the chemical formula of gypsum means there is water incorporated into the mineral structure.

carbonate, CaMg(CO$_3$)$_2$). Calcite-rich carbonate rocks are called **limestone**, whereas rocks rich in the mineral dolomite are also called dolomite. Many geologists use the term **dolostone** for the rock name to distinguish it from the mineral dolomite.

Seawater contains large quantities of dissolved calcium carbonate (CaCO$_3$). Clams, oysters, corals, some types of algae, and a variety of other marine organisms convert dissolved calcium carbonate to shells and other hard body parts. When these organisms die, waves and ocean currents break the shells into small fragments, called **bioclastic sediment**. A rock formed by lithification of such sediment is called **bioclastic limestone**, indicating that it forms by both biological and clastic processes. Many limestones are bioclastic. The bits and pieces of shells appear as fossils in the rock (Fig. 7–14).

Organisms that form limestone thrive and multiply in warm, shallow seas because the sun shines directly on the ocean floor, where most of them live. Therefore, bioclastic limestone typically forms in shallow water along coastlines at low and middle latitudes. It also forms on continents when rising sea level floods land with shallow seas.

Coquina is bioclastic limestone consisting wholly of coarse shell fragments cemented together. **Chalk** is a very fine-grained, soft, white bioclastic limestone made of the shells and skeletons of microorganisms that float near the surface of the oceans. When they die, their remains sink to the bottom and accumulate to form chalk. The pale-yellow chalks of Kansas, the off-white chalks of Texas, and the gray chalks of Alabama remind us that all of these areas once lay beneath the sea (Fig. 7–15).

Dolomite composes more than half of all carbonate rocks that are over a billion years old and a smaller, although substantial, proportion of younger carbonate rocks. Because it is so abundant, we would expect to see dolomite forming today; yet today there is no place in the world where dolomite is forming in large amounts. This dilemma is known as **the dolomite problem**.

The general consensus among geologists is that most dolomite did not form as a primary sediment or rock. Instead, it formed as magnesium-rich solutions derived from seawater percolated through limestone beds. Magnesium ions replaced half of the calcium in the calcite, converting the limestone beds to dolostone.

▶ 7.6 SEDIMENTARY STRUCTURES

Nearly all sedimentary rocks contain **sedimentary structures**, features that developed during or shortly after deposition of the sediment. These structures help us

(a)

(b)

Figure 7–14 Most limestone is lithified shell fragments and other remains of marine organisms. (a) A limestone mountain in British Columbia, Canada. (b) A close-up of shell fragments in limestone.   (© Breck P. Kent)

understand how the sediment was transported and deposited.

The most obvious and common sedimentary structure is **bedding**, or **stratification**—layering that develops as sediment is deposited (Fig. 7–16). Bedding forms because sediment accumulates layer by layer. Nearly all

Figure 7–15 The Niobrara chalk of western Kansas consists of the remains of tiny marine organisms. (David Schwimmer)

sedimentary beds were originally horizontal because most sediment accumulates on nearly level surfaces.

Cross-bedding consists of small beds lying at an angle to the main sedimentary layering (Fig. 7–17a). Cross-bedding forms in many environments where wind or water transports and deposits sediment. For example, wind heaps sand into parallel ridges called **dunes**, and flowing water forms similar features called **sand waves**. Figure 7–17b shows that cross-beds are the layers formed by sand grains tumbling down the steep downstream face of a dune or sand wave. Cross-bedding is common in sands deposited by wind, streams, ocean currents, and waves on beaches.

Ripple marks are small, nearly parallel sand ridges and troughs that are also formed by moving water or wind. They are like dunes and sand waves, but smaller. If the water or wind flows in a single direction, the ripple marks become asymmetrical, like miniature dunes. In other cases, waves move back and forth in shallow water, forming symmetrical ripple marks in bottom sand (Fig. 7–18). Ripple marks are often preserved in sandy sedimentary rocks (Fig. 7–19).

In **graded bedding**, the largest grains collect at the bottom of a layer and the grain size decreases toward the top (Fig. 7–20). Graded beds commonly form when some violent activity, such as a major flood or submarine land-

Figure 7–16 Sedimentary bedding shows clearly in the walls of the Grand Canyon. (Donovan Reese/Tony Stone Images)

(a)

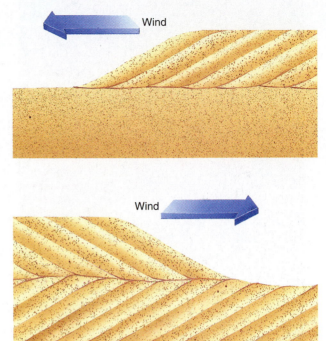

(b)

Figure 7–17 (a) Cross-bedding preserved in lithified ancient sand dunes in Arches National Park, Utah. (b) The development of cross-bedding in sand as a dune migrates.

slide, mixes a range of grain sizes together in water. The larger grains settle rapidly and concentrate at the base of the bed. Finer particles settle more slowly and accumulate in the upper parts of the bed.

Mud cracks are polygonal cracks that form when mud shrinks as it dries (Fig. 7–21). They indicate that the mud accumulated in shallow water that periodically dried up. For example, mud cracks are common on intertidal mud flats where sediment is flooded by water at high tide and exposed at low tide. The cracks often fill with

sediment carried in by the next high tide and are commonly well preserved in rocks.

Occasionally, very delicate sedimentary structures are preserved in rocks. Geologists have found imprints of raindrops that fell on a muddy surface about 1 billion years ago (Fig. 7–22) and imprints of salt crystals that formed as a puddle of salt water evaporated. Like mud cracks, raindrop and salt imprints show that the mud must have been deposited in shallow water that intermittently dried up.

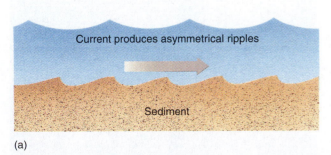

(a)

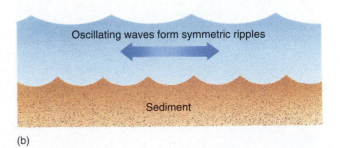

(b)

Figure 7–18 (a) Asymmetric ripple marks form when wind or currents move continuously in the same direction. (b) Symmetric ripple marks form when waves oscillate back and forth.

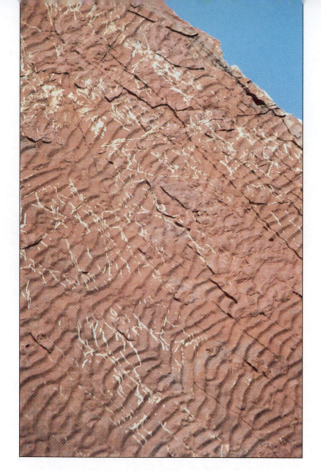

Figure 7–19 Ripple marks in billion-year-old mud rocks in eastern Utah.

Figure 7–20 A graded bed in Tonga, southwestern Pacific. Larger grains collected near the bottom, and smaller particles settled near the top of the bed. (Peter Ballance)

Figure 7–21 Mud cracks form when wet mud dries and shrinks.

Fossils are any remains or traces of a plant or animal preserved in rock—any evidence of past life. Fossils include remains of shells, bones, or teeth; whole bodies preserved in amber or ice; and a variety of tracks, burrows, and chemical remains. Fossils are discussed further in Chapter 9.

▶ **7.7 INTERPRETING SEDIMENTARY ROCKS: DEPOSITIONAL ENVIRONMENTS**

Imagine that you encounter a limestone outcrop as you walk in the hills. Entombed in the limestone you find

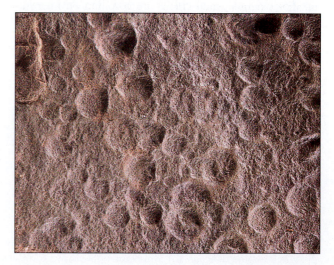

Figure 7–22 Delicate raindrop imprints formed by rain that fell about a billion years ago on a mudflat. (© Breck P. Kent)

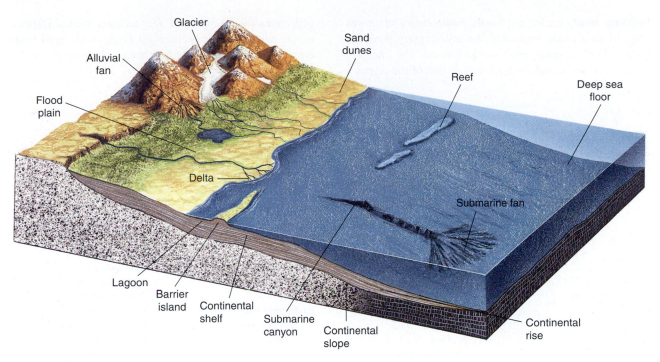

Figure 7–23 Common depositional environments.

fossils of marine clams that lived in shallow water. Therefore, you infer that the limestone must have formed in a shallow sea. Further, since the limestone is now well above sea level, you infer that tectonic forces have lifted this portion of the sea bed to form the hills.

Geologists study sedimentary rocks to help us understand the past. When geologists study sedimentary rocks, they ask questions such as: Where did the sediment originate? Was the sediment transported by a stream, wind, or a glacier? In what environment did the sediment accumulate? If it was deposited in the sea, was it on a beach or in deep water? If it was deposited on land, was it in a lake, a stream bed, or a flood plain?

Geologists answer these questions by analyzing the minerals, textures, and structures of sedimentary rocks. Additionally, the size and shape of a sedimentary rock layer contain clues to its depositional environment. Accurate interpretations of depositional environments are often rewarding because valuable concentrations of oil and gas, coal, evaporites, and metals form in certain types of environments.

Depositional environments vary greatly in scale, from an entire ocean basin to a 3-meter-long sand bar in a stream. Many small-scale environments may be active within a single large-scale depositional system (Fig. 7–23).

SUMMARY

Sedimentary rocks cover about three fourths of the Earth's land surface. **Clastic sediment** is sediment composed of fragments of weathered rock called **clasts**. Clastic sediment is **rounded** and **sorted** during transport and then deposited. Most sediment becomes lithified by **compaction** and **cementation.**

Clastic sedimentary rocks are composed of lithified clastic sediment and are named and classified primarily according to the size of the clastic grains. Common types are **conglomerate, sandstone, siltstone, shale, claystone,** and **mudstone. Organic sedimentary**

rocks are made up of the remains of organisms. **Coal** and **chert** are common organic sedimentary rocks. **Chemical sedimentary rocks** include **evaporites**, rocks that precipitate directly from solution as lake water or seawater evaporates. Most **limestone** is **bioclastic** and forms from broken shell fragments. **Dolostone** is a carbonate rock in which half of the calcium in calcite has been replaced by magnesium.

Sedimentary structures are features that develop during or shortly after sediment is deposited. They include **bedding, ripple marks, cross-bedding, graded**

bedding, **mud cracks**, and **fossils**. Sedimentary structures contain vital clues regarding the sedimentary environments in which sedimentary rocks formed. The interpretation of depositional environments is one of the primary objectives of the study of sedimentary rocks. Depositional environments include all large- and small-scale environments in which sediments are deposited.

Important Sedimentary Rocks

Conglomerate	Breccia	Sandstone	Arkose	Graywacke
Claystone	Shale	Mudstone	Siltstone	Chert
Coal	Limestone	Dolostone	Coquina	Chalk

KEY WORDS

clast   *110*
bioclastic   *110*
clastic sediment   *110*
gravel   *110*
rubble   *110*
sand   *110*
silt   *110*
clay   *110*
mud   *110*
rounding   *111*

sorting   *111*
viscosity   *111*
lithification   *112*
pore   *112*
compaction   *112*
cementing   *112*
quartz sandstone   *113*
fissility   *114*
nodule   *115*

peat   *116*
chemical sedimentary
  rock   *116*
carbonate rock   *116*
bioclastic sediment   *117*
bioclastic limestone   *117*
dolomite problem   *117*
sedimentary structures
  117

bedding   *117*
stratification   *117*
cross-bedding   *118*
dune   *118*
sand wave   *118*
ripple mark   *118*
graded bedding   *118*
mud crack   *119*
fossil   *120*

REVIEW QUESTIONS

1. Why do sedimentary rocks cover more than 75 percent of the Earth's land surface when they compose only 5 percent of the volume of the continental crust?

2. List the five stages in the formation of sedimentary rocks.

3. How do clastic sediments differ from dissolved sediment and chemical sediment?

4. Define bioclastic sediment.

5. In what ways are clastic sediments modified during transport?

6. Why is the maximum size of particles transported by wind finer than the maximum size transported by streams?

7. Why is the maximum size of particles transported by glaciers coarser than the maximum size transported by streams?

8. Describe how loose clastic sediment becomes lithified to form hard rock.

9. What is pore space in a clastic sediment? How is it modified during lithification?

10. What is the difference between conglomerate and breccias?

11. Why are most sandstones made up predominantly of quartz?

12. In what geologic environment does arkose form?

13. How do shale, sandstone, and limestone differ from one another?

14. How do shales acquire fissility? Why do mudstones lack that property?

15. How do limestones form?

16. What is bioclastic limestone?

17. How do dolomites form? What is the dolomite problem?

18. How does coal form?

19. How do evaporites form?

20. What does cross-bedding in a sandstone tell you about depositional environment?

21. What do the presence of mud cracks in a mudstone tell you about the depositional environment?

DISCUSSION QUESTIONS

1. Field geologists sometimes come upon large sections of sedimentary rocks that have been turned upside down by tectonic activities. How would you use sedimentary structures to determine whether a sequence of sedimentary rocks is upright or overturned?

2. On a field trip you discover a sequence of sedimentary rocks composed of thin black shales containing marine fossils interbedded with layers of gypsum and halite. What can you deduce about the depositional environment of these rocks?

3. Large portions of the Canadian Rockies are composed of limestone and shale. From this information alone, what can you tell about the geologic history of the region?

4. All the large-scale depositional environments are marine, while small-scale depositional environments can be either marine or terrestrial. Explain.

5. What types of sedimentary structures would you expect to find under the following circumstances?   a. A catastrophic flood washes a huge amount of mixed sediment into a lake.   b. Sand accumulates in a dry, windy environment. The prevailing wind direction shifts periodically, over a few million years.   c. Mud collects on the bottom of a large, shallow, inland sea.

6. Why is shale the most abundant sedimentary rock?

7. Would you expect to find large quantities of sedimentary rocks on the Moon? Why or why not? If you do expect to find them, what types would you expect?

Metamorphic Rocks

A potter forms a delicate vase from moist clay. She places the soft piece in a kiln and slowly heats it to 1000°C. As temperature rises, the clay minerals decompose. Atoms from the clay then recombine to form new minerals that make the vase strong and hard. The breakdown of the clay minerals, growth of new minerals, and hardening of the vase all occur without melting. The reactions in a potter's kiln are called **solid-state reactions** because they occur in solid materials.

Chemical reactions occur more rapidly in liquid or gas than in a solid because atoms and molecules are more mobile in a fluid. However, with enough time and elevated temperature, atoms in solid rock also react. Small amounts of fluid, such as the water in the potter's clay, increase the mobility of atoms and speed the reactions, but the reactions take place in solid materials.

Metamorphism (from the Greek words for "changing form") is the process by which rising temperature—and changes in other environmental conditions—transforms rocks and minerals. Metamorphism occurs in solid rock—like the transformations in the vase as the potter fires it in her kiln. Metamorphism can change any type of parent rock: sedimentary, igneous, or even another metamorphic rock.

Many metamorphic rocks show evidence of high temperature and plastic deformation. (J. M. Harrison/Geological Survey of Canada)

▶ 8.1 MINERAL STABILITY AND METAMORPHISM

A mineral that does not decompose or change in other ways, no matter how much time passes, is a **stable** mineral. Millions of years ago, weathering processes may have formed the clay minerals used by the potter to create her vase. They were stable and had remained unchanged since they formed. A stable mineral can become **unstable** when environmental conditions change. Three types of environmental change cause metamorphism: rising temperature, rising pressure, and changing chemical composition.

For example, when the potter put the clay in her kiln and raised the temperature, the clay minerals decomposed because they became unstable at the higher temperature. The atoms from the clay then recombined to form new minerals that were stable at the higher temperature. Like the clay, every mineral is stable only within a certain temperature range. In a similar manner, each mineral is stable only within a certain pressure range.

In addition, a mineral is stable only in a certain chemical environment. If fluids transport new chemicals to a rock, those chemicals may react with the original minerals to form new ones that are stable in the altered chemical environment. If fluids remove chemical components from a rock, new minerals may form for the same reason.

Metamorphism occurs because each mineral is stable only within a certain range of temperature, pressure, and chemical environment. If temperature or pressure rises above that range, or if chemicals are added to or removed from the rock, the rock's original minerals may decompose and their components recombine to form new minerals that are stable under the new conditions.

▶ 8.2 METAMORPHIC CHANGES

Metamorphism commonly alters both the texture and mineral content of a rock.

TEXTURAL CHANGES

As a rock undergoes metamorphism, some mineral grains grow larger and others shrink. The shapes of the grains may also change. For example, fossils give fossiliferous limestone its texture (Fig. 8–1). Both the fossils and the cement between them are made of small calcite crystals. If the limestone is buried and heated, some of the calcite grains grow larger at the expense of others. In the process, the fossiliferous texture is destroyed.

Figure 8–1 Fossils give this limestone its fossiliferous texture.

Metamorphism transforms limestone into a metamorphic rock called **marble** (Fig. 8–2). Like the fossiliferous limestone, the marble is composed of calcite, but the texture is now one of large interlocking grains, and the fossils have vanished.

Figure 8–2 Metamorphism has destroyed the fossiliferous texture of the limestone in Figure 8–1 and replaced it with the large, interlocking calcite grains of marble.

Figure 8–3 Shale is a very fine-grained sedimentary rock, containing clay, quartz, and feldspar.

Figure 8–4 Hornfels forms by metamorphism of shale. The white spots are metamorphic minerals. (Geoffrey Sutton)

MINERALOGICAL CHANGES

As a general rule, when a parent rock (the original rock) contains only one mineral, metamorphism transforms the rock into one composed of the same mineral but with a coarser texture. The mineral content does not change because no other chemical components are available during metamorphism. Limestone converting to marble is one example of this generalization. Another is the metamorphism of quartz sandstone to **quartzite**, a rock composed of recrystallized quartz grains.

In contrast, metamorphism of a parent rock containing several minerals usually forms a rock with new and different minerals *and* a new texture. For example, a typical shale contains large amounts of clay, as well as quartz and feldspar (Fig. 8–3). When heated, some of those minerals decompose, and their atoms recombine to form new minerals such as mica, garnet, and a different kind of feldspar. Figure 8–4 shows a rock called **hornfels** that formed when metamorphism altered both the texture and minerals of shale.

If migrating fluids change the chemical composition of a rock, new minerals invariably form. These effects are discussed further in Section 8.3.

DEFORMATION AND FOLIATION

Changes in temperature, pressure, or the chemical environment alter a rock's texture during metamorphism. But another factor also causes profound textural changes. Metamorphic rocks commonly form in large regions of the Earth's crust near a subduction zone, where two tectonic plates converge. The tectonic forces crush, break, and bend rocks in this environment as the rocks are undergoing metamorphism. This combination of metamorphism and **deformation** creates layering in the rocks.

Micas are common metamorphic minerals; they form as many different parent rocks undergo metamorphism. Recall from Chapter 3 that micas are shaped like pie plates. When metamorphism occurs without deformation, the micas grow with random orientations, like pie plates flying through the air (Fig. 8–5). However, when metamorphism and deformation occur together, the

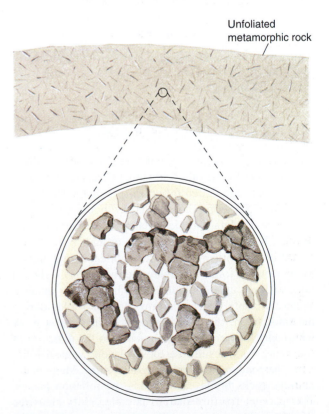

Unfoliated metamorphic rock

Figure 8–5 When metamorphism occurs without deformation, platy micas grow with random orientations.

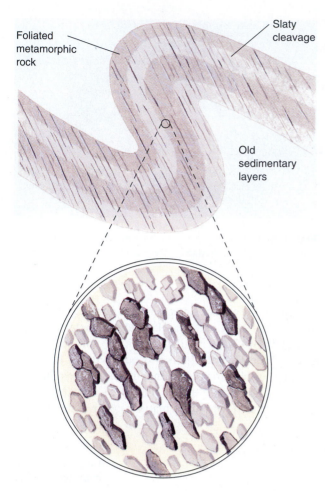

Figure 8–6   When deformation accompanies metamorphism, platy micas orient in a parallel manner to produce metamorphic layering called foliation.

Figure 8–7   Horizontal compression formed this tight fold in interbedded shale and sandstone. Slaty cleavage developed in the shale but not in the sandstone.   (Karl Mueller)

micas develop a parallel orientation. This parallel alignment of micas (and other minerals) produces the metamorphic layering called **foliation** (Fig. 8–6). The layers range from a fraction of a millimeter to a meter or more in thickness. Metamorphic foliation can resemble sedimentary bedding but is different in origin.

Micas and other platy minerals orient at right angles to the tectonic force squeezing the rocks. Pencil-shaped minerals such as amphiboles align in a similar manner. When horizontal forces deform shale into folds during metamorphism, the clays decompose and micas grow with their flat surfaces perpendicular to the direction of squeezing. As a result, the rock develops vertical foliation—perpendicular to the horizontal force. Many metamorphic rocks break easily along the foliation planes. This parallel fracture pattern is called **slaty cleavage** (Fig. 8–7). In most cases, slaty cleavage cuts across the original sedimentary bedding.

METAMORPHIC GRADE

Metamorphic grade expresses the intensity of metamorphism that affected a rock. Because temperature is the most important factor in metamorphism, metamorphic grade closely reflects the highest temperature attained during metamorphism. Geologists can interpret the metamorphic grade of most rocks because many metamorphic minerals form only within certain temperature ranges.

The temperature in shallow parts of the Earth's crust rises by an average of 30°C for each kilometer of depth. It continues to rise in deeper parts of the crust and in the mantle, but at a lesser rate. The rate at which temperature increases with depth is called the **geothermal gradient**. Consequently, the metamorphic grade of many rocks is related to the depth to which they were buried (Fig. 8–8). Low-grade metamorphism occurs at shallow depths, less than 10 to 12 kilometers beneath the surface, where temperature is below 350°C. High-grade conditions are found deep within continental crust and in the upper mantle, 40 or more kilometers below the Earth's surface. The temperature in these regions is 600°C or hotter and is near the melting point of rock. High-grade metamorphism can occur at shallower depths, where magma rises to a shallow level of the Earth's crust.

THE RATE OF METAMORPHISM

A rule of thumb among laboratory chemists is that the speed of a chemical reaction doubles with every 10°C

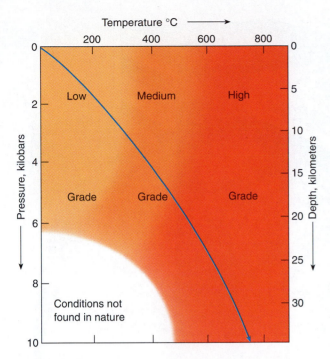

Figure 8–8 Metamorphic grade commonly increases with depth because the Earth becomes hotter with increasing depth. The blue line traces an average geothermal gradient: the path of temperature and pressure in an average part of continental crust.

rise in temperature. Thus, reactions occur slowly in a cold environment, but rapidly in a hot one. For the same reason, metamorphic changes occur slowly at low temperature, but much faster at high temperature. For example, clay minerals in 20-million-year-old shale buried to a depth of 2 to 3 kilometers in the Mississippi River delta show mineralogical changes at 50°C, about the temperature of a cup of hot coffee. Elsewhere, similar clays at the same temperature, but only 1 million years old, show no changes. Thus, metamorphism can occur at temperatures as low as 50°C, but the reactions require millions of years.[1] In contrast, geologists routinely produce metamorphic reactions in the laboratory at temperatures above 500°C in a few days.

The upper limit of metamorphism is the point at which rocks melt to form magma. That temperature varies depending on rock composition, pressure, and amount of water present, but it is between 600° and 1200°C for most rocks. A rock heated to its melting point creates magma, which forms igneous rocks when it solidifies.

---

[1]Many geologists call mineral reactions that occur between 50°C and 250°C *diagenesis* and reserve the term *metamorphism* for changes that occur at temperatures above about 250°C.

Metamorphism refers only to changes that occur *without* melting.

▶ 8.3   TYPES OF METAMORPHISM AND METAMORPHIC ROCKS

Recall that three conditions cause metamorphism: rising temperature, rising pressure, and changing chemical environment. In addition, tectonic deformation develops foliation and thus strongly affects the texture of a metamorphic rock. These conditions occur in four geologic environments.

CONTACT METAMORPHISM

Contact metamorphism occurs where hot magma intrudes cooler country rock. The country rock may be of any type—sedimentary, metamorphic, or igneous. The highest-grade metamorphic rocks form at the contact, closest to the magma. Lower-grade rocks develop farther out (Fig. 8–9). A metamorphic halo around a pluton can range in width from less than a meter to hundreds of meters, depending on the size and temperature of the intrusion and the effects of water or other fluids.

Contact metamorphism commonly occurs without deformation. As a result, the metamorphic minerals grow with random orientations—like the pie plates flying through the air—and the rocks develop no metamorphic layering.

Common Contact Metamorphic Rocks

The hornfels shown in Figure 8–4 is a hard, dark, fine-grained rock usually formed by contact metamorphism of shale. Mica and chlorite are common in the cooler, outer parts of a hornfels halo. Hornblende and other amphiboles occur in the middle of the halo, and pyroxenes can form next to the pluton, in the highest-temperature zone. Quartz and feldspar are common throughout the halo, because they are stable over a wide temperature range.

BURIAL METAMORPHISM

Burial metamorphism results from deep burial of rocks in a sedimentary basin. A large river carries massive amounts of sediment to the ocean every year, where it accumulates on a delta. Over tens or even hundreds of millions of years, the weight of the sediment becomes so great that the entire region sinks isostatically, just as a canoe sinks when you climb into it. Younger sediment may bury the oldest layers to a depth of more than 10 kilometers in a large basin.

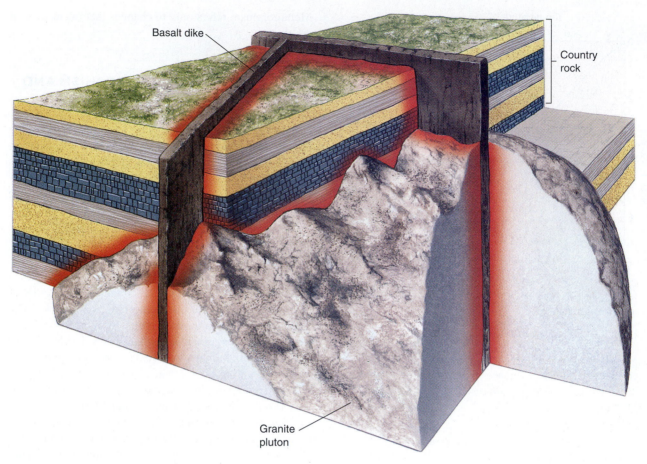

Basalt dike

Country rock

Granite pluton

Figure 8–9 A halo of contact metamorphism in red surrounds a pluton. The later intrusion of the basalt dike metamorphosed both the pluton and the sedimentary rock.

Pressure is commonly expressed in kilobars. One kilobar is approximately equal to 1000 times the pressure of the atmosphere at sea level.[2] Because rocks are heavy, pressure within the Earth increases rapidly with depth, at a rate of about 0.3 kilobar per kilometer. Over time, temperature and pressure increase within the deeper layers until burial metamorphism begins.

Geologists have studied sedimentary basins in detail because most of the world's oil and gas form in them. Thousands of wells have been drilled into the Mississippi River basin, where burial metamorphism is occurring today. Temperature measurements made in the wells (the deepest of which reaches a depth of about 8 kilometers), combined with rock samples recovered as the wells were drilled, allow geologists to identify the mineralogical changes that occur with increasing depth and temperature.

No metamorphic changes occur in the upper 2 kilometers of the Mississippi delta sediment. In this zone, pressure compacts the clay-rich mud and squeezes most of the water from the sediment. The water rises and returns to the ocean.

At about 2 kilometers, however, where the temperature reaches 50°C,[3] the original clay minerals decompose, and their components recombine to form new, different clays. At greater depths and higher temperatures, the clays continue to react and change character. At the greatest depths attained in the basin, corresponding to temperatures of about 250° to 300°C, the clay minerals have completely transformed to mica and chlorite. Similar metamorphic reactions are occurring today in the sediments underlying many large deltas, including the Amazon Basin on the east coast of South America and the Niger River delta on the west coast of Africa.

[2]By international agreement, geologists now express pressure in units of *gigapascals* (Gpa). One Gpa is equal to 10 kilobars. We continue to use kilobars in this text because it is a more familiar unit of pressure.

[3]50°C does not correspond to a depth of 2 kilometers in a region with a normal geothermal gradient. The gradient in these sediments is abnormally low because shallow parts of the delta are filled with young, cold sediment.

Figure 8–10 Argillite forms this cliff in western Montana.

Burial metamorphism occurs without tectonic deformation. Consequently, metamorphic minerals grow with random orientations and, like contact metamorphic rocks, burial metamorphic rocks are unfoliated.

Common Burial Metamorphic Rocks

Because of the lack of deformation, rocks formed by burial metamorphism often retain sedimentary structures. Shale and siltstone become harder and better lithified to form **argillite** (Fig. 8–10), which looks like the parent rock although new minerals have replaced the original ones. Quartz sandstone becomes quartzite. When sandstone is broken, the fractures occur in the cement *between* the sand grains. In contrast, quartzite becomes so firmly cemented during metamorphism that the rock fractures *through* the grains. Burial metamorphism converts limestone and dolomite to marble.

REGIONAL METAMORPHISM

Regional metamorphism occurs in and near a subduction zone, where tectonic forces build mountains and deform rocks. It is the most common and widespread type of metamorphism and affects broad regions of the Earth's crust.

Figure 8–11 shows magma forming in a subduction zone, where oceanic lithosphere is sinking beneath a continent. As the magma rises, it heats large regions of the crust. The high temperatures cause new metamorphic minerals to form throughout the region. At the same time, the tectonic forces squeeze and deform rocks. The

rising magma further deforms the hot, plastic country rock as it forces its way upward. As a result of all of these processes acting together, regionally metamorphosed rocks are strongly foliated and are typically associated with mountains and igneous rocks. Regional metamorphism produces zones of foliated metamorphic rock tens to hundreds of kilometers across.

Common Rocks Formed by Regional Metamorphism

Shale consists of clay minerals, quartz, and feldspar and is the most abundant sedimentary rock. The mineral grains are too small to be seen with the naked eye and can barely be seen with a microscope. Shale undergoes a sequence of changes as metamorphic grade increases.

Figure 8–12 shows the temperatures at which certain metamorphic minerals are stable. Thus, it shows the sequence in which minerals appear, and then decompose, as metamorphic grade increases. As regional metamorphism begins, the clay minerals break down and are replaced by mica and chlorite. These new, platy minerals grow perpendicular to the direction of tectonic squeezing. As a result, the rock develops slaty cleavage and is called **slate** (Fig. 8–13b). With rising temperature and continued deformation, the micas and chlorite grow larger, and wavy or wrinkled surfaces replace the flat, slaty cleavage, giving **phyllite** a silky appearance (Fig. 8–13c).

As temperature continues to rise, the mica and chlorite grow large enough to be seen by the naked eye, and foliation becomes very well developed. Rock of this type is called **schist** (Fig. 8–13d). Schist forms approximately

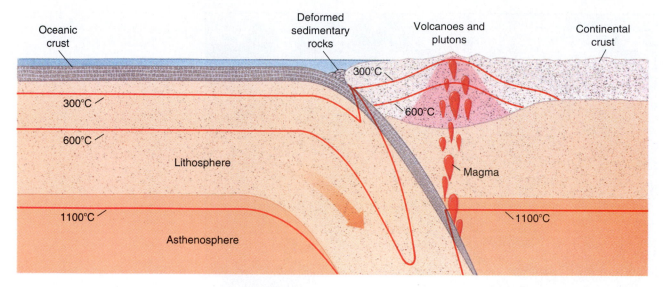

Figure 8–11 Regional metamorphism is common near a subduction zone. The pink shaded area is a zone where rising magma and tectonic force cause abnormally high temperatures and regional metamorphism. The red lines connect points of equal temperature and are called isotherms.

at the transition from low to intermediate metamorphic grades. In some schists, crystals of nonplaty minerals such as garnet, quartz, and feldspar give the rock a knotty appearance.

At high metamorphic grades, light- and dark-colored minerals often separate into bands that are thicker than the layers of schist to form a rock called **gneiss** (pronounced "nice") (Fig. 8–13e). At the highest metamorphic grade, the rock begins to melt, forming small veins of granitic magma. When metamorphism wanes and the rock cools, the magma veins solidify to form **migmatite**, a mixture of igneous and metamorphic rock (Fig. 8–13f).

Under conditions of regional metamorphism, quartz sandstone and limestone transform to foliated quartzite and foliated marble, respectively.

HYDROTHERMAL METAMORPHISM

Water is a chemically active fluid; it attacks and dissolves many minerals. If the water is hot, it attacks minerals even more rapidly. **Hydrothermal metamorphism** (also called *hydrothermal alteration* and *metasomatism*) occurs when hot water and ions dissolved in the hot water react with a rock to change its chemical composition and minerals. In some hydrothermal environments, water reacts with sulfur minerals to form sulfuric acid, making the solution even more corrosive.

The water responsible for hydrothermal metamorphism can originate from three sources. **Magmatic water** is given off by a cooling magma. **Metamorphic water**

is released from rocks during metamorphism. Most hydrothermal alteration, however, is caused by circulating ground water—the water that saturates soil and bedrock. Cold ground water sinks through bedrock fractures to depths of a few kilometers, where it is heated by the hotter rocks at depth or, in some cases, by a hot, shallow pluton. Upon heating, the water expands and rises back toward the surface through other fractures (Fig. 8–14). As it rises, it alters the country rock through which it flows.

Rocks Formed by Hydrothermal Metamorphism

Hydrothermal metamorphism is like an accelerated form of weathering. As in weathering, feldspars and many other minerals of the parent rock dissolve. The hot water carries away soluble components, such as potassium, sodium, calcium, and magnesium. Aluminum and silicon remain because they have low solubilities. They combine with oxygen and water to form clay minerals. Hydrothermally metamorphosed rocks often have a white, bleached appearance and a soft consistency because the clays are white and soft.

Most rocks and magma contain low concentrations of metals such as copper, gold, lead, zinc, and silver. For example, gold makes up 0.0000002 percent of average crustal rock, while copper makes up 0.0058 percent and lead 0.0001 percent. Although the metals are present in very low concentrations, hydrothermal solutions sweep slowly through vast volumes of country rock, dissolving

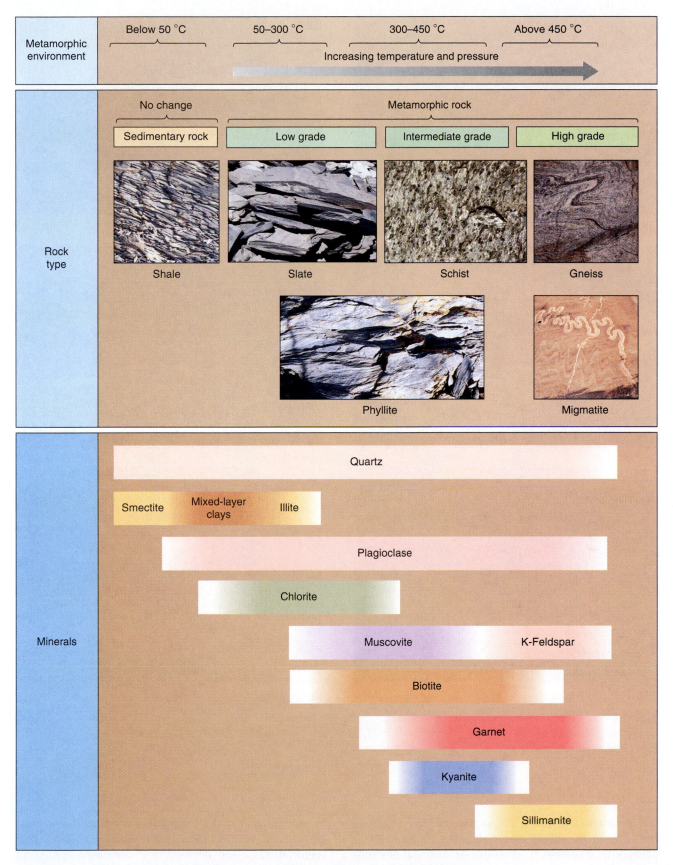

Figure 8–12 Shale changes in both texture and minerals as metamorphic grade increases. The lower part of the figure shows the stability ranges of common metamorphic minerals.

(a)

(b)

(c)

(d)

(e)

(f)

Figure 8–13 (a) Shale is the most common sedimentary rock. Regional metamorphism progressively converts shale to slate (b), phyllite (c), schist (d), and gneiss (e). Migmatite (f) forms when gneiss begins to melt.

and accumulating the metals as they go. The solutions then deposit the dissolved metals when they encounter changes in temperature, pressure, or chemical environment (Fig. 8–15). In this way, hydrothermal solutions scavenge and concentrate metals from average crustal rocks and then deposit them locally to form ore. Hydrothermal ore deposits are discussed further in Chapter 19.

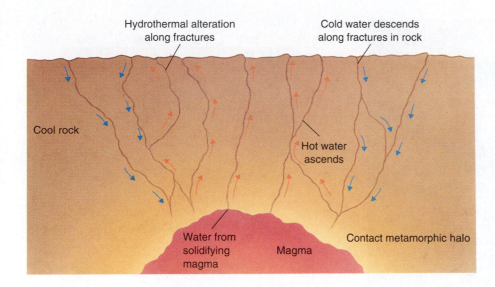

Hydrothermal alteration
along fractures

Cold water descends
along fractures in rock

Cool rock

Hot water
ascends

Water from
solidifying
magma

Magma

Contact metamorphic halo

Figure 8–14 Ground water descending through fractured rock is heated by magma and rises through other cracks, causing hydrothermal metamorphism in nearby rock.

► 8.4 MEASURING METAMORPHIC GRADE

If you are studying a metamorphic rock exposed on the Earth's surface, it is impossible to measure the temperature and pressure at which it formed because the rock may have formed several kilometers beneath the surface and millions or even a few billion years ago. However, scientists estimate the temperature and pressure at which the rock formed using an experimental approach. They heat and apply pressure to chemical compounds similar to the composition of the rock until new minerals form. They then repeat the experiment at different temperatures and pressures until they duplicate the mineral con-

tent of the real rock. Thus, by comparing natural rocks with experimental results, scientists determine the temperature and pressure of metamorphism within 10° or 20°C and a fraction of a kilobar. This experimental approach is not reliable for the slow reactions that form low-grade metamorphic rocks, but it works well for determining the temperature and pressure at which higher-grade rocks formed.

METAMORPHIC FACIES

Imagine that you are studying the outcrop of metamorphic rock shown in Figure 8–16. One striking feature of this outcrop is that it contains two very different rocks,

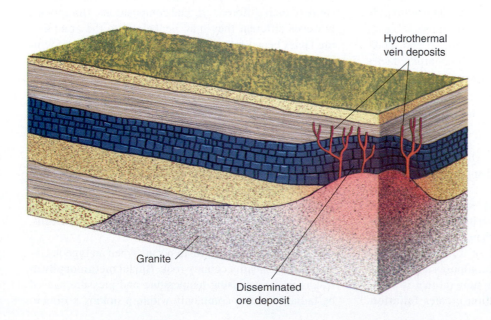

Hydrothermal
vein deposits

Granite

Disseminated
ore deposit

Figure 8–15 Hydrothermal ore deposits form when hot water deposits metals in fractures and surrounding country rock.

Figure 8–16 The same metamorphic conditions have converted limestone to white marble and shale to dark schist in this outcrop in Connecticut.

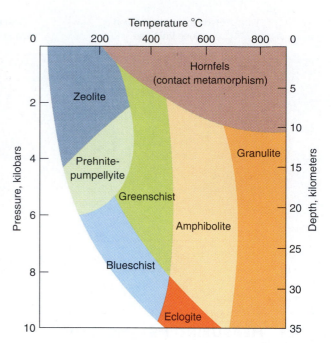

Figure 8–17 The names and metamorphic conditions of the metamorphic facies.

one consisting mostly of black minerals and the other of white ones. Recall that temperature, pressure, and composition control the mineral content of a metamorphic rock. You know that the metamorphic temperature and pressure must have been identical for the two rocks because they are so close together. Therefore, the difference in mineral content must result from a compositional contrast between the two rocks.

Ideally, all metamorphic rocks that formed under identical temperature and pressure conditions are grouped together into a single category called a **metamorphic facies**. Each rock differs from others in the same facies by having a different chemical composition and therefore a different mineral assemblage. Metamorphic facies differ from one another in that they form under different

conditions of temperature and pressure. Each facies is given a name derived from a mineral and/or texture commonly found in rocks of that facies (Fig. 8–17).

Think of the white and black rocks shown in Figure 8–16. Initially, the white rock layers were limestone, and the black layers were shale. Limestone has a different composition from that of shale. Both rocks were metamorphosed at the temperature and pressure of the amphibolite facies. The limestone became marble. The shale converted to schist. Each contains different minerals because of their different original compositions. Both rocks, however different they may be, belong to the amphibolite facies because they both formed under the same temperature and pressure conditions.

SUMMARY

Metamorphism is the process by which solid rocks and minerals change in response to changing environmental conditions.

Most metamorphic reactions occur because each mineral is stable only within a certain range of temperature, pressure, and chemical environment. If temperature or pressure rises above that range, or if the chemical environment changes, the mineral decomposes and its components recombine to form a new mineral that is stable at the new conditions. Deformation creates **foliation**.

Both the texture and the minerals can change as a rock is metamorphosed. The mineralogy of a metamorphic rock reflects its **metamorphic grade**, the temperature and pressure at which it formed. Metamorphic grade is often expressed by the relative terms **low-**, **medium-**, and **high-grade metamorphism**.

Contact metamorphism occurs when an igneous intrusion heats nearby country rock. **Burial metamorphism** results from increasing temperature and pressure caused by burial of rocks, commonly within a sinking sedimentary basin. Both types of metamorphism produce **nonfoliated** metamorphic rocks. **Regional metamorphism** forms **foliated** rocks. It is the most common type of metamorphism and is caused by rising temperature and pressure accompanied by tectonic deformation, commonly near a subduction zone. **Hydrothermal meta-**

morphism occurs when hot, migrating fluids change the chemical composition of country rock.

All metamorphic rocks that formed under identical temperature and pressure are grouped into a **metamorphic facies**.

Important Metamorphic Rocks

Marble	**Quartzite**	**Hornfels**	**Argillite**	**Slate**
Phyllite	**Schist**	**Gneiss**	**Migmatite**	

KEY WORDS

solid-state reaction *124*
metamorphism *124*
stable *126*
unstable *126*
deformation *127*
foliation *128*
slaty cleavage *128*

metamorphic grade *128*
geothermal gradient *128*
contact metamorphism *129*
burial metamorphism *129*

regional metamorphism *131*
hydrothermal metamorphism *132*
metasomatism *132*
magmatic water *132*

metamorphic water *132*
metamorphic facies *136*

REVIEW QUESTIONS

1. Describe the two general kinds of changes that a rock undergoes during metamorphism.

2. Describe four main factors that cause and control metamorphism.

3. How does the nature of parent rock affect the products of metamorphism?

4. What is the approximate temperature range over which metamorphism occurs? What factors define the upper and lower temperatures of metamorphism?

5. Explain how deformation produces foliated textures in metamorphic rocks.

6. Name and briefly describe each of the different types of metamorphism.

7. Describe and name a rock that might result from contact metamorphism of a shale.

8. What rock types might form by contact metamorphism of limestone?

9. Describe and name the succession of metamorphic rocks that form as shale experiences progressively higher grades of regional dynamothermal metamorphism.

10. Where does the water responsible for hydrothermal alteration originate?

11. How does contact metamorphism differ from regional metamorphism?

12. What is a metamorphic facies?

DISCUSSION QUESTIONS

1. Discuss processes that might cause the composition of a rock to change during metamorphism. What would be the effects of a change in composition?

2. Referring to Figure 8–8, what metamorphic grades would be expected to occur in rocks exposed to the following conditions?   a. 400°C and 3 kilobars;   b. 400°C and a depth of 12 kilometers;   c. 600°C at the Earth's surface;   d. 200°C and a depth of 15 kilometers. Referring to Figure 8–17, what metamorphic facies would be expected to form under the same sets of conditions?

3. What types of metamorphic rocks would you expect to find in the following environments?   a. adjacent to a hot spring in Yellowstone Park;   b. in the Appalachian Mountains, which is an old region of mountain building caused by collision of two tectonic plates;   c. at a depth of 6000 meters beneath southern Louisiana.

4. Referring to Figure 8–17, what metamorphic facies would you expect to find at 600°C and at a depth of 20 kilometers?

Geologic Time: A Story in the Rocks

Geologists commonly study events that occurred in the past. They observe rocks and landforms and ask questions such as "What geologic processes shaped that mountain range?" "When did the mountains rise and erode?" For example, compare the Appalachians, a low, rounded mountain range, with the Tetons, whose rocky peaks rise precipitously from the valley floor. We might ask, "Do the two ranges seem so different because the Appalachians are older and have been eroding for a longer time? Were the Appalachians once as steep as the Tetons are today? If so, when did their rocky summits rise, and when did they become rounded?"

Fossil trilobites. Trilobites dominated the seas during Cambrian time and survived for about 300 million years, until the end of the Paleozoic Era. (© John Cancalosi/OKAPIA 1991)

Figure 9–1 Limestone was deposited in a shallow sea and then uplifted and folded in the Canadian Rockies, Alberta.

▶ 9.1 GEOLOGIC TIME

While most of us think of time in terms of days or years, geologists commonly refer to events that happened millions or billions of years ago. In Chapter 1 you learned that the Earth is approximately 4.6 billion years old. Yet humans and our human-like ancestors have existed for 4 million years, and recorded history is only a few thousand years old. How do geologists measure the ages of rocks and events that occurred millions or billions of years ago?

Geologists measure geologic time in two different ways. **Relative age** lists the order in which events occurred. Determination of relative age is based on a simple principle: In order for an event to affect a rock, the rock must exist first. Thus, the rock must be older than the event. This principle seems obvious, yet it is the basis of much geologic work. For example, consider the rocks shown in Figure 9–1. Sediment normally accumulates in horizontal layers. If you observe a fold in the layers, you can deduce that the folding occurred after the sediment was deposited. The order in which rocks and

geologic features formed can almost always be interpreted by observation and logic.

Absolute age is age in years. Dinosaurs became extinct 65 million years ago. The Teton Range in Wyoming began rising 6 million years ago. Absolute age tells us both the order in which events occurred and the amount of time that has passed since they occurred.

▶ 9.2 RELATIVE GEOLOGIC TIME

Absolute age measurements have become common only in the second half of this century. Prior to that time, geologists used field observations to determine relative ages. Even today, with sophisticated laboratory processes available, most field geologists routinely use relative ages. Geologists use a combination of common sense and a few simple principles to determine the order in which rocks formed and changed over time.

The **principle of original horizontality** is based on observation that sediment usually accumulates in horizontal layers (Fig. 9–2a). If sedimentary rocks lie at an

(a)

(b)

Figure 9–2 (a) The principle of original horizontality tells us that most sedimentary rocks are deposited with horizontal bedding (San Juan River, Utah). (b) When we see tilted rocks, we infer that they were tilted after they were deposited (Connecticut).

Older

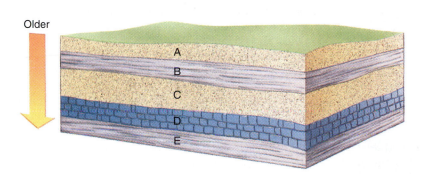

Figure 9–3 In a sequence of sedimentary beds, the oldest bed is the lowest, and the youngest is on top. These beds become older in the order A, B, C, D, E.

angle, as in Figure 9–2b, we can infer that tectonic forces tilted them after they formed.

The **principle of superposition** states that sedimentary rocks become younger from bottom to top (as long as tectonic forces have not turned them upside down). This is because younger layers of sediment always accumulate on top of older layers. In Figure 9–3, the sedimentary layers become progressively younger in the order E, D, C, B, and A.

The **principle of crosscutting relationships** is based on the obvious fact that a rock must first exist before anything can happen to it. Figure 9–4 shows light granite dikes cutting through older country rock. Clearly, the country rock must be older than the dikes. Figure 9–5 shows sedimentary rocks intruded by three granite dikes. Dike B cuts dike C, and dike A cuts dike B, so dike C is older than B, and dike A is the youngest. The sedimentary rocks must be older than all of the dikes.

Figure 9–4 Light granitic dikes cutting across older country rock along the coast in southeast Alaska.

Figure 9–5 Three granite dikes cutting sedimentary rocks. The dikes become younger in the order C, B, A. The sedimentary rocks must be older than all three dikes.

▶ 9.3   UNCONFORMITIES

The 2-kilometer-high walls of Grand Canyon are composed of sedimentary rocks lying on older igneous and metamorphic rocks. Their ages range from about 200 million years to nearly 2 billion years. The principle of superposition tells us that the deepest rocks are the oldest and the rocks become progressively younger as we climb up the canyon walls. However, no principle assures us that the rocks formed continuously from 2 billion to 200 million years ago. Thus, the rock record may not be complete. Suppose that no sediment was deposited for a period of time, or erosion removed some sedimentary layers before younger layers accumulated. In either case a gap would exist in the rock record. We know that any rock layer is younger than the layer below it, but without more information we do not know how much younger.

Layers of sedimentary rocks are **conformable** if they were deposited without interruption. An **unconformity** represents an interruption in deposition, usually of long duration. During the interval when no sediment was

Figure 9–7 A disconformity separates horizontally layered sandstone and an overlying conglomerate layer in Wyoming. Some sandstone layers were eroded away before the conglomerate was deposited.

deposited, some rock layers may have been eroded. Thus, an unconformity represents a long time interval for which no geologic record exists in that place. The lost record may involve hundreds of millions of years.

Several types of unconformities exist. In a **disconformity**, the sedimentary layers above and below the unconformity are parallel (Figs. 9–6 and 9–7). A discon-

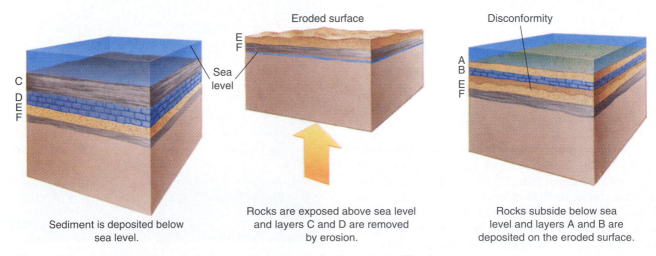

Sediment is deposited below sea level.

Rocks are exposed above sea level and layers C and D are removed by erosion.

Rocks subside below sea level and layers A and B are deposited on the eroded surface.

Figure 9–6 A sequence of events leading to development of a disconformity. The disconformity represents a gap in the rock record.

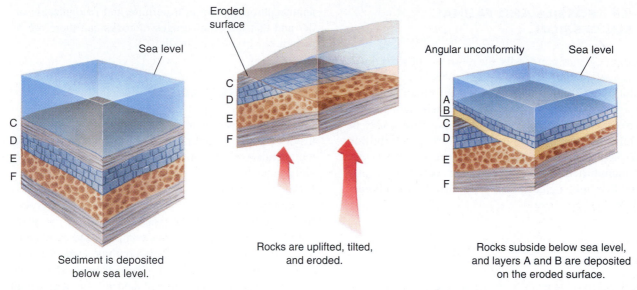

Figure 9–8   An angular unconformity develops when older sedimentary rocks are tilted and eroded before younger sediment accumulates.

formity may be difficult to recognize unless an obvious soil layer or erosional surface is developed. However, geologists identify disconformities by determining the ages of rocks using methods based on fossils and absolute dating, described later in this chapter.

In an **angular unconformity**, tectonic activity tilted older sedimentary rock layers before younger sediment accumulated (Figs. 9–8 and 9–9).

A **nonconformity** is an unconformity in which sedimentary rocks lie on igneous or metamorphic rocks. The nonconformity shown in Figure 9–10 represents a time gap of about 1 billion years.

Figure 9–10   A nonconformity in the cliffs of the Grand Canyon, Arizona. You can see sedimentary bedding in the red sandstone above the nonconformity but none in the igneous and metamorphic rocks below.

Figure 9–9   An angular unconformity near Capitol Reef National Park, Utah.

▶ 9.4 FOSSILS AND FAUNAL SUCCESSION

Paleontologists study **fossils**, the remains and other traces of prehistoric life, to understand the history of life and evolution. Fossils also provide information about the ages of sedimentary rocks and their depositional environments. The oldest known fossils are traces of bacteria-like organisms that lived about 3.5 billion years ago. A much younger fossil consists of the frozen and mummified remains of a Bronze Age man recently found frozen in a glacier near the Italian-Austrian border (Fig. 9–11).

INTERPRETING GEOLOGIC HISTORY FROM FOSSILS

Fossils allow geologists to interpret geologic history. For example, the remains of marine animals in rocks of the Canadian Rockies or near the top of Mount Everest tell us that these places once lay submerged beneath the sea. Therefore, we infer that later tectonic processes raised these regions to their present elevations.

The **theory of evolution** states that life forms have changed throughout geologic time. Fossils are useful in determining relative ages of rocks because different animals and plants lived at different times in the Earth's history. For example, trilobites lived from 535 million to 245 million years ago, and the first dinosaurs appeared about 220 million years ago.

In a sequence of sedimentary rocks that formed over a long time, different fossils appear and then vanish from bottom to top in the same order in which the organisms evolved and then became extinct through time. Rocks containing dinosaur bones must be younger than those containing trilobite remains. The **principle of faunal succession** states that fossil organisms succeeded one another through time in a definite and recognizable order and that the relative ages of rocks can therefore be recognized from their fossils.

▶ 9.5 CORRELATION

Ideally geologists would like to develop a continuous history for each region of the Earth by interpreting rocks that formed in that place throughout geologic time. Unfortunately, there is no single place on Earth where rocks formed and were preserved continuously. Erosion has removed some rock layers and at times no rocks formed. Consequently, the rock record in any one place is full of gaps. To assemble as complete and continuous a record as possible, geologists combine evidence from many localities. To do this, rocks of the same age from different localities must be matched in a process called **correlation**. But how do we correlate rocks over great distances?

If you follow a single continuous sedimentary bed from one place to another, then it is clearly the same layer in both localities. But this approach is impractical over long distances and where rocks are not exposed. Another problem arises when correlation is based on continuity of a sedimentary layer. When rocks are correlated for the purpose of building a geologic time scale, geologists want to show that certain rocks all formed at the same time. Suppose that you are attempting to trace a beach sandstone that formed as sea level rose. The beach would have migrated inland over time, and as a result, the sandstone would have become younger in a landward direction (Fig. 9–12).

Over a distance short enough to be covered in an hour or so of walking, the age difference may be unimportant. But if you trace a similar beach sand laterally over hundreds of kilometers, its age may vary by mil-

Figure 9–11 The mummified remains of this Bronze Age hunter were recently discovered preserved in glacial ice near the Austrian-Italian border. (Sygma)

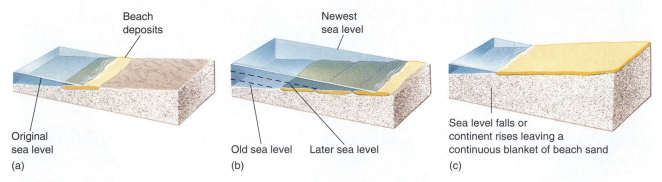

Figure 9–12 As sea level rises (a and b) or falls (c) slowly through time, a beach migrates laterally, forming a single sand layer that is of different ages in different geographic localities.

lions of years. Thus, there are two kinds of correlation: **Time correlation** means age equivalence, but **lithologic correlation** means continuity of a rock unit, such as the sandstone. The two are not always the same because some rock units, such as the sandstone, were deposited at different times in different places. To construct a record of Earth history and a geologic time scale, geologists must find other evidence of the geologic ages of the rocks.

INDEX FOSSILS

An **index fossil** indicates the age of rocks containing it. To be useful, an index fossil is produced by an organism that (1) is abundantly preserved in rocks, (2) was geographically widespread, (3) existed as a species or genus for only a relatively short time, and (4) is easily identified in the field. Floating or swimming marine animals that evolved rapidly make the best index fossils. A marine habitat allows rapid and widespread distribution of these organisms. If a species evolved rapidly and soon became extinct, then it existed for only a short time. The shorter the time span that a species existed, the more precisely the index fossil reflects the age of a rock.

In many cases, the presence of a single type of index fossil is sufficient to establish the age of a rock. More commonly, an assemblage of several fossils is used to date and correlate rocks. Figure 9–13 shows an example of how index fossils and fossil assemblages are used in correlation.

KEY BEDS

A **key bed** is a thin, widespread sedimentary layer that was deposited rapidly and synchronously over a wide area and is easily recognized. Many volcanic eruptions eject large volumes of fine, glassy volcanic ash into the atmosphere. Wind carries the ash over great distances before it settles. Some historic ash clouds have encircled the globe. When the ash settles, the glass rapidly crystallizes to form a pale clay layer that is incorporated into sedimentary rocks. Such volcanic eruptions occur at a precise point in time, so the ash is the same age everywhere.

▶ 9.6 ABSOLUTE GEOLOGIC TIME

How does a geologist measure the absolute age of an event that occurred before calendars and even before humans evolved to keep calendars? Think of how a calendar measures time. The Earth rotates about its axis at a constant rate, once a day. Thus, each time the Sun rises, you know that a day has passed and you check it off on your calendar. If you mark off each day as the Sun rises, you record the passage of time. To know how many days have passed since you started keeping time, you just count the check marks. Absolute age measurement depends on two factors: *a process that occurs at a constant rate* (e.g., the Earth rotates once every 24 hours) and some way to keep *a cumulative record of that process* (e.g., marking the calendar each time the Sun rises). Measurement of time with a calendar, a clock, an hourglass, or any other device depends on these two factors.

Geologists have found a natural process that occurs at a constant rate and accumulates its own record: It is the radioactive decay of elements that are present in many rocks. Thus, many rocks have built-in calendars. We must understand radioactivity to read the calendars.

RADIOACTIVITY

Recall from Chapter 3 that an atom consists of a small, dense nucleus surrounded by a cloud of electrons. A nu-

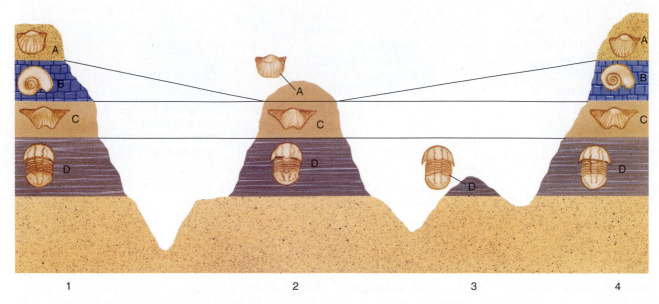

1   2   3   4

Figure 9–13 Schematic view showing the use of fossils and faunal succession to demonstrate age equivalency of sedimentary rocks from widely separated geographic localities. Sedimentary beds containing the same fossil assemblages are interpreted to be of the same age. The fossils show that at locality 2 the sedimentary beds of layer B are missing because layer A directly overlies layer C. Either layer B was deposited and then eroded away before A was deposited at locality 2, or layer B was never deposited here. At locality 3 all layers above D are missing, either because of erosion or because they were never deposited here.

cleus consists of positively charged **protons** and neutral particles called **neutrons**. All atoms of any given element have the same number of protons in the nucleus. However, the number of neutrons may vary. **Isotopes** are atoms of the same element with different numbers of neutrons. For example, all isotopes of potassium have 19 protons, but one isotope has 21 neutrons and another has 20 neutrons. Each isotope is given the name of the element followed by the total number of protons *plus* neutrons in its nucleus. Thus, potassium-40 contains 19 protons and 21 neutrons. Potassium-39 has 19 protons but only 20 neutrons.

Many isotopes are **stable** and do not change with time. If you studied a sample of potassium-39 for 10 billion years, all the atoms would remain unchanged. Other isotopes are **unstable** or radioactive. Given time, their nuclei spontaneously break apart. Potassium-40 decomposes naturally to form two other isotopes, argon-40 and calcium-40 (Fig. 9–14). A radioactive isotope such as potassium-40 is known as a **parent isotope**. An isotope created by radioactivity, such as argon-40 or calcium-40, is called a **daughter isotope**.

Many common elements, such as potassium, consist of a mixture of radioactive and nonradioactive isotopes. With time, the radioactive isotopes decay, but the nonra-

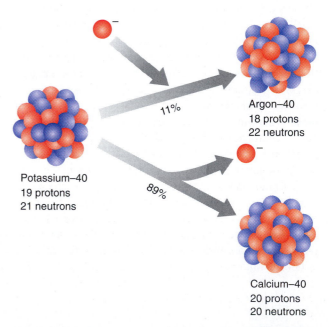

Potassium–40
19 protons
21 neutrons

11%

89%

Argon–40
18 protons
22 neutrons

Calcium–40
20 protons
20 neutrons

Figure 9–14 The radioactive decay of potassium-40 to argon-40 and calcium-40. Eleven percent of the potassium-40 atoms that decay convert to argon-40; a small negatively charged particle is added. The other 89 percent convert to calcium-40; a small negatively charged particle is released.

CARBON-14 DATING

Carbon-14 dating differs from the other parent–daughter pairs shown in Table 9–1 in two ways. First, carbon-14 has a half-life of only 5730 years, in contrast to the millions or billions of years of other isotopes used for radiometric dating. Second, accumulation of the daughter, nitrogen-14, cannot be measured. Nitrogen-14 is the most common isotope of nitrogen, and it is impossible to distinguish nitrogen-14 produced by radioactive decay from other nitrogen-14 in an object being dated.

The abundant stable isotope of carbon is carbon-12. Carbon-14 is continuously created in the atmosphere as cosmic radiation bombards nitrogen-14, converting it to carbon-14. The carbon-14 then decays back to nitrogen-14. Because it is continuously created and because it decays at a constant rate, the ratio of carbon-14 to carbon-12 in the atmosphere remains nearly constant. While an organism is alive, it absorbs carbon from the atmosphere. Therefore, the organism contains the same ratio of carbon-14 to carbon-12 as found in the atmosphere. However, when the organism dies, it stops absorbing new carbon. Therefore, the proportion of carbon-14 in the remains of the organism begins to diminish at death as the carbon-14 decays. Thus, carbon-14 age determinations are made by measuring the ratio of carbon-14 to carbon-12 in organic material. As time passes, the proportion of carbon-14 steadily decreases. By the time an organism has been dead for 50,000 years, so little carbon-14 remains that measurement is very difficult. After about 70,000 years, nearly all of the carbon-14 has decayed to nitrogen-14, and the method is no longer useful.

DISCUSSION QUESTION

Explain why carbon-14 dating would not be useful to date Mesozoic carbonized fossils.

---

dioactive ones do not. A few elements, such as uranium, consist only of radioactive isotopes. The amount of uranium on Earth slowly decreases as it decomposes to other elements, such as lead.

RADIOACTIVITY AND HALF-LIFE

If you watch a single atom of potassium-40, when will it decompose? This question cannot be answered because any particular potassium-40 atom may or may not decompose at any time. But recall from Chapter 3 that even a small sample of a material contains a huge number of atoms. Each atom has a certain probability of decaying at any time. Averaged over time, half of the atoms in any sample of potassium-40 will decompose in 1.3 billion years. The **half-life** is the time it takes for half of the atoms in a sample to decompose. The half-life of potassium-40 is 1.3 billion years. Therefore, if 1 gram of potassium-40 were placed in a container, 0.5 gram would remain after 1.3 billion years, 0.25 gram after 2.6 billion years, and so on. Each radioactive isotope has its own half-life; some half-lives are fractions of a second and others are measured in billions of years.

THE BASIS OF RADIOMETRIC DATING

Two aspects of radioactivity are essential to the calendars in rocks. First, the half-life of a radioactive isotope is constant. It is easily measured in the laboratory and is unaffected by geologic processes. So radioactive decay occurs at a known, constant rate. Second, as a parent isotope decays, its daughter accumulates in the rock. The longer the rock exists, the more daughter isotope accumulates. The accumulation of a daughter isotope is analogous to marking off days on a calendar. Because radioactive isotopes are widely distributed throughout the Earth, many rocks have built-in calendars that allow us to measure their ages. **Radiometric dating** is the process of determining the ages of rocks, minerals, and fossils by measuring their parent and daughter isotopes.

Figure 9–15 shows the relationships between age and relative amounts of parent and daughter isotopes. At the end of one half-life, 50 percent of the parent atoms have decayed to daughter. At the end of two half-lives, the mixture is 25 percent parent and 75 percent daughter. To determine the age of a rock, a geologist measures the proportions of parent and daughter isotopes in a sample and compares the ratio to a similar graph. Consider a hypothetical parent–daughter pair having a half-life of 1 million years. If we determine that a rock contains a mixture of 25 percent parent isotope and 75 percent daughter, Figure 9–15 shows that the age is two half-lives, or 2 million years.

If the half-life of a radioactive isotope is short, an isotope gives accurate ages for young materials. For example, carbon-14 has a half-life of 5730 years. Carbon-14 dating gives accurate ages for materials younger than 70,000 years. It is useless for older materials, because by

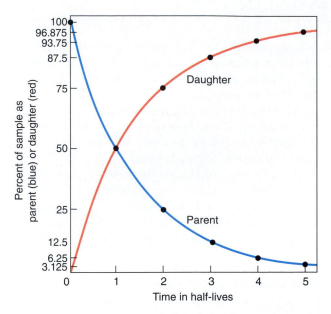

Figure 9–15 As a radioactive parent isotope decays to a daughter, the proportion of parent decreases (blue line), and the amount of daughter increases (red line). The half-life is the amount of time required for half of the parent to convert to daughter. At time zero, when the radiometric calendar starts, a sample is 100 percent parent. At the end of one half-life, 50 percent of the parent has converted to daughter. At the end of two half-lives, 25 percent of the sample is parent and 75 percent is daughter. Thus, by measuring the proportions of parent and daughter in a rock, its age in half-life can be obtained. Because the half-lives of all radioactive isotopes are well known, it is simple to convert age in half-life to age in years.

70,000 years virtually all the carbon-14 has decayed, and no additional change can be measured. Isotopes with long half-lives give good ages for old rocks, but not enough daughter accumulates in young rocks to be measured. For example, rubidium-87 has a half-life of 47 billion years. In a geologically short period of time—10 million years or less—so little of its daughter has accumulated that it is impossible to measure accurately. Therefore, rubidium-87 is not useful for rocks younger than about 10 million years. The six radioactive isotopes that are most commonly used for dating are summarized in Table 9–1.

WHAT IS MEASURED BY A RADIOMETRIC AGE DATE?

Biotite is rich in potassium, so the potassium-40/argon-40 (abbreviated K/Ar) parent–daughter pair can be used successfully. Suppose that we collect a fresh sample of biotite-bearing granite, separate out a few grams of biotite, and obtain a K/Ar date of 100 million years. What happened 100 million years ago to start the K/Ar "calendar" (Fig. 9–16)?

The granite started out as molten magma, which slowly solidified below the Earth's surface. But at this point, the granite was still buried several kilometers below the surface and was still hot. Later, perhaps over millions of years, the granite cooled slowly as tectonic forces pushed it toward the surface, where we collected our

---

Table 9–1 • THE MOST COMMONLY USED ISOTOPES IN RADIOMETRIC AGE DATING

ISOTOPES		HALF-LIFE OF PARENT (YEARS)	EFFECTIVE DATING RANGE (YEARS)	MINERALS AND OTHER MATERIALS THAT CAN BE DATED
Parent	Daughter			
Carbon-14	Nitrogen-14	5730 ± 30	100–70,000	Anything that was once alive: wood, other plant matter, bone, flesh, or shells; also, carbon in carbon dioxide dissolved in ground water, deep layers of the ocean, or glacier ice
Potassium-40	Argon-40 Calcium-40	1.3 billion	50,000–4.6 billion	Muscovite Biotite Hornblende Whole volcanic rock
Uranium-238	Lead-206	4.5 billion	10 million–4.6 billion	Zircon Uraninite and pitchblende
Uranium-235 Thorium-232	Lead-207 Lead-208	710 million 14 billion		
Rubidium-87	Strontium-87	47 billion	10 million–4.6 billion	Muscovite Biotite Potassium feldspar Whole metamorphic or igneous rock

Granite pluton intrudes country rock and solidifies 1 billion years ago. Radiometric calendar starts.

Basalt dike intrudes 0.6 billion years ago. Heat from basalt resets calendar to time zero in nearby granite.

Uplift and erosion expose granite at Earth's surface.

Geologist dates sample from here. Obtains date of 1 billion years.

Geologist dates sample from here. Obtains date of 0.6 billion years.

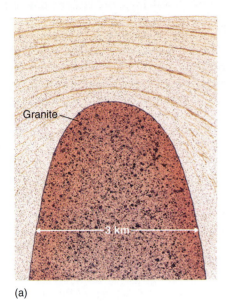

Granite

3 km

(a)

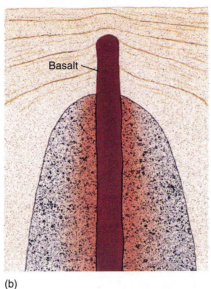

Basalt

(b)

(c)

Figure 9–16   A radiometric age is the time that has elapsed since a rock or mineral last cooled. For example, in (a) this granite magma solidified and cooled 1 billion years ago. It then began to accumulate argon-40 as potassium-40 decayed and the radiometric calendar started. (b) Later, 0.6 billion years ago, a large basalt dike intruded the granite, heating it up again. This heat allowed the argon to escape, resetting the calendar to time zero. (c) A geologist samples the granite and measures its potassium–argon age today at 0.6 billion years. This is the age of the contact metamorphic heating event, not of the original formation of the granite.

sample. So what does our 100-million-year radiometric date tell us: The age of formation of the original granite magma? The time when the magma became solid? The time of uplift?

We can measure time because potassium-40 decays to argon-40, which accumulates in the biotite crystal. As more time elapses, more argon-40 accumulates. Therefore, the biotite calendar does not start recording time until argon atoms begin to accumulate in biotite. Argon is an inert gas that is trapped in biotite as the argon forms. But if the biotite is heated above a certain temperature, the argon escapes. Biotite retains argon only when it cools below 350°C.

Thus, the 100-million-year age date on the granite is the time that has passed since the granite cooled below 350°C, probably corresponding to cooling that occurred during uplift and erosion. If a basalt dike heated 1-billion-year-old granite to a temperature above 350°C, argon would escape from the biotite and the radiometric calendar would be reset to zero (Fig. 9–16).

Now consider the problems in interpreting the radiometric age of a sedimentary rock such as sandstone. Most of the sand grains are quartz, which contains no radioactive isotopes and is useless for dating. However, some sandstones contain small amounts of potassium-bearing minerals that can be used for dating. But those minerals, like the quartz grains, were eroded from older rocks. Consequently, radiometric dating of those grains gives the age of the parent rock, not of the sandstone. However, in some sedimentary rocks, clays and feldspars crystallize as the sediment accumulates. These minerals can be separated and radiometrically dated to give the time that the rocks formed.

Two additional conditions must be met for a radiometric date to be accurate: First, when a mineral or rock cools, no original daughter isotope is trapped in the mineral or rock. Second, once the clock starts, no parent or daughter isotopes are added or removed from the rock or mineral. Metamorphism, weathering, and circulating fluids can add or remove parent or daughter isotopes. Sound geologic reasoning and discretion must be applied when choosing materials for radiometric dating and when interpreting radiometric age dates.

▶ 9.7   THE GEOLOGIC COLUMN AND TIME SCALE

As mentioned earlier, no single locality exists on Earth where a complete sequence of rocks formed continuously throughout geologic time. However, geologists have correlated rocks that accumulated continuously through portions of geologic time from many different localities around the world. The information has been combined to create the **geologic column**, which is a nearly complete composite record of geologic time (Table 9–2). The worldwide geologic column is frequently revised as geologic mapping continues.

Geologists divide all of geologic time into smaller units for convenience. Just as a year is subdivided into months, months into weeks, and weeks into days, large geologic time units are split into smaller intervals. The units are named, just as months and days are. The largest time units are **eons**, which are divided into **eras**. Eras are subdivided, in turn, into **periods**, which are further subdivided into **epochs**.

The geologic column and the geologic time scale were constructed on the basis of relative age determinations. When geologists developed radiometric dating, they added absolute ages to the column and time scale.

Today, geologists use this time scale to date rocks in the field. Imagine that you are studying a sedimentary sequence. If you find an index fossil or a key bed that has already been dated by other scientists, you know the age of the rock and you do not need to send the sample to a laboratory for radiometric dating.

GEOLOGIC TIME

Look again at the geologic time scale of Table 9–2. Notice that the Phanerozoic Eon, which comprises the most recent 538 million years of geologic time, takes up most of the table and contains all of the named subdivisions. The earlier eons—the Proterozoic, Archean, and Hadean—are often not subdivided at all, even though together they constitute a time interval of 4 billion years, almost eight times as long as the Phanerozoic. Why is Phanerozoic time subdivided so finely? Or, conversely, what prevents subdivision of the earlier eons?

Earliest Eons of Geologic Time: Precambrian Time

Geologic time units are based largely on fossils found in rocks. Only a few Earth rocks are known that formed during the **Hadean Eon** (Greek for "beneath the Earth"), the earliest time in Earth history. No fossils of Hadean age are known. It may be that erosion or metamorphism destroyed traces of Hadean life, or that Hadean time preceded the evolution of life. In any case, Hadean time is not amenable to subdivision based on fossils.

Most rocks of the **Archean Eon** (Greek for "ancient") are igneous or metamorphic, although a few Archean sedimentary rocks are preserved. Some contain microscopic fossils of single-celled organisms. Life on Earth apparently began sometime during the Archean Eon, although fossils are neither numerous nor well preserved enough to permit much fine tuning of Archean time.

Large and diverse groups of fossils have been found in sedimentary rocks of the **Proterozoic Eon** (Greek for "earlier life"). The most complex are not only multicellular but have different kinds of cells arranged into tissues and organs. Some of these organisms look so much like modern jellyfish, corals, and worms that some paleontologists view them as ancestors of organisms alive today. However, other paleontologists believe that the resemblance is only superficial. A few types of Proterozoic shell-bearing organisms have been identified, but shelled organisms did not become abundant until the Paleozoic Era.

Commonly, the Hadean, Archean, and Proterozoic Eons are collectively referred to by the informal term **Precambrian**, because they preceded the Cambrian Period, when fossil remains first became very abundant.

Phanerozoic Eon

The word *Phanerozoic* is from the Greek, meaning "visible life." Sedimentary rocks of the **Phanerozoic Eon** contain plentiful and easily discernible fossils. The beginning of Phanerozoic time marks a dramatic increase in the abundance and diversity of life.

Subdivision of the Phanerozoic Eon into three eras is based on the most common types of life during each era. Sedimentary rocks formed during the **Paleozoic era** (Greek for "old life") contain fossils of early life forms, such as invertebrates, fishes, amphibians, reptiles, ferns, and cone-bearing trees. Sedimentary rocks of the **Mesozoic era** (Greek for "middle life") contain new types of **phytoplankton**, microscopic plants that float at or near the sea surface, and beautiful, swimming cephalopods called **ammonoids**. However, the Mesozoic Era is most famous for the **dinosaurs** that dominated the land (Fig. 9–17). Mammals and flowering plants also evolved during this era. During the **Cenozoic era** (Greek for "recent life"), mammals and grasses became abundant.

The eras of Phanerozoic time are subdivided into **periods**, the time unit most commonly used by geolo-

Table 9–2 • THE GEOLOGIC COLUMN AND TIME SCALE*

TIME UNITS OF THE GEOLOGIC TIME SCALE

Eon	Era	Period		Epoch		DISTINCTIVE PLANTS AND ANIMALS
Phanerozoic Eon (*Phaneros* = "evident"; *Zoon* = "life")	Cenozoic Era	Quaternary		Recent or Holocene	"Age of Mammals"	Humans
				Pleistocene —2		Mammals develop and become dominant
		Tertiary	Neogene	Pliocene —5		
				Miocene —24		
			Paleogene	Oligocene —37		
				Eocene —58		Extinction of dinosaurs and many other species
				Paleocene		
		—66				
	Mesozoic Era	Cretaceous			"Age of Reptiles"	First flowering plants, greatest development of dinosaurs
		—144				
		Jurassic				First birds and mammals, abundant dinosaurs
		—208				
		Triassic				First dinosaurs
		—245				
	Paleozoic Era	Permian			"Age of Amphibians"	Extinction of trilobites and many other marine animals
		—286				
		Carboniferous	Pennsylvanian			Great coal forests; abundant insects, first reptiles
		—320				
			Mississippian			Large primitive trees
		—360				
		Devonian			"Age of Fishes"	First amphibians
		—408				
		Silurian				First land plant fossils
		—438				
		Ordovician			"Age of Marine Invertebrates"	First fish
		—505				
		Cambrian				First organisms with shells, trilobites dominant
		—538				
Proterozoic						First multicelled organisms
2500		Sometimes collectively called Precambrian				
Archean						First one-celled organisms
3800						Approximate age of oldest rocks
Hadean						Origin of the Earth
4600±						

*Time is given in millions of years (for example, 1000 stands for 1000 million, which is 1 billion). The table is *not* drawn to scale. We know relatively little about events that occurred during the early part of the Earth's history. Therefore, the first 4 billion years are given relatively little space on this chart, while the more recent Phanerozoic Eon, which spans only 538 million years, receives proportionally more space.

gist. Some of the periods are named after special characteristics of the rocks formed during that period. For example, the Cretaceous Period is named from the Latin word for "chalk" (creta) after chalk beds of this age in Africa, North America, and Europe. Other periods are named for the geographic localities where rocks of that age were first described. For example, the Jurassic Period is named for the Jura Mountains of France and Switzerland. The Cambrian Period is named for Cambria, the Roman name for Wales, where rocks of this age were first studied.

In addition to the abundance of fossils, another reason that details of Phanerozoic time are better known than those of Precambrian time is that many of the older rocks have been metamorphosed, deformed, and eroded. It is a simple matter of probability that the older a rock is, the greater the chance that metamorphism or erosion has obliterated fossils and other evidence of its history.

Figure 9–17 In this reconstruction, a mother duckbill dinosaur nurtures her babies in their nest on a mud flat, 100 million years ago in Montana. (Museum of the Rockies)

SUMMARY

Determinations of **relative time** are based on geologic relationships among rocks and the evolution of life forms through time. The criteria for relative dating are summarized in a few simple principles: the **principle of original horizontality**, the **principle of superposition**, the **principle of crosscutting relationships**, and the **principle of faunal succession**.

Layers of sedimentary rock are **conformable** if they were deposited without major interruptions. An **unconformity** represents a major interruption of deposition and a significant time gap between formation of successive layers of rock. In a **disconformity**, layers of sedimentary rock on either side of the unconformity are parallel. An **angular unconformity** forms when lower layers of rock are tilted prior to deposition of the upper beds. In a **nonconformity**, sedimentary layers overly an erosion surface developed on igneous or metamorphic rocks.

Fossils are used to date rocks according to the **principle of faunal succession**.

Correlation is the demonstration of equivalency of rocks that are geographically separated. **Index fossils**

and **key beds** are important tools in **time correlation**, the demonstration that sedimentary rocks from different geographic localities formed at the same time. Worldwide correlation of rocks of all ages has resulted in the **geologic column**, a composite record of rocks formed throughout the history of the Earth.

Absolute time is measured by **radiometric age dating**, which relies on the fact that **radioactive parent isotopes** decay to form **daughter isotopes** at a fixed, known rate as expressed by the **half-life** of the isotope. The cumulative effects of the radioactive decay process can be determined because the daughter isotopes accumulate in rocks and minerals.

The major units of the **geologic time scale** are **eons**, **eras**, **periods**, and **epochs**. The **Phanerozoic Eon** is finely and accurately subdivided because sedimentary rocks deposited at this time are often well preserved and they contain abundant well-preserved fossils. In contrast, **Precambrian** rocks and time are only coarsely subdivided because fossils are scarce and poorly preserved and the rocks are often altered.

KEY WORDS

relative age   *140*
absolute age   *140*
principle of original
    horizontality   *140*
principle of superposition
    141

principle of crosscutting
    relationships   *141*
conformable   *142*
unconformity   *142*
disconformity   *142*

angular unconformity
    143
nonconformity   *143*
fossil   *144*
evolution   *144*

principle of faunal
    succession   *144*
correlation   *144*
time correlation   *145*

REVIEW QUESTIONS

1. Describe the two ways of measuring geologic time. How do they differ?

2. Give an example of how the principle of original horizontality might be used to determine the order of events affecting a sequence of folded sedimentary rocks.

3. How does the principle of superposition allow us to determine the relative ages of a sequence of unfolded sedimentary rocks?

4. Explain the principle of crosscutting relationships and how it can be used to determine age relationships among sedimentary rocks.

5. Explain a conformable relationship in sedimentary rocks.

6. Explain the differences among unconformities, disconformities, angular unconformities, and nonconformities.

7. List five different types of fossils and how they are formed.

8. How does a trace fossil differ from other types of fossils?

9. Discuss the principle of faunal succession and the use of index fossils in time correlation.

10. What are the two different types of correlation of rock units? How do they differ?

11. What tools or principles are most commonly used in correlation?

12. Describe the similarities and differences between how a calendar records time and how minerals and rocks containing radioactive isotopes record time.

13. What is radioactivity?

14. What is a stable isotope? An unstable isotope?

15. What is the relationship between parent and daughter isotopes?

16. What is meant by the half-life of a radioactive isotope? How is the half-life used in radiometric age dating?

17. Some radioactive isotopes are useful for measuring relatively young ages, whereas others are useful for measuring older ages. Why is this true?

18. What geologic event is actually measured by a radiometric age determination of an igneous rock or mineral?

19. Why is the Phanerozoic Eon separated into so many subdivisions in contrast to much longer Precambrian time, which has few subdivisions?

20. Describe and discuss the events in Earth history responsible for rock features that allow us to divide geologic time into eras and periods. Describe and discuss the rock features as well.

21. What geologic events are recorded by an angular unconformity? A disconformity? A nonconformity? What can be inferred about the timing of each set of events?

DISCUSSION QUESTIONS

1. Devise two metaphors for the length of geologic time in addition to the metaphor used in Chapter 1. Locate some of the most important time boundaries in your analogy.

2. Suppose that you landed on the Moon and were able to travel in a vehicle that could carry you over the lunar surface to see a wide variety of rocks, but that you had no laboratory equipment to work with. What principles and tools could you use to determine relative ages of Moon rocks?

3. Imagine that one species lived between 538 and 505 million years ago and another lived between 520 and 245 million years ago. What can you say about the age of a rock that contains fossils of both species?

4. Imagine that someone handed you a sample of sedimentary rock containing abundant fossils. What could you tell about its age if you didn't use radiometric dating and you didn't know where it was collected from? What additional information could you determine if you studied the outcrop that it came from?

5. What geologic events are represented by a potassium–argon age from flakes of biotite in   a. granite, b. biotite schist, and   c. sandstone?

6. Suppose you were using the potassium–argon method to measure the age of biotite in granite. What would be the effect on the age measurement if the biotite had been slowly leaking small amounts of argon since it crystallized?

7. Some pyroxenes contain enough potassium to be dated by the potassium–argon method. However, pyroxenes are notorious scavengers of argon. That is, they absorb substantial amounts of argon from magma as they crystallize. How would this argon scavenging affect a potassium–argon age measurement done on a pyroxene?

8. Suppose that you measured and described a sequence of Middle Paleozoic sedimentary rocks in northern Ohio and later did the same with a sequence of the same age in Wyoming. How would you correlate the two sections?

Earthquakes and the Earth's Structure

Continents glide continuously around the globe, but we cannot feel the motion because it is too slow. Occasionally, however, the Earth trembles noticeably. The ground rises and falls and undulates back and forth, as if it were an ocean wave. Buildings topple, bridges fail, roadways and pipelines snap. An **earthquake** is a sudden motion or trembling of the Earth caused by the abrupt release of energy that is stored in rocks.

Before the plate tectonics theory was developed, geologists recognized that earthquakes occur frequently in some regions and infrequently in others, but they did not understand why. Modern geologists know that most earthquakes occur along plate boundaries, where huge tectonic plates separate, converge, or slip past one another.

The January 1995 Kobe earthquake destroyed this portion of the Kobe-Osaka Highway in western Japan and killed nearly 5000 people. (Atsushi Tsukada/AP Wide World)

▶ 10.1 WHAT IS AN EARTHQUAKE?

How do rocks store energy, and why do they suddenly release it as an earthquake?

Stress is a force exerted against an object.[1] You stress a cable when you use it to tow a neighbor's car. Tectonic forces stress rocks. The movement of lithospheric plates is the most common source of tectonic stress.

When an object is stressed, it changes volume and shape. If a solid object is stressed slowly, it first deforms in an elastic manner: When the stress is removed, the object springs back to its original size and shape. A rubber band exhibits elastic deformation. The energy used to stretch a rubber band is stored in the elongated rubber. When the stress is removed, the rubber band springs back to its initial size and shape and releases the stored energy. Rocks also deform elastically when tectonic stress is applied (Fig. 10–1).

Every rock has a limit beyond which it cannot deform elastically. Under certain conditions, when its elastic limit is exceeded, a rock continues to deform like putty. This behavior is called **plastic deformation**. A rock that has deformed plastically retains its new shape when the stress is released (Fig. 10–2). Earthquakes do not occur when rocks deform plastically.

Under other conditions, an elastically stressed rock may rupture by **brittle fracture** (Fig. 10–3). The fracture releases the elastic energy, and the surrounding rock springs back to its original shape. This rapid motion creates vibrations that travel through the Earth and are felt as an earthquake.

Earthquakes also occur when rock slips along previously established faults. Tectonic plate boundaries are huge faults that have moved many times in the past and will move again in the future (Fig. 10–4).

Although tectonic plates move at rates between 1 and 16 centimeters per year, friction prevents the plates from slipping past one another continuously. As a result, rock near a plate boundary stretches or compresses. When its accumulated elastic energy overcomes the friction that binds plates together, the rock suddenly slips along

---

[1]More precisely, stress is defined as force per unit area and is measured in units of newtons per square meter (N/m²).

Figure 10–1 The behavior of a rock as stress increases in graphical form (a), in schematic form (b). At first the rock deforms by elastic deformation in which the amount of deformation is directly proportional to the amount of stress. Beyond the elastic limit, the rock deforms plastically and a small amount of additional stress causes a large increase in distortion. Finally, at the yield point, the rock fractures. Many stressed rocks deform elastically and then rupture, with little or no intermediate plastic deformation. The factors that control rock behavior are discussed further in Chapter 12.

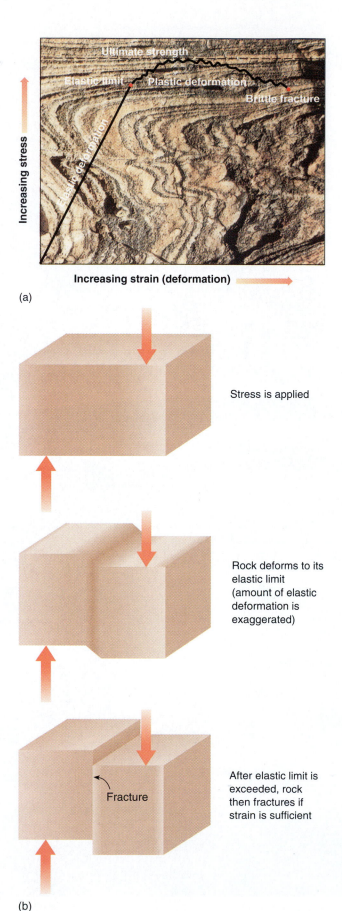

(a)

Increasing stress →

Ultimate strength

Elastic limit   Plastic deformation

Brittle fracture

Elastic deformation

Increasing strain (deformation) →

Stress is applied

Rock deforms to its elastic limit (amount of elastic deformation is exaggerated)

Fracture

After elastic limit is exceeded, rock then fractures if strain is sufficient

(b)

Figure 10–2 Rocks may deform plastically when stressed. Plastic deformation contorted the layering in metamorphic rocks in Connecticut.

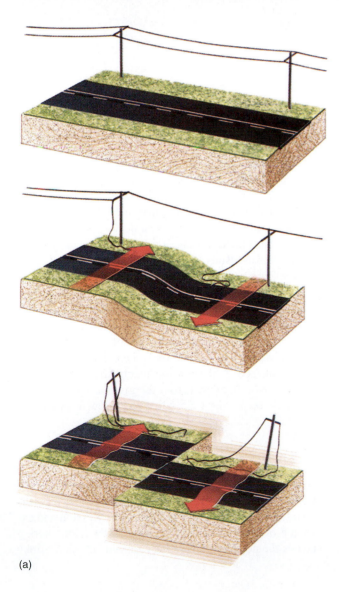

(a)

(b)

Figure 10–3 (a) A rock stores elastic energy when it is distorted by a tectonic force. When the rock fractures, it snaps back to its original shape, creating an earthquake. In the process, the rock moves along the fracture. (b) Moving rock and soil fractured and displaced this roadway during the Loma Prieta earthquake in 1989.

Figure 10–4 California's San Andreas fault, the source of many earthquakes, is the boundary between the Pacific plate, on the left in this photo, and the North American plate, on the right. (R.E. Wallace/USGS)

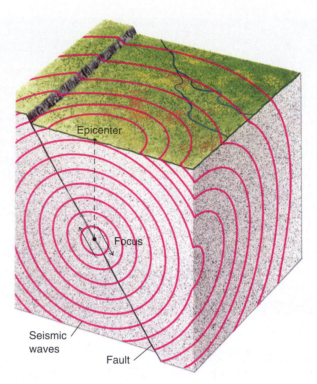

Figure 10–5 Body waves radiate outward from the focus of an earthquake.

the fault, generating an earthquake. The rocks may move from a few centimeters to a few meters, depending on the amount of stored energy.

▶ 10.2 EARTHQUAKE WAVES

If you have ever bought a watermelon, you know the challenge of picking out a ripe, juicy one without being able to look inside. One trick is to tap the melon gently with your knuckle. If you hear a sharp, clean sound, it is probably ripe; a dull thud indicates that it may be over-ripe and mushy. The watermelon illustrates two points that can be applied to the Earth: (1) The energy of your tap travels through the melon, and (2) the nature of the melon's interior affects the quality of the sound.

A wave transmits energy from one place to another. Thus, a drumbeat travels through air as a sequence of

waves, the Sun's heat travels to Earth as waves, and a tap travels through a watermelon in waves. Waves that travel through rock are called **seismic waves**. Earthquakes and explosions produce seismic waves. **Seismology** is the study of earthquakes and the nature of the Earth's interior based on evidence from seismic waves.

An earthquake produces several different types of seismic waves. **Body waves** travel through the Earth's interior. They radiate from the initial rupture point of an earthquake, called the **focus** (Fig. 10–5).

The point on the Earth's surface directly above the focus is the **epicenter**. During an earthquake, body waves carry some of the energy from the focus to the surface. **Surface waves** then radiate from the epicenter along the Earth's surface. Although the mechanism is different, surface waves undulate across the ground like the waves that ripple across the water after you throw a rock into a calm lake.

BODY WAVES

Two main types of body waves travel through the Earth's interior. A **P wave** (also called a compressional wave) is an elastic wave that causes alternate compression and expansion of the rock (Fig. 10–6). Consider a long spring such as the popular Slinky™ toy. If you stretch a Slinky

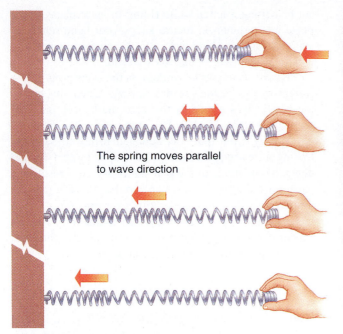

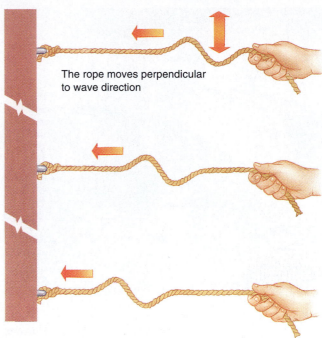

Figure 10–6 Model of a P wave; a compressional wave. The wave is propagated along the spring. The particles in the spring move parallel to the direction of wave propagation.

Figure 10–7 Model of an S wave; a shear wave. The wave is propagated along the rope. The particles in the rope move perpendicular to the direction of wave propagation.

and strike one end, a compressional wave travels along its length. P waves travel through air, liquid, and solid material. Next time you take a bath, immerse your head until your ears are under water and listen as you tap the sides of the tub with your knuckles. You are hearing P waves.

P waves travel at speeds between 4 and 7 kilometers per second in the Earth's crust and at about 8 kilometers per second in the uppermost mantle. As a comparison, the speed of sound in air is only 0.34 kilometer per second, and the fastest jet fighters fly at about 0.85 kilometer per second. P waves are called primary waves because they are so fast that they are the first waves to reach an observer.

A second type of body wave, called an **S wave**, is a **shear wave**. An S wave can be illustrated by tying a rope to a wall, holding the end, and giving it a sharp up-and-down jerk (Fig. 10–7). Although the wave travels parallel to the rope, the individual particles in the rope move at right angles to the rope length. A similar motion in an S wave produces shear stress in rock and gives the wave its name. S waves are slower than P waves and travel at speeds between 3 and 4 kilometers per second in the crust. As a result, S waves arrive after P waves and are the secondary waves to reach an observer.

Unlike P waves, S waves move only through solids. Because molecules in liquids and gases are only weakly

bound to one another, they slip past each other and thus cannot transmit a shear wave.

SURFACE WAVES

Surface waves travel more slowly than body waves. Two types of surface waves occur simultaneously in the Earth (Fig. 10–8). A **Rayleigh wave** moves with an up-and-down rolling motion like an ocean wave. **Love waves** produce a side-to-side vibration. Thus, during an earth-

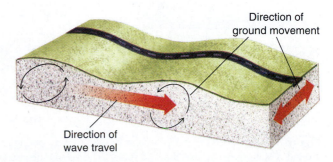

Figure 10–8 Surface waves. Surface motion includes up-and-down movement like that of an ocean wave and also a side-to-side sway.

Figure 10–9 Surface waves cause a large proportion of earthquake damage. Collapse of Interstate Highway 880 during the 1989 Loma Prieta, California, earthquake. (Paul Scott/Sygma)

quake, the Earth's surface rolls like ocean waves and writhes from side to side like a snake (Fig. 10–9).

MEASUREMENT OF SEISMIC WAVES

A **seismograph** is a device that records seismic waves. To understand how a seismograph works, consider the act of writing a letter while riding in an airplane. If the plane hits turbulence, inertia keeps your hand relatively stationary as the plane moves back and forth beneath it, and your handwriting becomes erratic.

Early seismographs worked on the same principle. A weight was suspended from a spring. A pen attached to the weight was aimed at the zero mark on a piece of graph paper (Fig. 10–10). The graph paper was mounted on a rotary drum that was attached firmly to bedrock. During an earthquake, the graph paper jiggled up and down, but inertia kept the pen stationary. As a result, the paper moved up and down beneath the pen. The rotating drum recorded earthquake motion over time. This record of Earth vibration is called a **seismogram** (Fig. 10–11). Modern seismographs use electronic motion detectors which transmit the signal to a computer.

MEASUREMENT OF EARTHQUAKE STRENGTH

Over the past century, geologists have devised several scales to express the size of an earthquake. Before seismographs were in common use, earthquakes were evaluated on the basis of structural damage. One that destroyed many buildings was rated as more intense than one that destroyed only a few. This system did not accurately measure the energy released by a quake, however, because structural damage depends on distance from the focus, the rock or soil beneath the structure, and the quality of construction.

In 1935 Charles Richter devised the **Richter scale** to express earthquake magnitude. Richter magnitude is

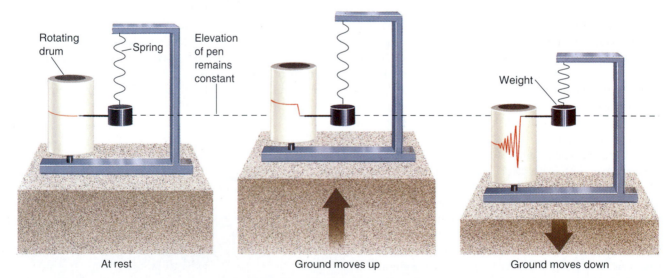

Rotating drum — Spring — Elevation of pen remains constant — Weight

At rest   Ground moves up   Ground moves down

Figure 10–10 A seismograph records ground motion during an earthquake. When the ground is stationary, the pen draws a straight line across the rotating drum. When the ground rises abruptly during an earthquake, it carries the drum up with it. But the spring stretches, so the weight and pen hardly move. Therefore, the pen marks a line lower on the drum. Conversely, when the ground sinks, the pen marks a line higher on the drum. During an earthquake, the pen traces a jagged line as the drum rises and falls.

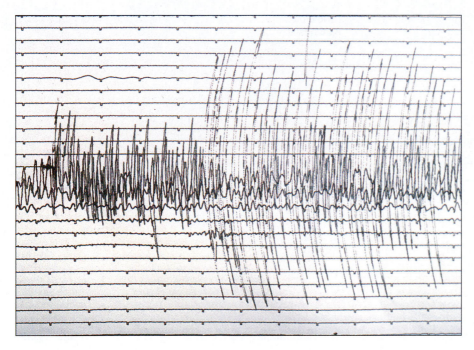

Figure 10–11 This seismogram records north-south ground movements during the October 1989 Loma Prieta earthquake. (Russell D. Curtis/USGS)

calculated from the height of the largest earthquake body wave recorded on a specific type of seismograph. The Richter scale is more quantitative than earlier intensity scales, but it is not a precise measure of earthquake energy. A sharp, quick jolt would register as a high peak on a Richter seismograph, but a very large earthquake can shake the ground for a long time without generating extremely high peaks. In this way, a great earthquake can release a huge amount of energy that is not reflected in the height of a single peak, and thus is not adequately expressed by Richter magnitude.

Modern equipment and methods enable seismologists to measure the amount of slip and the surface area of a fault that moved during a quake. The product of these two values allows them to calculate the **moment magnitude**. Most seismologists now use moment magnitude rather than Richter magnitude because it more closely reflects the total amount of energy released during an earthquake. An earthquake with a moment magnitude of 6.5 has an energy of about 10^{25} (10 followed by 25 zeros) ergs.[2] The atomic bomb dropped on the Japanese city of Hiroshima at the end of World War II released about that much energy.

On both the moment magnitude and Richter scales, the energy of the quake increases by about a factor of 30 for each successive increment on the scale. Thus, a magnitude 6 earthquake releases roughly 30 times more energy than a magnitude 5 earthquake.

The largest possible earthquake is determined by the strength of rocks. A strong rock can store more elastic energy before it fractures than a weak rock. The largest earthquakes ever measured had magnitudes of 8.5 to 8.7, about 900 times greater than the energy released by the Hiroshima bomb.

LOCATING THE SOURCE OF AN EARTHQUAKE

If you have ever watched an electrical storm, you may have used a simple technique for estimating the distance between you and the place where the lightning strikes. After the flash of a lightning bolt, count the seconds that pass before you hear thunder. Although the electrical discharge produces thunder and lightning simultaneously, light travels much faster than sound. Therefore, light reaches you virtually instantaneously, whereas sound travels much more slowly, at 340 meters per second. If the time interval between the flash and the thunder is 1 second, then the lightning struck 340 meters away and was very close.

The same principle is used to determine the distance from a recording station to both the epicenter and focus of an earthquake. Recall that P waves travel faster than

[2]An erg is the standard unit of energy in scientific usage. One erg is a small amount of energy. Approximately 3×10^{12} ergs are needed to light a 100-watt light bulb for 1 hour. However, 10^{25} is a very large number, and 10^{25} ergs represents a considerable amount of energy.

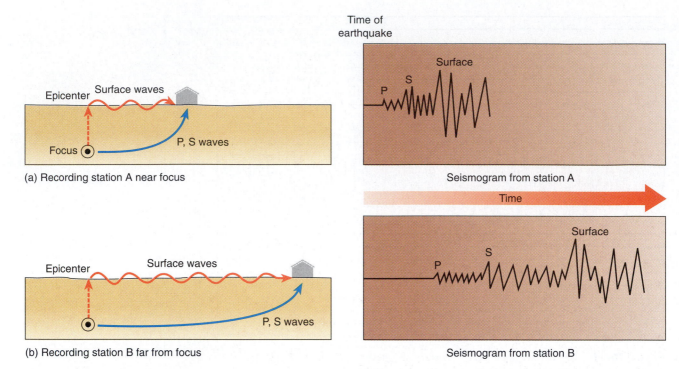

Figure 10–12 The time intervals between arrivals of P, S, and L waves at a recording station increase with distance from the focus of an earthquake.

S waves and that surface waves are slower yet. If a seismograph is located close to an earthquake epicenter, the different waves will arrive in rapid succession for the same reason that the thunder and lightning come close together when a storm is close. On the other hand, if a seismograph is located far from the epicenter, the S waves arrive at correspondingly later times after the P waves arrive, and the surface waves are even farther behind, as shown in Figure 10–12.

Geologists use a **time-travel curve** to calculate the distance between an earthquake epicenter and a seismograph. To make a time-travel curve, a number of seismic stations at different locations record the times of arrival of seismic waves from an earthquake with a known epicenter and occurrence time. Then a graph such as Figure 10–13 is drawn. This graph can then be used to measure the distance between a recording station and an earthquake whose epicenter is unknown.

Time-travel curves were first constructed from data obtained from natural earthquakes. However, scientists do not always know precisely when and where an earthquake occurred. In the 1950s and 1960s, geologists studied seismic waves from atomic bomb tests to improve the time-travel curves because they knew both the locations and timing of the explosions.

Figure 10–13 shows us that if the first P wave arrives 3 minutes before the first S wave, the recording sta-

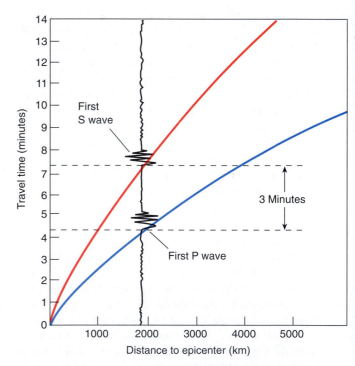

Figure 10–13 A time-travel curve. With this graph you can calculate the distance from a seismic station to the source of an earthquake. In the example shown, a 3-minute delay between the first arrivals of P waves and S waves corresponds to an earthquake with an epicenter 1900 kilometers from the seismic station.

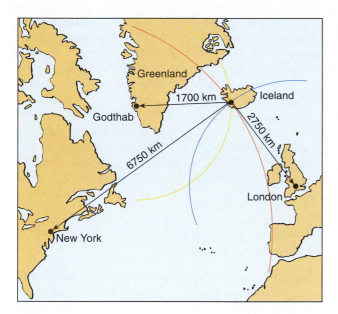

Figure 10–14 Locating an earthquake. The distance from each of three seismic stations to the earthquake is determined from time-travel curves. The three arcs are drawn. They intersect at only one point, which is the epicenter of the earthquake.

Figure 10–15 Jon Turk stands in an area of permanent ground displacement caused by the Loma Prieta, California, earthquake of 1989. (Christine Seashore)

tion is about 1900 kilometers from the epicenter. But this distance does not indicate whether the earthquake originated to the north, south, east, or west. To pinpoint the location of an earthquake, geologists compare data from three or more recording stations. If a seismic station in New York City records an earthquake with an epicenter 6750 kilometers away, geologists know that the epicenter lies somewhere on a circle 6750 kilometers from New York City (Fig. 10–14). The same epicenter is reported to be 2750 kilometers from a seismic station in London and 1700 kilometers from one in Godthab, Greenland. If one circle is drawn for each recording station, the arcs intersect at the epicenter of the quake.

▶ 10.3 EARTHQUAKE DAMAGE

Large earthquakes can displace rock and alter the Earth's surface (Fig. 10–15). The New Madrid, Missouri, earthquake of 1811 changed the course of the Mississippi River. During the 1964 Alaskan earthquake, some beaches rose 12 meters, leaving harbors high and dry, while other beaches sank 2 meters, causing coastal flooding.

Most earthquake fatalities and injuries occur when falling structures crush people. Structural damage, injury, and death depend on the magnitude of the quake, its proximity to population centers, rock and soil types, topography, and the quality of construction in the region.

HOW ROCK AND SOIL INFLUENCE EARTHQUAKE DAMAGE

In many regions, bedrock lies at or near the Earth's surface and buildings are anchored directly to the rock. Bedrock vibrates during an earthquake and buildings may fail if the motion is violent enough. However, most bedrock returns to its original shape when the earthquake is over, so if structures can withstand the shaking, they will survive. Thus, bedrock forms a desirable foundation in earthquake hazard areas.

In many places, structures are built on sand, clay, or silt. Sandy sediment and soil commonly settle during an earthquake. This displacement tilts buildings, breaks pipelines and roadways, and fractures dams. To avert structural failure in such soils, engineers drive steel or concrete pilings through the sand to the bedrock below. These pilings anchor and support the structures even if the ground beneath them settles.

Mexico City provides one example of what can happen to clay-rich soils during an earthquake. The city is built on a high plateau ringed by even higher mountains. When the Spaniards invaded central Mexico, lakes dotted the plateau and the Aztec capital lay on an island at the end of a long causeway in one of the lakes. Over the following centuries, European settlers drained the lake and built the modern city on the water-soaked, clay-rich lake-bed sediment. On September 19, 1985, an earthquake with a magnitude of 8.1 struck about 500 kilometers west of the city. Seismic waves shook the wet clay beneath the city and reflected back and forth between the bedrock sides and bottom of the basin, just as waves in a bowl of Jell-O™ bounce off the side and bottom of the bowl. The reflections amplified the waves, which destroyed more than 500 buildings and killed between 8000 and 10,000 people (Fig. 10–16). Meanwhile, there was comparatively little damage in Acapulco, which was much closer to the epicenter but is built on bedrock.

If soil is saturated with water, the sudden shock of an earthquake can cause the grains to shift closer together, expelling some of the water. When this occurs, increased stress is transferred to the pore water, and the pore pressure may rise sufficiently to suspend the grains in the water. In this case, the soil loses its shear strength and behaves as a fluid. This process is called **liquefaction.** When soils liquefy on a hillside, the slurry flows downslope, carrying structures along with it. During the 1964 earthquake near Anchorage, Alaska, a clay-rich bluff 2.8 kilometers long, 300 meters wide, and 22 meters high liquefied. The slurry carried houses into the ocean and buried some so deeply that bodies were never recovered.

CONSTRUCTION DESIGN AND EARTHQUAKE DAMAGE

A magnitude 6.4 earthquake struck central India in 1993, killing 30,000 people. In contrast, the 1994 magnitude 6.6 quake in Northridge (near Los Angeles) killed only 55. The tremendous mortality in India occurred because buildings were not engineered to withstand earthquakes.

Figure 10–16 The 1985 Mexico City earthquake had a magnitude of 8.1 and killed between 8000 and 10,000 people. Earthquake waves amplified within the soil so that the effects were greater in Mexico City than they were in Acapulco, which was much closer to the epicenter. (Wide World Photos)

Some common framing materials used in buildings, such as wood and steel, bend and sway during an earthquake but resist failure. However, brick, stone, concrete, adobe (dried mud), and other masonry products are brittle and likely to fail during an earthquake. Although masonry can be reinforced with steel, in many regions of the world people cannot afford such reinforcement.

FIRE

Earthquakes commonly rupture buried gas pipes and electrical wires, leading to fire, explosions, and electrocutions (Fig. 10–17). Water pipes may also break, so fire fighters cannot fight the blazes effectively. Most of the damage from the 1906 San Francisco earthquake resulted from fires.

LANDSLIDES

Landslides are common when the Earth trembles. Earthquake-related landslides are discussed in more detail in Chapter 13.

TSUNAMIS

When an earthquake occurs beneath the sea, part of the sea floor rises or falls (Fig. 10–18). Water is displaced in

response to the rock movement, forming a wave. Sea waves produced by an earthquake are often called tidal waves, but they have nothing to do with tides. Therefore, geologists call them by their Japanese name, **tsunami**.

In the open sea, a tsunami is so flat that it is barely detectable. Typically, the crest may be only 1 to 3 meters high, and successive crests may be more than 100 to 150 kilometers apart. However, a tsunami may travel at 750 kilometers per hour. When the wave approaches the shallow water near shore, the base of the wave drags against the bottom and the water stacks up, increasing the height of the wave. The rising wall of water then flows inland. A tsunami can flood the land for as long as 5 to 10 minutes.

▶ 10.4 EARTHQUAKES AND TECTONIC PLATE BOUNDARIES

Although many faults are located within tectonic plates, the largest and most active faults are the boundaries between tectonic plates. Therefore, as Figure 10–19 shows, earthquakes occur most frequently along plate boundaries.

EARTHQUAKES AT A TRANSFORM PLATE BOUNDARY: THE SAN ANDREAS FAULT ZONE

The populous region from San Francisco to San Diego straddles the San Andreas fault zone, which is a transform boundary between the Pacific plate and the North American plate (Fig. 10–20). The fault itself is vertical and the rocks on opposite sides move horizontally. A fault of this type is called a **strike–slip fault** (Fig. 10–21). Plate motion stresses rock adjacent to the fault, generating numerous smaller faults, shown by the solid lines in the figure. The San Andreas fault and its satellites form a broad region called the **San Andreas fault zone**.

In the past few centuries, hundreds of thousands of earthquakes have occurred in this zone. Geologists of the United States Geological Survey recorded 10,000 earthquakes in 1984 alone, although most could be detected only with seismographs. Severe quakes occur periodically. One shook Los Angeles in 1857, and another destroyed San Francisco in 1906. A large quake in 1989 occurred south of San Francisco, and another rocked Northridge, just outside Los Angeles, in January 1994. The fact that the San Andreas fault zone is part of a major plate boundary tells us that more earthquakes are inevitable.

The plates move past one another in three different ways along different segments of the San Andreas fault zone:

Figure 10–17 Ruptured gas and electric lines often cause fires during earthquakes in urban areas. This blaze followed the 1989 San Francisco earthquake. (Michael Williamson/Sygma)

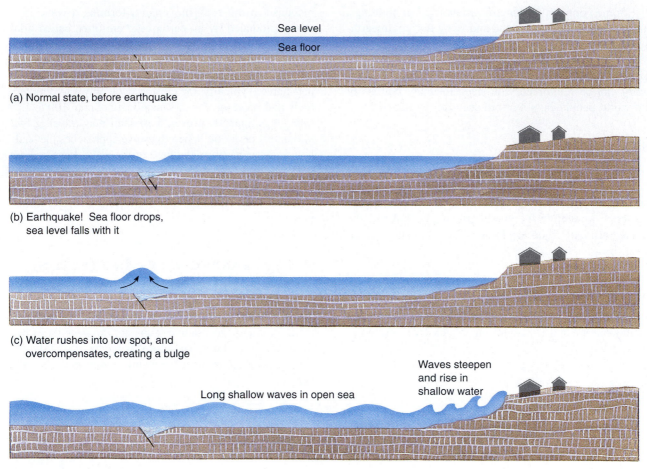

(a) Normal state, before earthquake

(b) Earthquake! Sea floor drops, sea level falls with it

(c) Water rushes into low spot, and overcompensates, creating a bulge

(d) Tsunami generated

Figure 10–18 Formation of a tsunami. If a portion of the sea floor drops during an earthquake, the sea level falls with it. Water rushes into the low spot and overcompensates, creating a bulge. The long, shallow waves build up when they reach land.

1. Along some portions of the fault, rocks slip past one another at a continuous, snail-like pace called **fault creep**. The movement occurs without violent and destructive earthquakes because the rocks move continuously and slowly.

2. In other segments of the fault, the plates pass one another in a series of small hops, causing numerous small, nondamaging earthquakes.

3. Along the remaining portions of the fault, friction prevents slippage of the fault although the plates continue to move past one another. In this case, rock near the fault deforms and stores elastic energy. Because the plates move past one another at 3.5 centimeters per year, 3.5 meters of elastic deformation accumulate over a period of 100 years. When the accumulated elastic energy exceeds friction, the rock suddenly slips along the fault and snaps back

to its original shape, producing a large, destructive earthquake.

The Northridge Earthquake of January 1994

In January 1994, a magnitude 6.6 earthquake struck Northridge in the San Fernando Valley just north of Los Angeles (Fig. 10–22). Fifty-five people died and property damage was estimated at $8 billion.

As explained earlier, the San Andreas fault is a strike–slip fault that is part of a transform plate boundary. In the mid-1980s, geologists also discovered buried thrust faults in Southern California. A **thrust fault** is one in which rock on one side of the fault slides up and over the rock on the other side (Fig. 10–23). While the San

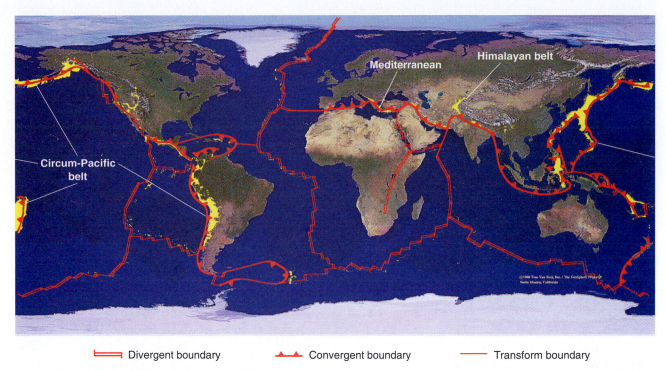

⊨⊨ Divergent boundary   ▲▲ Convergent boundary   — Transform boundary

Figure 10–19   The Earth's major earthquake zones coincide with tectonic plate boundaries. Each yellow dot represents an earthquake that occurred between 1961 and 1967.   (Tom Van Sant, Geosphere Project)

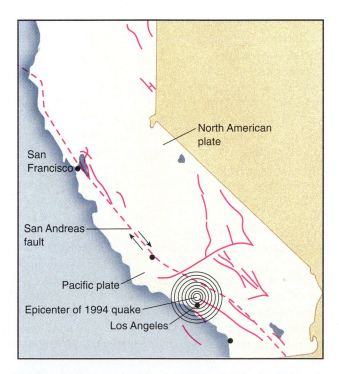

Figure 10–20   Faults and earthquakes in California. The dashed line is the San Andreas fault, and solid lines are related faults. Blue dots are epicenters of recent earthquakes.   (Redrawn from USGS data)

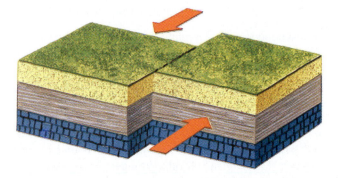

Figure 10–21   A strike–slip fault is vertical, and the rock on opposite sides of the fracture moves horizontally.

Andreas fault lies east of metropolitan Los Angeles, the Santa Monica thrust fault lies directly under the city (Fig. 10–24). A major quake on one of these faults can be more disastrous than one on the main San Andreas fault. The Northridge earthquake occurred when one of these thrust faults slipped. According to geologist James Dolan of the California Institute of Technology, "There's a whole seismic hazard from buried thrust faults that we didn't even appreciate until six years ago."

The existence of thrust faults indicates that in Southern California, both the direction of tectonic forces

Figure 10–22 The January 1994 Northridge, California, earthquake killed 55 people and caused eight billion dollars in damage. (Earthquake Engineering Institute)

and the manner in which stress is relieved are more complicated than a simple model expresses. Although many disastrous and expensive earthquakes have shaken southern California in the past few decades, none of them has been the Big One that seismologists still fear.

EARTHQUAKES AT SUBDUCTION ZONES

In a subduction zone, a relatively cold, rigid lithospheric plate dives beneath another plate and slowly sinks into the mantle. In most places, the subducting plate sinks with intermittent slips and jerks, giving rise to numerous earthquakes. The earthquakes concentrate along the upper part of the sinking plate, where it scrapes past the opposing plate (Fig. 10–25). This earthquake zone is called the **Benioff zone**, after the geologist who first recognized it. Many of the world's strongest earthquakes occur in subduction zones.

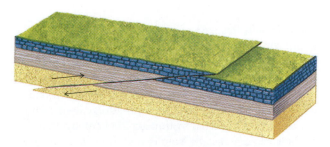

Figure 10–23 A thrust fault is a low-angle fault in which rock on one side of the fault slides up and over rock on the other side.

 CASE STUDY Earthquake Activity in the Pacific Northwest

The small Juan de Fuca plate, which lies off the coasts of Oregon, Washington, and southern British Columbia, is diving beneath North America at a rate of 3 to 4 centimeters per year. Thus, the region should experience subduction zone earthquakes. Yet although small earthquakes occasionally shake the Pacific Northwest, no large ones have occurred in the past 150 to 200 years.

Why are earthquakes relatively uncommon in the Pacific Northwest? Geologists have suggested two possible answers to that question. Subduction may be occurring slowly and continuously by fault creep. If this is the case, elastic energy would not accumulate in nearby rocks, and strong earthquakes would be unlikely. Alternatively, rocks along the fault may be locked together by friction, accumulating a huge amount of elastic energy that will be released in a giant, destructive quake sometime in the future.

Recently geologists have discovered probable evidence of great prehistoric earthquakes in the Pacific Northwest. A major coastal earthquake commonly creates violent sea waves, which deposit a layer of sand along the coast. Geologists have found several such sand layers, each burying a layer of peat and mud that accumulated in coastal swamps during the quiet intervals between earthquakes. In addition, they have found submarine landslide deposits lying on the deep sea floor off the coast that formed when earthquakes triggered submarine landslides that carried sand and mud from the coast to the sea floor. These deposits show that 13 major earth-

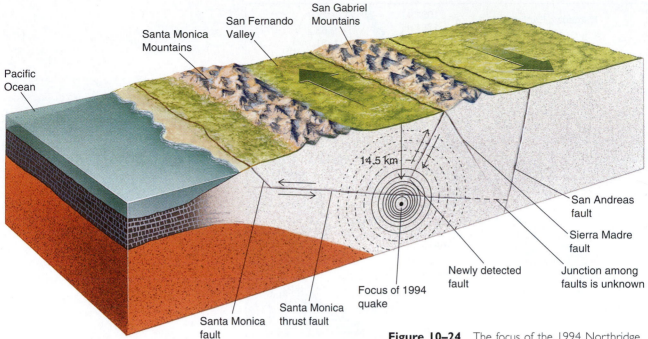

Figure 10–24   The focus of the 1994 Northridge quake was on a previously undetected thrust fault west of the San Andreas fault.

quakes, separated by 300 to 900 years, struck the coast during the past 7700 years. There is also evidence for one major historic earthquake. Oral accounts of the native inhabitants chronicle the loss of a small village in British Columbia and a significant amount of ground shaking in northern California. The same earthquake may have caused a 2-meter-high tsunami in Japan in January of 1700. Thus, many geologists anticipate another major, destructive earthquake in the Pacific Northwest during the next 600 years.

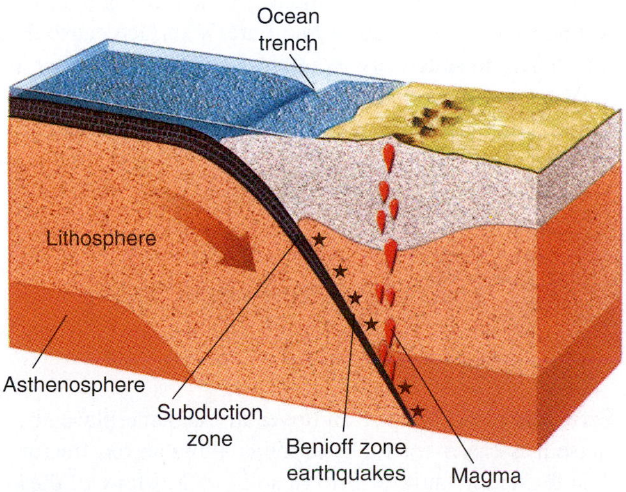

EARTHQUAKES AT DIVERGENT PLATE BOUNDARIES

Earthquakes frequently shake the mid-oceanic ridge system as a result of faults that form as the two plates separate. Blocks of oceanic crust drop downward along most mid-oceanic ridges, forming a rift valley in the center of the ridge. Only shallow earthquakes occur along the mid-oceanic ridge because here the asthenosphere rises to within 20 to 30 kilometers of the Earth's surface and is too hot and plastic to fracture.

EARTHQUAKES IN PLATE INTERIORS

No major earthquakes have occurred in the central or eastern United States in the past 100 years, and no lithospheric plate boundaries are known in these regions. Therefore, one might infer that earthquake danger is insignificant. However, the largest historical earthquake sequence in the contiguous 48 states occurred near New Madrid, Missouri. In 1811 and 1812, three shocks with estimated magnitudes between 7.3 and 7.8 altered the course of the Mississippi River and rang church bells 1500 kilometers away in Washington, D.C.

Figure 10–25   A descending lithospheric plate generates magma and earthquakes in a subduction zone. Earthquakes concentrate along the upper portion of the subducting plate, called the Benioff zone.

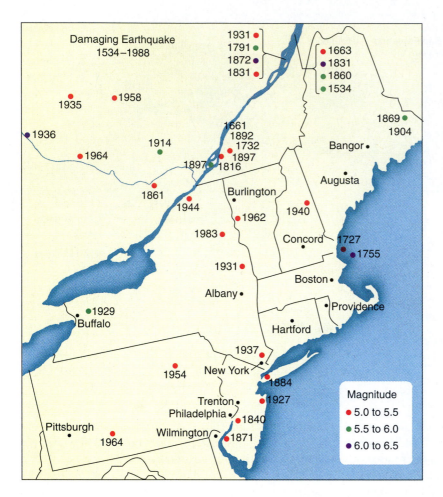

Figure 10–26 Earthquake activity in the northeastern United States and southeastern Canada between 1534 and 1988. (Redrawn from *Geotimes*, May 1991, p. 6)

Geologists have remeasured distances between old survey bench marks near New York City and found that the marks have moved significantly during the past 50 to 100 years. This motion indicates that the crust in this region is being deformed. Historical reviews show that, although earthquakes are infrequent in the Northeast, they do occur (Fig. 10–26). If a major quake were centered near New York City or Boston today, the consequences could be disastrous.

Earthquakes in plate interiors are not as well understood as those at plate boundaries, but modern research is revealing some clues. The New Madrid region lies in an extinct continental rift zone bounded by deep faults. Although the rift failed to develop into a divergent plate boundary, the deep faults remain a weakness in the lithosphere. As the North America plate glides over the asthenosphere, it may pass over irregularities, or "bumps," in that plastic zone, causing slippage along the deep faults near New Madrid.

Some intraplate earthquakes occur where thick piles of sediment have accumulated on great river deltas such as the Mississippi River delta. The underlying litho-sphere cannot support the weight of sediment, and the lithosphere fractures as it settles. Human activity may also induce intraplate earthquakes when changes in the weight of water or rock on the Earth's surface cause the lithosphere to settle. For example, water accumulating in a new reservoir is thought to have caused a magnitude 6 earthquake that killed more than 12,000 people in central India in September 1993.

▶ 10.5 EARTHQUAKE PREDICTION

LONG-TERM PREDICTION

Earthquakes occur over and over in the same places because it is easier for rocks to move along an old fracture than for a new fault to form in solid rock. Many of these faults lie along tectonic plate boundaries. Therefore, long-term earthquake prediction recognizes that earthquakes have recurred many times in a specific place and will probably occur there again (Fig. 10–27).

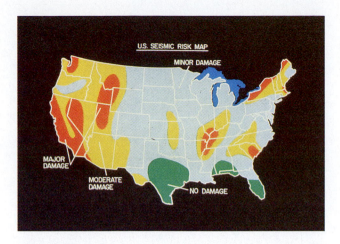

Figure 10–27 This map shows potential earthquake damage in the United States. The predictions are based on records of frequency and magnitude of historical earthquakes. (Ward's Natural Science Establishment, Inc.)

SHORT-TERM PREDICTION

Short-term predictions are forecasts that an earthquake may occur at a specific place and time. Short-term prediction depends on signals that immediately precede an earthquake.

Foreshocks are small earthquakes that precede a large quake by a few seconds to a few weeks. The cause of foreshocks can be explained by a simple analogy. If you try to break a stick by bending it slowly, you may hear a few small cracking sounds just before the final snap. If foreshocks consistently preceded major earthquakes, they would be a reliable tool for short-term prediction. However, foreshocks preceded only about half of a group of recent major earthquakes. At other times, swarms of small shocks were not followed by a large quake.

Another approach to short-term earthquake prediction is to measure changes in the land surface near an active fault zone. Seismologists monitor unusual Earth movements with tiltmeters and laser surveying instruments because distortions of the crust may precede a major earthquake. This method has successfully predicted some earthquakes, but in other instances predicted quakes did not occur or quakes occurred that had not been predicted.

When rock is deformed to near its rupture point prior to an earthquake, microscopic cracks may form. In some cases, the cracks release radon gas previously trapped in rocks and minerals. In addition, the cracks may fill with water and cause the water levels in wells to fluctuate. Furthermore, air-filled cracks do not conduct electricity as well as solid rock, so the electrical conductivity of rock decreases as cracks form.

Chinese scientists reported that, just prior to the 1975 quake in the city of Haicheng, snakes crawled out of their holes, chickens refused to enter their coops, cows broke their halters and ran off, and even well-trained police dogs became restless and refused to obey commands. Some researchers in the United States have attempted to quantify the relationship between animal behavior and earthquakes, but without success.

In January 1975, Chinese geophysicists recorded swarms of foreshocks and unusual bulges near the city of Haicheng, which had a previous history of earthquakes. When the foreshocks became intense on February 1, authorities evacuated portions of the city. The evacuation was completed on the morning of February 4, and in the early evening of the same day, an earthquake destroyed houses, apartments, and factories but caused few deaths.

After that success, geologists hoped that a new era of earthquake prediction had begun. But a year later, Chinese scientists failed to predict an earthquake in the adjacent city of Tangshan. This major quake was not preceded by foreshocks, so no warning was given, and at least 250,000 people died.[3] Over the past few decades, short-term prediction has not been reliable.

▶ 10.6 STUDYING THE EARTH'S INTERIOR

Recall from Chapter 2 that the Earth is composed of a thin crust, a thick mantle, and a core. The three layers are distinguished by different chemical compositions. In turn, both the mantle and core contain finer layers based on changing physical properties. Scientists have learned a remarkable amount about the Earth's structure even though the deepest well is only a 12-kilometer hole in northern Russia. Scientists deduce the composition and properties of the Earth's interior by studying the behavior of seismic waves. Some of the principles necessary for understanding the behavior of seismic waves are as follows:

1. In a uniform, homogeneous medium, a wave radiates outward in concentric spheres and at constant velocity.
2. The velocity of a seismic wave depends on the nature of the material that it travels through. Thus, seismic waves travel at different velocities in different types of rock. In addition, wave velocity varies with changing rigidity and density of a rock.

[3]Accurate reports of the death toll are unavailable. Published estimates range from 250,000 to 650,000.

3. When a wave passes from one material to another, it refracts (bends) and sometimes reflects (bounces back). Both **refraction** and **reflection** are easily seen in light waves. If you place a pencil in a glass half filled with water, the pencil appears bent. Of course the pencil does not bend; the light rays do. Light rays slow down when they pass from air to water, and as the velocity changes, the waves refract (Fig. 10–28). If you look in a mirror, the mirror reflects your image. In a similar manner, boundaries between the Earth's layers refract and reflect seismic waves.

4. P waves are compressional waves and travel through all gases, liquids, and solids, whereas S waves travel only through solids.

DISCOVERY OF THE CRUST–MANTLE BOUNDARY

Figure 10–29 shows that some waves travel directly through the crust to a nearby seismograph. Others travel downward into the mantle and then refract back upward to the same seismograph. The route through the mantle is longer than that through the crust. However, seismic waves travel faster in the mantle than they do in the crust. Over a short distance (less than 300 kilometers), waves traveling through the crust arrive at a seismograph before those following the longer route through the mantle. However, for longer distances, the longer route

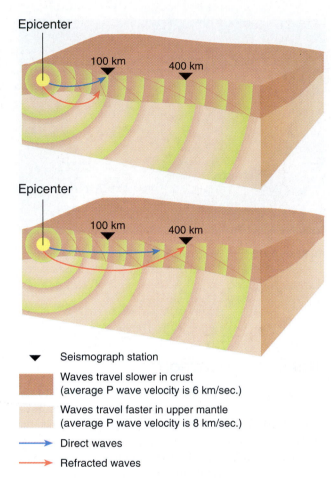

▼ Seismograph station

☐ Waves travel slower in crust
(average P wave velocity is 6 km/sec.)

☐ Waves travel faster in upper mantle
(average P wave velocity is 8 km/sec.)

→ Direct waves

→ Refracted waves

Figure 10–29 The travel path of a seismic wave. The closer station receives the direct waves first because they travel the least distance. However, a station 400 kilometers away receives the refracted wave first. Even though it travels a longer distance, most of its path is in the denser mantle, so it travels fast enough to reach the seismic station first. Think of a commuter who takes a longer route on the interstate highway rather than going via a shorter road that is choked with heavy traffic.

Figure 10–28 If you place a pencil in water, the pencil appears bent. It actually remains straight, but our eyes are fooled because light rays bend, or refract, as they cross the boundary between air and water.

through the mantle is faster because waves travel more quickly in the mantle.

The situation is analogous to the two different routes you may use to travel from your house to a friend's. The shorter route is a city street where traffic moves slowly. The longer route is an interstate highway, but you have to drive several kilometers out of your way to get to the highway and then another few kilometers from the highway to your friend's house. If your friend lives nearby, it is faster to take the city street. But if your friend lives far away, it is faster to take the longer route and make up time on the highway.

In 1909, Andrija Mohorovičić discovered that seismic waves from a distant earthquake traveled more rapidly than those from a nearby earthquake. By analyzing the arrival times of earthquake waves to many different seismographs, Mohorovičić identified the boundary between the crust and the mantle. Today, this boundary is called the **Mohorovičić discontinuity**, or the **Moho**, in honor of its discoverer.

The Moho lies at a depth ranging from 5 to 70 kilometers. Oceanic crust is thinner than continental crust, and continental crust is thicker under mountain ranges than it is under plains.

THE STRUCTURE OF THE MANTLE

The mantle is almost 2900 kilometers thick and comprises about 80 percent of the Earth's volume. Much of our knowledge of the composition and structure of the mantle comes from seismic data. As explained earlier, seismic waves speed up abruptly at the crust-mantle boundary (Fig. 10–30). Between 75 and 125 kilometers, at the base of the lithosphere, seismic waves slow down again because the high temperature at this depth causes solid rock to become plastic. Recall that the plastic layer is called the asthenosphere. The plasticity and partially melted character of the asthenosphere slow down the seismic waves. At the base of the asthenosphere 350 kilometers below the surface, seismic waves speed up again because increasing pressure overwhelms the tem-

perature effect, and the mantle becomes less plastic.

At a depth of about 660 kilometers, seismic wave velocities increase again because pressure is great enough that the minerals in the mantle recrystallize to form denser minerals. The zone where the change occurs is called the **660-kilometer discontinuity**. The base of the mantle lies at a depth of 2900 kilometers.

DISCOVERY OF THE CORE

Using a global array of seismographs, seismologists detect direct P and S waves up to 105° from the focus of an earthquake. Between 105° and 140° is a "shadow zone" where no direct P waves arrive at the Earth's surface. This shadow zone is caused by a discontinuity, which is the mantle-core boundary. When P waves pass from the mantle into the core, they are refracted, or bent, as shown in Figure 10–31. The refraction deflects the P waves away from the shadow zone.

No S waves arrive beyond 105°. Their absence in this region shows that they do not travel through the outer core. Recall that S waves are not transmitted through liquids. The failure of S waves to pass through the outer core indicates that the outer core is liquid.

Refraction patterns of P waves, shown in Figure 10–31, shows that another boundary exists within the core. It is the boundary between the liquid outer core and

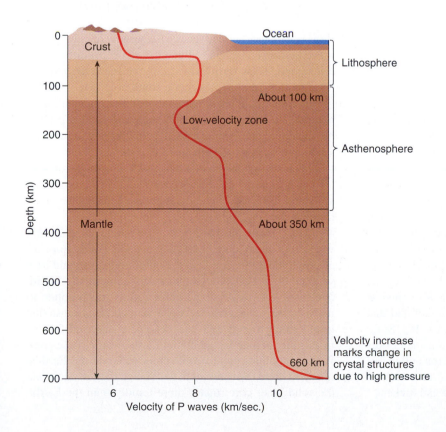

Figure 10–30 Velocities of P waves in the crust and the upper mantle. As a general rule, the velocity of P waves increases with depth. However, in the asthenosphere, the temperature is high enough that rock is plastic. As a result, seismic waves slow down in this region, called the low-velocity zone. Wave velocity increases rapidly at the 660-kilometer discontinuity, the boundary between the upper and the lower mantle, probably because of a change in mineral content due to increasing pressure.

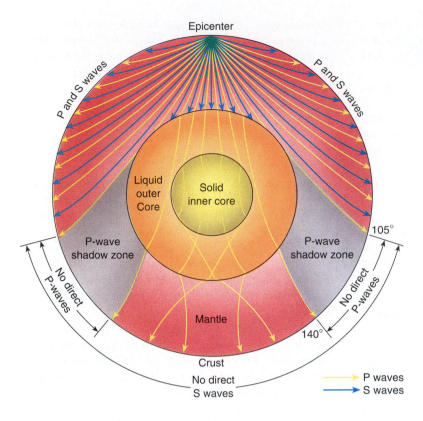

Figure 10–31 Cross section of the Earth showing paths of seismic waves. They bend gradually because of increasing pressure with depth. They also bend sharply where they cross major layer boundaries in the Earth's interior. Note that S waves do not travel through the liquid outer core, and therefore direct S waves are only observed within an arc of 105° of the epicenter. P waves are refracted sharply at the core–mantle boundary, so there is a shadow zone of no direct P waves from 105° to 140°.

the solid inner core. Although seismic waves tell us that the outer core is liquid and the inner core is solid, other evidence tells us that the core is composed of iron and nickel.

DENSITY MEASUREMENTS

The overall density of the Earth is 5.5 grams per cubic centimeter (g/cm^3); but both crust and mantle have average densities less than this value. The density of the crust ranges from 2.5 to 3 g/cm^3, and the density of the mantle varies from 3.3 g/cm^3 to 5.5 g/cm^3. Since the mantle and crust account for slightly more than 80 percent of the Earth's volume, the core must be very dense to account for the average density of the Earth. Calculations show that the density of the core must be 10 to 13 g/cm^3, which is the density of many metals under high pressure.

Many meteorites are composed mainly of iron and nickel. Cosmologists think that meteorites formed at about the same time that the Solar System did, and that they reflect the composition of the primordial Solar System. Because the Earth coalesced from meteorites and similar objects, scientists believe that iron and nickel must be abundant on Earth. Therefore, they conclude that the metallic core is composed of iron and nickel.

▶ 10.7 THE EARTH'S MAGNETISM

Early navigators learned that no matter where they sailed, a needle-shaped magnet aligned itself in a north–south orientation. Thus, they learned that the Earth has a magnetic north pole and a magnetic south pole (Fig. 10–32).

The Earth's interior is too hot for a permanent magnet to exist. Instead, the Earth's magnetic field is probably electromagnetic in origin. If you wrap a wire around a nail and connect the ends of the wire to a battery, the nail becomes magnetized and can pick up small iron objects. The battery causes electrons to flow through the wire, and this flow of electrical charges creates the electromagnetic field.

Most likely, the Earth's magnetic field is generated within the outer core. Metals are good conductors of electricity and the metals in the outer core are liquid and very mobile. Two types of motion occur in the liquid outer core. (1) Because the outer core is much hotter at its base than at its surface, convection currents cause the liquid metal to rise and fall. (2) The rising and falling metals are then deflected by the Earth's spin. These convecting, spinning liquid conductors generate the Earth's magnetic field. New research has shown that, in addition, the solid inner core rotates more rapidly than the Earth.

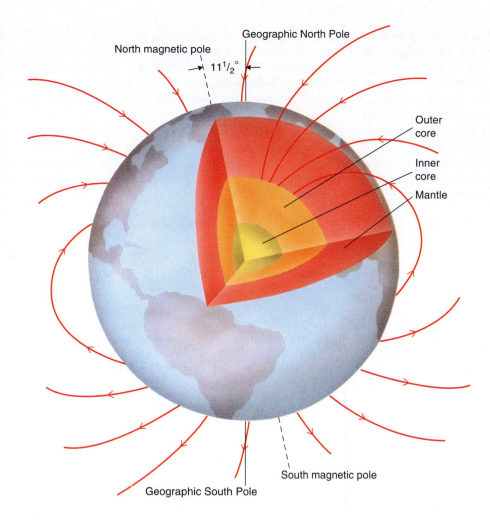

North magnetic pole

Geographic North Pole

$11\frac{1}{2}°$

Outer core

Inner core

Mantle

South magnetic pole

Geographic South Pole

Figure 10–32   The magnetic field of the Earth. Note that the magnetic north pole is 11.5° offset from the geographic pole.

Magnetic fields are common in planets, stars, and other objects in space. Our Solar System almost certainly possessed a weak magnetic field when it first formed. The flowing metals of the liquid outer core amplified some of this original magnetic force. If we observe this magnetic field over thousands of years, its axis is approximately lined up with the Earth's rotational axis, because the Earth's spin affects the flow of metal in the outer core.

SUMMARY

An **earthquake** is a sudden motion or trembling of the Earth caused by the abrupt release of slowly accumulated energy in rocks. Most earthquakes occur along tectonic plate boundaries. Earthquakes occur either when the elastic energy accumulated in rock exceeds the friction that holds rock along a fault or when the elastic energy exceeds the strength of the rock and the rock breaks by **brittle fracture**.

An earthquake starts at the initial point of rupture, called the **focus**. The location on the Earth's surface directly above the focus is the **epicenter**. **Seismic waves** include **body waves**, which travel through the interior of the Earth, and **surface waves**, which travel on the surface. **P waves** are compressional body waves. **S waves** are body waves that travel slower than P waves. They consist of a shearing motion and travel through solids but not liquids. Surface waves travel more slowly than either type of body wave. Seismic waves are recorded on a **seismograph**. Modern geologists use the **moment magnitude** scale to record the energy released during an earthquake. The distance from a seismic station to an earthquake is calculated by recording the time between the arrival of P and S waves. The epicenter can be located by measuring the distance from three or more seismic stations. Earthquake damage is influenced by rock and soil type, construction design, and the likelihood of fires, landslides, and tsunamis.

Earthquakes are common at all three types of plate boundaries. The San Andreas fault zone is an example of a transform plate boundary, where two plates slide past one another. Subduction zone earthquakes occur when the subducting plate slips suddenly. Earthquakes occur at divergent plate boundaries as blocks of lithosphere along the fault drop downward. Earthquakes occur in plate interiors along old faults or where sediment depresses the lithosphere.

Long-term earthquake prediction is based on the observation that most earthquakes occur at tectonic plate boundaries. Short-term prediction is based on occurrences of **foreshocks**, release of radon gas, changes in the land surface, the water table, electrical conductivity, and erratic animal behavior.

The Earth's internal structure and properties are known by studies of earthquake wave velocities and **refraction** and **reflection** of seismic waves as they pass through the Earth. Flowing metal in the outer core generates the Earth's magnetic field.

KEY WORDS

earthquake *154*	**epicenter** *158*	**seismogram** *160*	**San Andreas fault**
stress *156*	**surface wave** *158*	**Richter scale** *160*	**zone** *165*
plastic deformation *156*	**P wave** *158*	**moment magnitude**	**fault creep** *166*
brittle fracture *156*	**S wave** *159*	**scale** *161*	**thrust fault** *166*
seismic wave *158*	**shear wave** *159*	**time-travel curve** *162*	**Benioff zone** *168*
seismology *158*	**Rayleigh wave** *159*	**liquefaction** *164*	**foreshock** *171*
body wave *158*	**Love wave** *159*	**tsunami** *165*	**Mohorovičić discontinuity**
focus *158*	**seismograph** *160*	**strike–slip fault** *165*	**(Moho)** *173*

REVIEW QUESTIONS

1. Explain how energy is stored prior to and then released during an earthquake.

2. Describe the behavior of rock during elastic deformation, plastic deformation, and brittle fracture.

3. Give two mechanisms that can release accumulated elastic energy in rocks.

4. Why do most earthquakes occur at the boundaries between tectonic plates? Are there any exceptions?

5. Define *focus* and *epicenter*.

6. Discuss the differences between P waves, S waves, and surface waves.

7. Explain how a seismograph works. Sketch what an imaginary seismogram would look like before and during an earthquake.

8. Describe the similarities and differences between the Richter and moment magnitude scales. What is actually measured, and what information is obtained?

9. Describe how the epicenter of an earthquake is located.

10. List five different factors that affect earthquake damage. Discuss each briefly.

11. Discuss earthquake mechanisms at the three different types of tectonic plate boundaries.

12. Briefly discuss major faults close to Los Angeles.

13. Discuss earthquake mechanisms at plate interiors.

14. Discuss the scientific reasoning behind long-term and short-term earthquake prediction.

15. Outline the seismic gap hypothesis. Discuss modern objections to the theory.

16. What is the Moho? How was it discovered?

17. Explain how geologists learned that the core is composed of iron and nickel.

18. Briefly discuss the theories for the existence of the Earth's magnetic field.

DISCUSSION QUESTIONS

1. Using the graph in Figure 10–13, determine how far away from an earthquake you would be if the first P wave arrived 5 minutes before the first S wave.

2. Using a map of the United States, locate an earthquake that is 1000 kilometers from Seattle, 1300 kilometers from San Francisco, and 700 kilometers from Denver.

3. Mortality was high in the India earthquake in 1993 because the quake occurred at night when people were sleeping in their homes. However, mortality in the Northridge earthquake was low because it occurred early in the morning rather than during rush hour. Is there a contradiction in these two statements?

4. Explain why the existence of thrust faults west of the San Andreas fault complicates attempts at earthquake prediction near Los Angeles.

5. Argue for or against placing stringent building codes for earthquake-resistant design in the Seattle area.

6. Imagine that geologists predict a major earthquake in a densely populated region. The prediction may be right or it may be wrong. City planners may heed it and evacuate the city or ignore it. The possibilities lead to four combinations of predictions and responses, which can be set out in a grid as follows:

Will the predicted earthquake really occur?

		Yes	No
Is the city evacuated?	Yes		
	No		

For example, the space in the upper left corner of the grid represents the situation in which the predicted earthquake occurs and the city is evacuated. For each space in the square, outline the consequences of the sequence of events.

7. Engineers know that if a major quake were to strike a large California city, many structures would fail and people would die. Furthermore, it is possible, but expensive, to construct homes, commercial buildings, and bridges that would not fail even in a major quake. Discuss the tradeoff between money and human lives in construction in earthquake-prone zones. Would you be willing to pay twice as much for a house that was earthquake proof, as compared with a normal house? Would you be willing to pay more taxes for safer highway bridges? Do you feel that different types of structures (i.e., residential homes, commercial buildings, nuclear power plants) should be built to different safety standards?

8. From Figure 10–30, what is the speed of P waves at a depth of 25 kilometers, 200 kilometers, 500 kilometers?

9. Give two reasons why the Earth's magnetic field cannot be formed by a giant bar magnet within the Earth's core.

CHAPTER 11

Ocean Basins

If you were to ask most people to describe the difference between a continent and an ocean, they would almost certainly reply, "Why, obviously a continent is land and an ocean is water!" This observation is true, of course, but to a geologist another distinction is more important. He or she would explain that sea-floor rocks are different from those of a continent. The accumulation of seawater in the world's ocean basins is a *result* of that difference.

Recall that the Earth's lithosphere floats on the asthenosphere and that the upper part of the lithosphere is either oceanic or continental crust. Oceanic crust is dense basalt and varies from 5 to 10 kilometers thick. In contrast, continental crust is made of less dense granite and averages 20 to 40 kilometers thick. Thus, continental lithosphere is both thicker and less dense than oceanic lithosphere. The thick, less dense continental lithosphere floats isostatically at high elevations, whereas that of the ocean basins sinks to low elevations. Most of the Earth's water collects in the depressions formed by oceanic lithosphere. Even if no water existed on the Earth's surface, oceanic crust would form deep basins and continental crust would rise to higher elevations.

Reefs, such as this one in Papua New Guinea, grow abundantly in the shallow waters of tropical oceans.   (Mark J. Thomas/Dembinski Photo Assoc.)

▶ 11.1 THE EARTH'S OCEANS

Oceans cover about 71 percent of the Earth's surface. The Pacific Ocean is the largest and deepest ocean. It covers one third of Earth's surface, more than all land combined. The Atlantic Ocean covers about half the area of the Pacific. The Indian Ocean is the smallest of the three ocean basins and lies primarily in the Southern Hemisphere. Other oceans, seas, and gulfs are portions of those three major ocean basins. For example, the Arctic Ocean is the northern extension of the Atlantic.

The sea floor is about 5 kilometers deep in most parts of the ocean basins, although it is only 2 to 3 kilometers deep above the mid-oceanic ridges and it can plunge to 10 kilometers deep in an oceanic trench. As a result of their area and depth, the ocean basins contain 1.4 billion cubic kilometers of water—18 times more than the volume of all land above sea level.

The size of an ocean basin changes over geologic time because new oceanic crust forms at spreading centers, and old sea floor is consumed at subduction zones. At present, the Atlantic Ocean is growing wider at the mid-Atlantic ridge, while the Pacific is shrinking at subduction zones around its edges—that is, the Atlantic Ocean basin is increasing in size at the expense of the Pacific.

Oceans profoundly affect the Earth's climate. Some marine currents carry heat from the equator toward the poles, while others carry cold water toward the equator. Without this heat exchange, the equator would be unbearably hot, and high latitudes would be colder than at present. Because plate tectonic activity alters the distribution of continents and oceans, it also alters ocean circulation and thereby causes long-term climate change.

▶ 11.2 STUDYING THE SEA FLOOR

Seventy-five years ago, scientists had better maps of the Moon than of the sea floor. The Moon is clearly visible in the night sky, and we can view its surface with a telescope. The sea floor, on the other hand, is deep, dark, and inhospitable to humans. Modern oceanographers use a variety of techniques to study the sea floor, including several types of sampling and remote sensing.

SAMPLING

Several devices collect sediment and rock directly from the ocean floor. A **rock dredge** is an open-mouthed steel net dragged along the sea floor behind a research ship. The dredge breaks rocks from submarine outcrops and hauls them to the surface. Sediment near the surface of the sea floor can be sampled by a weighted, hollow steel

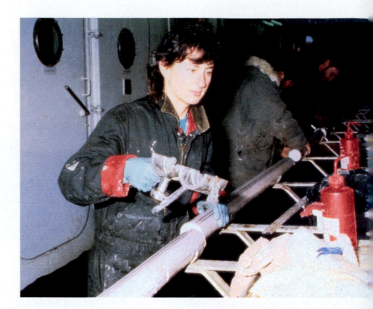

Figure 11–1 An oceanographer extracts sediment from a core retrieved from the sea floor. (Ocean Drilling Program, Texas A&M University)

pipe lowered on a cable from a research vessel. The weight drives the pipe into the soft sediment, which is forced into the pipe. The sediment **core** is retrieved from the pipe after it is winched back to the surface. If the core is removed from the pipe carefully, even the most delicate sedimentary layering is preserved (Fig. 11–1).

Sea-floor drilling methods developed for oil exploration also take core samples from oceanic crust. Large drill rigs are mounted on offshore platforms and on research vessels. The drill cuts cylindrical cores from both sediment and rock, which are then brought to the surface for study. Although this type of sampling is expensive, cores can be taken from depths of several kilometers into oceanic crust.

A number of countries, including France, Japan, Russia, and the United States, have built small submarines to carry oceanographers to the sea floor, where they view, photograph, and sample sea-floor rocks, sediment, and life (Fig. 11–2). More recently, scientists have used deep-diving robots and laser imagers to sample and photograph the sea floor. A robot is cheaper and safer than a submarine, and a laser imager penetrates up to eight times farther through water than a conventional camera.

REMOTE SENSING

Remote sensing methods do not require direct physical contact with the ocean floor, and for some studies this

Figure 11–2 *Alvin* is a research submarine capable of diving to the sea floor. Scientists on board control robot arms to collect sea-floor rocks and sediment. (Rod Catanach, Woods Hole Oceanographic Institution)

approach is both effective and economical. The **echo sounder** is commonly used to map sea-floor topography. It emits a sound signal from a research ship and then

records the signal after it bounces off the sea floor and travels back up to the ship (Fig. 11–3a). The water depth is calculated from the time required for the sound to make the round trip. A topographic map of the sea floor is constructed as the ship steers a carefully navigated course with the echo sounder operating continuously. Modern echo sounders transmit up to 1000 signals at a time to create more complete and accurate maps.

The **seismic profiler** works in the same way but uses a higher-energy signal that penetrates and reflects from layers in the sediment and rock. This gives a picture of the layering and structure of oceanic crust, as well as the sea-floor topography (Fig. 11–3b).

A **magnetometer** is an instrument that measures a magnetic field. Magnetometers towed behind research ships measure the magnetism of sea-floor rocks.

▶ 11.3 SEA-FLOOR MAGNETISM

Recall from Chapter 2 that Alfred Wegener proposed that continents had migrated across the globe, but because he was unable to explain *how* continents moved, his theory was largely ignored. Then, in the 1960s, 30 years after Wegener's death, geologists studying the magnetism of sea-floor rocks detected odd magnetic patterns on the sea floor. Their interpretations of those patterns quickly led to the development of the plate tectonics theory and

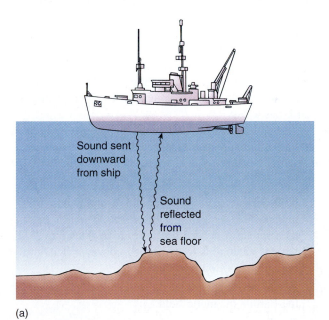

(a)

Figure 11–3 (a) Mapping the topography of the sea floor with an echo sounder. A sound signal generated by the echo sounder bounces off the sea floor and back up to the ship, where its travel time is recorded. (b) A seismic profiler record of sediment layers and basaltic ocean crust in the Sea of Japan. (Ocean Drilling Program, Texas A&M University)

(b)

proved that Wegener's hypothesis of continental drift had been correct.

To understand how magnetic patterns on the sea floor led to the plate tectonics theory, we must consider the relationships between the Earth's magnetic field and magnetism in rocks. Many iron-bearing minerals are permanent magnets. Their magnetism is much weaker than that of a magnet used to stick cartoons on your refrigerator door, but it is strong enough to measure with a magnetometer.

When magma solidifies, certain iron-bearing minerals crystallize and become permanent magnets. When such a mineral cools within the Earth's magnetic field, the mineral's magnetic field aligns parallel to the Earth's field just as a compass needle does (Fig. 11–4). Thus, minerals in an igneous rock record the orientation of the Earth's magnetic field at the time the rock cooled.

Many sedimentary rocks also preserve a record of the orientation of the Earth's magnetic field at the time the sediment was deposited. As sediment settles through water, magnetic mineral grains tend to settle with their magnetic axes parallel to the Earth's field. Even silt particles settling through air orient parallel to the magnetic field.

MAGNETIC REVERSALS

The **polarity** of a magnetic field is the orientation of its positive, or north, end and its negative, or south, end. Because many rocks record the orientation of the Earth's magnetic field at the time the rocks formed, we can construct a record of the Earth's polarity by studying magnetic orientations in rocks from many different ages and places. When geologists constructed such a record, they discovered, to their amazement, that the Earth's magnetic field has reversed polarity throughout geologic history. When a **magnetic reversal** occurs, the north magnetic pole becomes the south magnetic pole, and vice versa. The orientation of the Earth's field at present is referred to as normal polarity, and that during a time of opposite polarity is called reversed polarity. The Earth's polarity has reversed about 130 times during the past 65 million years. But polarity reversals do not occur on a regular schedule. A period of normal polarity during the Mesozoic Era lasted for 40 million years.

As the Earth's magnetic field is about to reverse, it becomes progressively weaker, but its orientation remains constant. Then the magnetic field collapses to zero or close to zero. Soon a new magnetic field with a polarity opposite from the previous field begins to grow. The reversal takes 3000 to 5000 years, a very short time compared with many other geologic changes.

SEA-FLOOR SPREADING: THE BEGINNING OF THE PLATE TECTONICS THEORY

As you learned in Chapter 4, most oceanic crust is basalt that formed as magma erupted onto the sea floor from the mid-oceanic ridge system. Basalt contains iron-bearing minerals that record the orientation of the Earth's magnetic field at the time the basalt cooled.

Figure 11–5 shows magnetic orientations of the oceanic crust along a portion of the mid-oceanic ridge known as the Reykjanes ridge near Iceland. The black stripes represent rocks with normal polarity, and the intervening stripes represent rocks with reversed polarity. Notice that the stripes form a pattern of alternating normal and reversed polarity, and that the stripes are arranged symmetrically about the axis of the ridge. The central stripe is black, indicating that the rocks of the ridge axis have the same magnetic orientation that the Earth has today.

Shortly after this discovery, geologists suggested that a particular sequence of events created the alternating stripes of normal and reversed polarity in sea-floor rocks:

1. New oceanic crust forms continuously as basaltic magma rises beneath the ridge axis. The new crust then spreads outward from the ridge. This movement is analogous to two broad conveyor belts moving away from one another.

2. As the new crust cools, it acquires the orientation of the Earth's magnetic field.

3. The Earth's magnetic field reverses orientation on an average of every half-million years.

4. Thus, the magnetic stripes on the sea floor record a succession of reversals in the Earth's magnetic field that occurred as the sea floor spread away from the

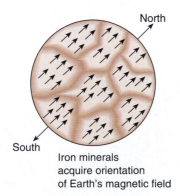

Figure 11–4 An iron-bearing mineral in an igneous rock acquires a permanent magnetic orientation parallel to that of the Earth's field, as the rock cools.

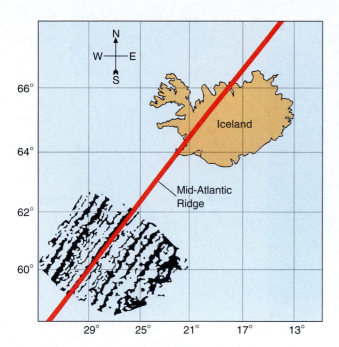

Figure 11–5 The mid-Atlantic ridge, shown in red, runs through Iceland. Magnetic orientations of sea-floor rocks near the ridge are shown in the lower left portion of the map. The black stripes represent sea-floor rocks with normal magnetic polarity, and the intervening stripes represent rocks with reversed polarity. The stripes form a symmetrical pattern of alternating normal and reversed polarity on each side of the ridge. (After Heirtzler et al., 1966, *Deep-Sea Research*, Vol. 13.)

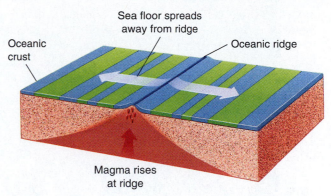

Figure 11–6 As new oceanic crust cools at the mid-oceanic ridge, it acquires the magnetic orientation of the Earth's field. Alternating stripes of normal (blue) and reversed (green) polarity record reversals in the Earth's magnetic field that occurred as the crust spread outward from the ridge.

gin of all oceanic crust.[1] In a short time, geologists combined Wegener's continental drift hypothesis and the newly developed sea-floor spreading hypothesis to develop the plate tectonics theory. You read in Chapter 2 that this theory explains how and why continents move, mountains rise, earthquakes shake our planet, and volcanoes erupt. It also explains the origin and features of the Earth's largest mountain chain: the mid-oceanic ridge.

▶ **11.4 THE MID-OCEANIC RIDGE**

The extensive use of submarines during World War II made it essential to have topographic maps of the sea floor. Those maps, made with early versions of the echo sounder, were kept secret by the military. When they became available to the public after peace was restored, scientists were surprised to learn that the ocean floor has at least as much topographic diversity and relief as the continents (Fig. 11–7a). Broad plains, high peaks, and deep valleys form a varied and fascinating submarine landscape, but the mid-oceanic ridge is the most impressive feature of the deep sea floor.

The mid-oceanic ridge system is a continuous submarine mountain chain that encircles the globe (Fig. 11–7b). Its total length exceeds 80,000 kilometers and it is more than 1500 kilometers wide in places. The ridge rises an average of 3 kilometers above the surrounding

ridge (Fig. 11–6). To return to our analogy, imagine that a can of white spray paint were mounted above two black conveyor belts moving apart. If someone sprayed paint at regular intervals as the conveyor belts moved, symmetric white and black stripes would appear on both belts.

At about the same time that oceanographers discovered the magnetic stripes on the sea floor, they also began to sample the mud lying on the sea floor. They discovered that the mud is thinnest at the mid-Atlantic ridge and becomes progressively thicker at greater distances from the ridge. Mud falls to the sea floor at about the same rate everywhere in the open ocean. It is thinnest at the ridge because the sea floor is youngest there; the mud thickens with increasing distance from the ridge because the sea floor becomes progressively older away from the ridge.

Oceanographers soon recognized similar magnetic stripes and sediment thickness trends along other portions of the mid-oceanic ridge. As a result, they proposed the hypothesis of **sea-floor spreading** to explain the ori-

[1]Hess and Dietz proposed the sea-floor spreading hypothesis in 1960, prior to Vine and Matthews's 1963 interpretation of sea-floor magnetic stripes, but the hypothesis received widespread attention only after 1963.

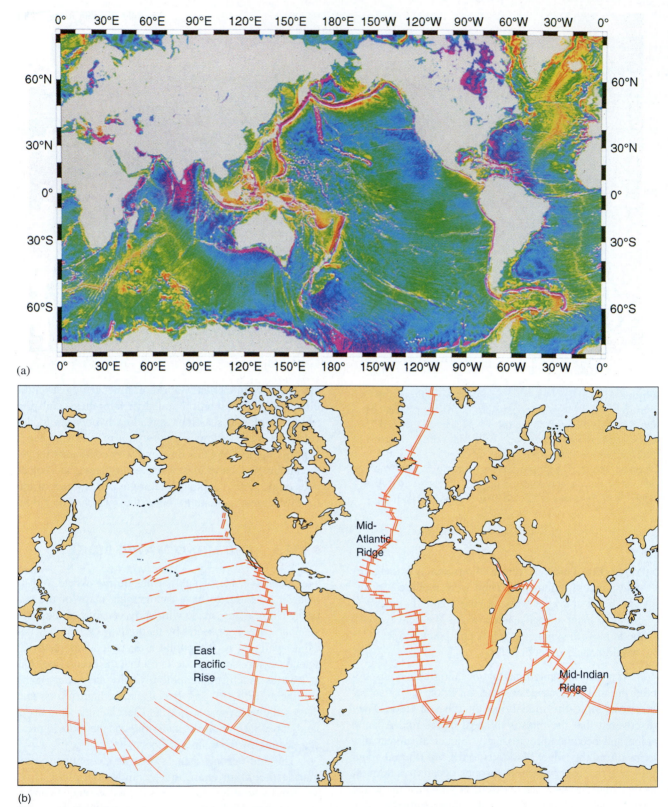

(a)

(b)

Figure 11–7 (a) Sea-floor topography is dominated by undersea mountain chains called mid-oceanic ridges and deep trenches called subduction zones. The green areas represent the relatively level portion of the ocean floor; the yellow-orange-red hues are mountains, and the blue-violet-magenta areas are trenches. (Scripps Institution of Oceanography, University of California, San Diego.) (b) A map of the ocean floor showing the mid-oceanic ridge in red. Double lines show the ridge axis; single lines are transform faults. Note that the deep sea mountain chain shown in (a) corresponds to the mid-oceanic ridge shown in (b).

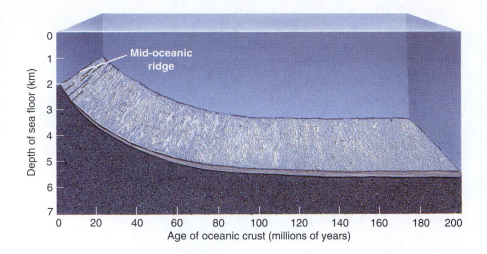

Figure 11–8 The sea floor sinks as it grows older. At the mid-oceanic ridge, new lithosphere is buoyant because it is hot and of low density. It ages, cools, thickens, and becomes denser as it moves away from the ridge and consequently sinks. The central portion of the sea floor lies at a depth of about 5 kilometers.

deep sea floor. It covers more than 20 percent of the Earth's surface—nearly as much as all continents combined—and is made up primarily of undeformed basalt.

On the mid-Atlantic ridge, which is a segment of the mid-oceanic ridge, a **rift valley** 1 to 2 kilometers deep and several kilometers wide splits the ridge crest. In 1974, French and American scientists used small research submarines to dive into the rift valley. They saw gaping vertical cracks up to 3 meters wide on the floor of the rift. Nearby were basalt flows so young that they were not covered by any mud. Recall that the mid-oceanic ridge is a spreading center. The cracks formed when brittle oceanic crust separated at the ridge axis. Basaltic magma then rose through the cracks and flowed onto the floor of the rift valley. This basalt became new oceanic crust as two lithospheric plates spread outward from the ridge axis. Not all spreading centers have rift valleys as deep and as wide as those of the mid-Atlantic ridge.

The new crust (and the underlying lithosphere) at the ridge axis is hot and therefore of relatively low density. Its buoyancy causes it to float high above the surrounding sea floor, forming the submarine mountain chain called the mid-oceanic ridge system. The new lithosphere cools as it spreads away from the ridge. As a result of cooling, it becomes thicker and denser and sinks to lower elevations, forming the deeper sea floor on both sides of the ridge (Fig. 11–8).

The **heat flow** (the rate at which heat flows outward from the Earth's surface) at the ridge is several times greater than that in other parts of the ocean basins. A great deal of Earth heat escapes from the ridge axis because the spreading lithosphere is stretched thin under the ridge, and the hot asthenosphere bulges toward the surface. Rising magma carries additional heat upward.

Shallow earthquakes are common at the mid-oceanic ridge because oceanic crust fractures as the two plates

separate (Fig. 11–9). Blocks of crust drop downward along the sea-floor cracks, forming the rift valley.

Hundreds of fractures called **transform faults** cut across the rift valley and the ridge (Fig. 11–10). These fractures extend through the entire thickness of the lithosphere. They develop because the mid-oceanic ridge actually consists of many short segments. Each segment is slightly offset from adjacent segments by a transform fault. Transform faults are original features of the mid-oceanic ridge; they develop when lithospheric spreading begins.

Some transform faults displace the ridge by less than a kilometer, but others offset the ridge by hundreds

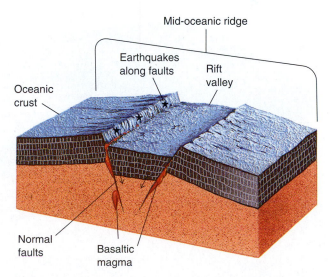

Figure 11–9 A cross-sectional view of the central rift valley in the mid-oceanic ridge. As the plates separate, blocks of rock drop down along the fractures to form the rift valley. The moving blocks cause earthquakes.

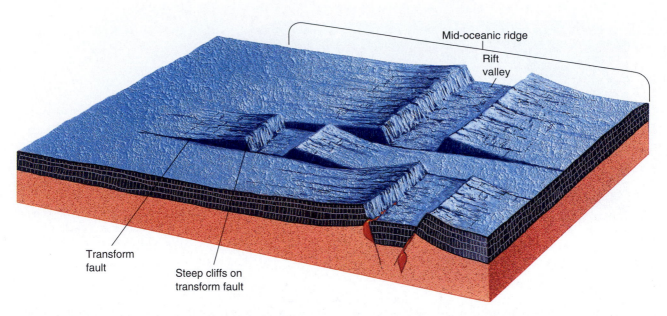

Figure 11–10 Transform faults offset segments of the mid-oceanic ridge. Adjacent segments of the ridge may be separated by steep cliffs 3 kilometers high. Note the flat abyssal plain far from the ridge.

of kilometers. In some cases, a transform fault can grow so large that it forms a transform plate boundary. The San Andreas fault in California is a transform plate boundary.

Shallow earthquakes occur along a transform fault between offset segments of the ridge crest. Here the ocean floor on each side of the fault moves in opposite directions (Fig. 11–11). Earthquakes rarely occur on transform faults beyond the ridge crest, however, because there the sea floor on both sides of the fault moves in the same direction.

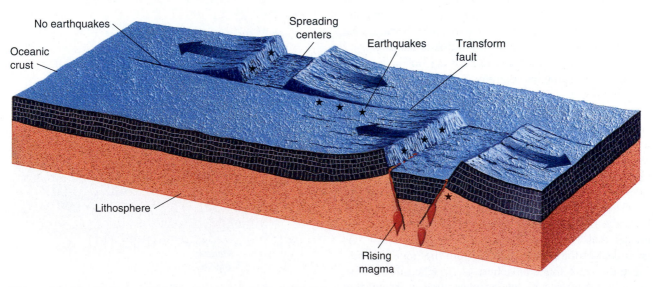

Figure 11–11 Shallow earthquakes (stars) occur along the ridge axis and on the transform faults *between* ridge segments, where plates move in opposite directions; no earthquakes occur on the transform faults *beyond* the ridge segments, where plates move in the same direction.

GLOBAL SEA-LEVEL CHANGES AND THE MID-OCEANIC RIDGE

As explained in Chapter 7, a thin layer of marine sedimentary rocks blankets the interior regions of most continents. These rocks tell us that those places must have been below sea level when the sediment accumulated.

Tectonic activity can cause a continent to sink, allowing the sea to flood a large area. However, at particular times in the past (most notably during the Cambrian, Carboniferous, and Cretaceous periods), seas flooded low-lying portions of all continents simultaneously. Although our plate tectonics model explains the sinking of individual continents, or parts of continents, it does not explain why all continents should sink at the same time. Therefore, we need to explain how sea level could rise globally by hundreds of meters to flood all continents simultaneously.

The alternating growth and melting of glaciers during the Pleistocene Epoch caused sea level to fluctuate by as much as 200 meters. However, the ages of most marine sedimentary rocks on continents do not coincide with times of glacial melting. Therefore, we must look for a different cause to explain continental flooding.

Recall from Section 11.4 that the new, hot lithosphere at a spreading center is buoyant, causing the mid-oceanic ridge to rise above the surrounding sea floor. This submarine mountain chain displaces a huge volume of seawater. If the mid-oceanic ridge were smaller, it would displace less seawater and sea level would fall. If it were larger, sea level would rise.

The mid-oceanic ridge stands highest at the spreading center, where new rock is hottest and has the lowest density. The elevation of the ridge decreases on both sides of the spreading center because the lithosphere cools and shrinks as it moves outward. Now consider a spreading center where spreading is very slow (e.g., 1 to 2 centimeters per year). At such a slow rate of spreading, the newly formed lithosphere would cool before it migrated far from the spreading center. This slow rate of spreading would produce a narrow, low-volume ridge, as shown in Figure 1. In contrast, rapid sea-floor spreading, on the order of 10 to 20 centimeters per year, would create a high-volume ridge because the newly formed, hot lithosphere would be carried a considerable distance away from the spreading center before it cooled and shrunk. This high-volume ridge would displace con-

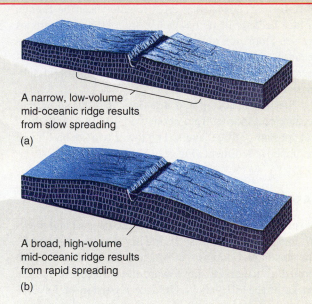

A narrow, low-volume mid-oceanic ridge results from slow spreading

(a)

A broad, high-volume mid-oceanic ridge results from rapid spreading

(b)

Figure 1   (a) Slow sea-floor spreading creates a narrow, low-volume mid-oceanic ridge that displaces less seawater and lowers sea level. (b) Rapid sea-floor spreading creates a wide, high-volume ridge that displaces more seawater and raises sea level.

siderably more seawater than a low-volume ridge and would cause a global sea level rise.

Sea-floor age data indicate that the rate of sea-floor spreading has varied from about 2 to 16 centimeters per year since Jurassic time, about 200 million years ago. Sea-floor spreading was unusually rapid during Late Cretaceous time, between 110 and 85 million years ago. That rapid spreading should have formed an unusually high-volume mid-oceanic ridge and resulted in flooding of low-lying portions of continents. Geologists have found Upper Cretaceous marine sedimentary rocks on nearly all continents, indicating that Late Cretaceous time was, in fact, a time of abnormally high global sea level. Unfortunately, because no oceanic crust is older than about 200 million years, the hypothesis cannot be tested for earlier times when extensive marine sedimentary rocks accumulated on continents.

DISCUSSION QUESTION

What factors other than variations in sea-floor spreading rates and growth and melting of glaciers might cause global sea-level fluctuations?

► II.5   SEDIMENT AND ROCKS OF THE DEEP SEA FLOOR

The Earth is 4.6 billion years old, and rocks as old as 3.96 billion years have been found on continents. Once formed, continental crust remains near the Earth's surface because of its buoyancy. In contrast, no parts of the sea floor are older than about 200 million years because oceanic crust forms continuously at the mid-oceanic ridge and then recycles into the mantle at subduction zones.

Seismic profiling and sea-floor drilling show that oceanic crust varies from about 5 to 10 kilometers thick and consists of three layers. The lower two are basalt and the upper is sediment (Fig. 11–12).

BASALTIC OCEANIC CRUST

Layer 3, 4 to 5 kilometers thick, is the deepest and thickest layer of oceanic crust. It directly overlies the mantle. The upper part consists of vertical basalt dikes, which formed as magma oozing toward the surface froze in the cracks of the rift valley. The lower portion of Layer 3 consists of horizontally layered gabbro, the coarse-grained equivalent of basalt. The gabbro forms as pools of magma cool beneath the basalt dikes.

Layer 2 lies above Layer 3 and is about 1 to 2 kilometers thick. It consists mostly of **pillow basalt**, which forms as hot magma oozes onto the sea floor, where contact with cold seawater causes the molten lava to contract into pillow-shaped spheroids (Fig. 11–13).

The basalt crust of layers 2 and 3 forms at the mid-oceanic ridge. However, these rocks make up the foundation of all oceanic crust because all oceanic crust forms at the ridge axis and then spreads outward. In some places, chemical reactions with seawater have altered the basalt of layers 2 and 3 to a soft, green rock that contains up to 13 percent water.

OCEAN FLOOR SEDIMENT

The uppermost layer of oceanic crust, called **Layer 1**, consists of two different types of sediment. **Terrigenous sediment** is sand, silt, and clay eroded from the continents and carried to the deep sea floor by submarine currents. Most of this sediment is found close to the continents. **Pelagic sediment**, on the other hand, collects even on the deep sea floor far from continents. It is a gray and red-brown mixture of clay, mostly carried from continents by wind, and the remains of tiny plants and animals that live in the surface waters of the oceans (Fig. 11–14). When these organisms die, their remains slowly settle to the ocean floor.

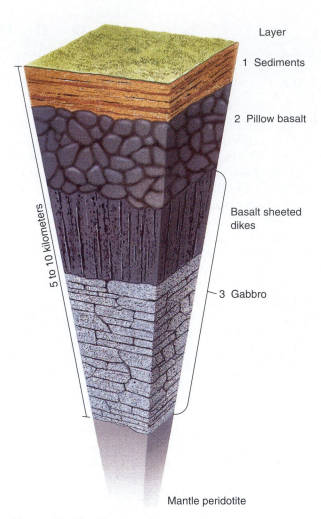

Layer

1 Sediments

2 Pillow basalt

Basalt sheeted dikes

3 Gabbro

5 to 10 kilometers

Mantle peridotite

Figure II–12   The three layers of oceanic crust. Layer I consists of mud. Layer 2 is pillow basalt. Layer 3 consists of vertical dikes overlying gabbro. Below Layer 3 is the upper mantle.

Figure II–13   Sea-floor pillow basalt in the Cayman trough.   (Woods Hole Oceanographic Institution)

Figure 11–14 A scanning electron microscope photo of pelagic foraminifera, tiny organisms that float near the surface of the seas. (Ocean Drilling Program, Texas A&M University)

Pelagic sediment accumulates at a rate of about 2 to 10 millimeters per 1000 years. As mentioned earlier, its thickness increases with distance from the ridge because the sea floor becomes older as it spreads away from the ridge (Fig. 11–15). Close to the ridge there is virtually no sediment. Close to shore, pelagic sediment gradually merges with the much thicker layers of terrigenous sediment, which can be 3 kilometers thick.

Parts of the ocean floor beyond the mid-oceanic ridge are flat, level, featureless submarine surfaces called the **abyssal plains** (Fig. 11–7). They are the flattest surfaces on Earth. Seismic profiling shows that the basaltic crust is rough and jagged throughout the ocean. On the

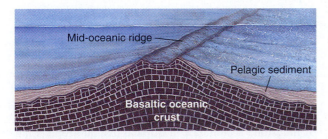

Figure 11–15 Deep sea mud becomes thicker with increasing distance from the mid-oceanic ridge.

abyssal plains, however, pelagic sediment buries this rugged profile, forming the smooth abyssal plains. If you were to remove all of the sediment, you would see rugged topography similar to that of the mid-oceanic ridge.

▶ 11.6 CONTINENTAL MARGINS

A **continental margin** is a place where continental crust meets oceanic crust. Two types of continental margins exist. A **passive margin** occurs where continental and oceanic crust are firmly joined together. Because it is not a plate boundary, little tectonic activity occurs at a passive margin. Continental margins on both sides of the Atlantic Ocean are passive margins. In contrast, an **active continental margin** occurs at a convergent plate boundary, where oceanic lithosphere sinks beneath the continent in a subduction zone. The west coast of South America is an active margin.

PASSIVE CONTINENTAL MARGINS

Consider the passive margin of eastern North America. Recall from Chapter 2 that, about 200 million years ago, all of the Earth's continents were joined into the supercontinent called Pangea. Shortly thereafter, Pangea began to rift apart into the continents as we know them today. The Atlantic Ocean opened as the east coast of North America separated from Europe and Africa.

As Pangea broke up, the crust of North America's east coast stretched and thinned near the fractures (Fig. 11–16). Basaltic magma rose at the spreading center, forming new oceanic crust between the separating continents. All tectonic activity then centered at the spreading mid-Atlantic ridge, and no further tectonic activity occurred at the continental margins; hence the term *passive* continental margin.

The Continental Shelf

On all continents, streams and rivers deposit sediment on coastal deltas, like the Mississippi River delta. Then ocean currents redistribute the sediment along the coast. The sediment forms a shallow, gently sloping submarine surface called a **continental shelf** on the edge of the continent (Fig. 11–17). As sediment accumulates on a continental shelf, the edge of the continent sinks isostatically because of the added weight. This effect keeps the shelf slightly below sea level.

Over millions of years, thick layers of sediment accumulated on the passive east coast of North America. The depth of the shelf increases gradually from the shore to about 200 meters at the outer shelf edge. The average inclination of the continental shelf is about 0.1°. A con-

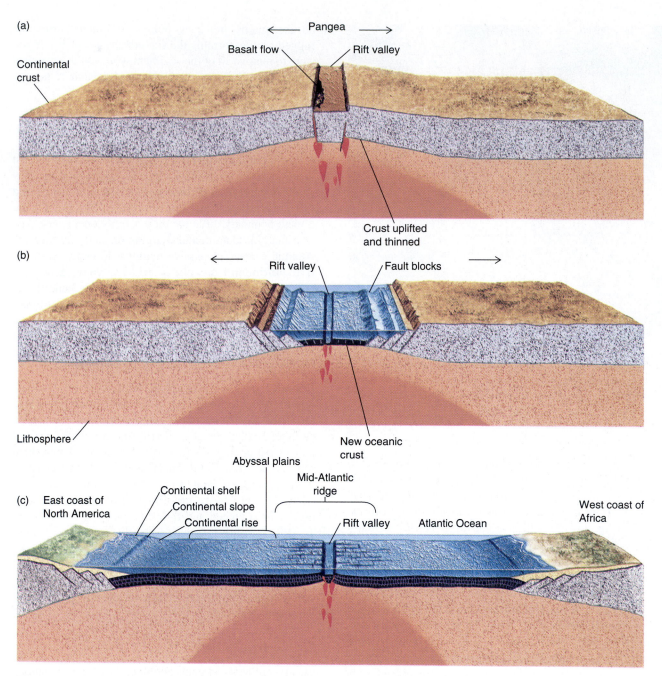

Figure 11–16   A passive continental margin developed on North America's east coast as Pangea rifted apart and the Atlantic Ocean basin began to open. (a) A mantle plume forces central Pangea upward. (b) Faulting and erosion thin the uplifted part of Pangea, and the crust begins to rift apart. Rising basalt magma forms new oceanic crust in the rift zone. (c) Sediment eroded from the continent forms a broad continental shelf–slope–rise complex.

tinental shelf on a passive margin can be a large feature. The shelf off the coast of southeastern Canada is about 500 kilometers wide, and parts of the shelves of Siberia and northwestern Europe are even wider.

In some places, a supply of sediment may be lacking, either because no rivers bring sand, silt, or clay to the shelf or because ocean currents bypass that area. In warm regions where sediment does not muddy the water, reef-building organisms thrive. As a result, thick beds of limestone accumulate in tropical and subtropical latitudes where clastic sediment is lacking. Limestone accumulations of this type may be hundreds of meters thick

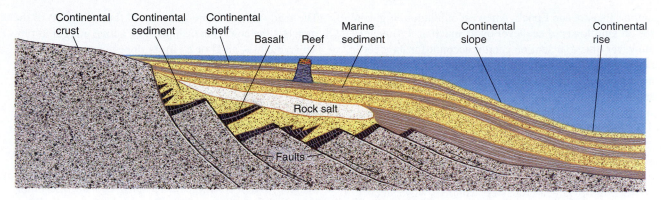

Figure 11–17 A passive continental margin consists of a broad continental shelf, slope, and rise formed by accumulation of sediment eroded from the continent. Salt deposits, reef limestone, and basalt sills are common in the shelf sedimentary rocks.

and hundreds of kilometers across and are called **carbonate platforms**. The Florida Keys and the Bahamas are modern-day examples of carbonate platforms on continental shelves.

Some of the world's richest petroleum reserves occur on the continental shelves of the North Sea between England and Scandinavia, in the Gulf of Mexico, and in the Beaufort Sea on the northern coast of Alaska and western Canada. In recent years, oil companies have explored and developed these offshore reserves. Deep drilling has revealed that granitic continental crust lies beneath the sedimentary rocks, confirming that the continental shelves are truly parts of the continents despite the fact that they are covered by seawater.

The Continental Slope and Rise

At the outer edge of a shelf, the sea floor suddenly steepens to about 4° to 5° as it falls away from 200 meters to about 5 kilometers in depth. This steep region of the sea floor averages about 50 kilometers wide and is called the **continental slope**. It is a surface formed by sediment accumulation, much like the shelf. Its steeper angle is due primarily to gradual thinning of continental crust in a transitional zone where it nears the junction with oceanic crust. Seismic profiler exploration shows that the sedimentary layering is commonly disrupted where sediment has slumped and slid down the steep incline.

A continental slope becomes less steep as it gradually merges with the deep ocean floor. This region, called the **continental rise**, consists of an apron of terrigenous sediment that was transported across the continental shelf and deposited on the deep ocean floor at the foot of the slope. The continental rise averages a few hundred kilometers wide. Typically, it joins the deep sea floor at a depth of about 5 kilometers.

In essence, then, the shelf–slope–rise complex is a smoothly sloping surface on the edge of a continent, formed by accumulation of sediment eroded from the continent.

Submarine Canyons and Abyssal Fans

In many places, sea-floor maps show deep valleys called **submarine canyons** eroded into the continental shelf and slope. They look like submarine stream valleys. A canyon typically starts on the outer edge of a continental shelf and continues across the slope to the rise (Fig. 11–18). At its lower end, a submarine canyon commonly leads into an **abyssal fan** (sometimes called a **submarine fan**), a large, fan-shaped pile of sediment lying on the continental rise.

Most submarine canyons occur where large rivers enter the sea. When they were first discovered, geologists thought the canyons had been eroded by rivers dur-

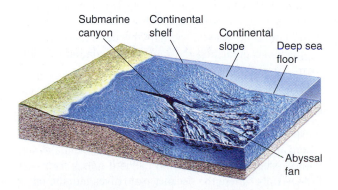

Figure 11–18 Turbidity currents cut submarine canyons into the continental shelf and slope and deposit sediment to form a submarine fan.

ing the Pleistocene Epoch, when accumulation of glacial ice on land lowered sea level by as much as 130 meters. However, this explanation cannot account for the deeper portions of submarine canyons cut into the lower continental slopes at depths of a kilometer or more. Therefore, submarine canyons must have formed under water, and a submarine mechanism must be found to explain them.

Geologists subsequently discovered that submarine canyons are cut by **turbidity currents**. A turbidity current develops when loose, wet sediment tumbles down the slope in a submarine landslide. The movement may be triggered by an earthquake or simply by oversteepening of the slope as sediment accumulates. When the sediment starts to move, it mixes with water. Because the mixture of sediment and water is denser than water alone, it flows as a turbulent, chaotic avalanche across the shelf and slope. A turbidity current can travel at speeds greater than 100 kilometers per hour and for distances up to 700 kilometers.

Sediment-laden water traveling at such speed has tremendous erosive power. Once a turbidity current cuts a small channel into the shelf and slope, subsequent currents follow the same channel, just as an intermittent surface stream uses the same channel year after year. Over time, the currents erode a deep submarine canyon into the shelf and slope. Turbidity currents slow down when they reach the deep sea floor. The sediment accumulates there to form an abyssal fan. Most submarine canyons and fans form near the mouths of large rivers because the rivers supply the great amount of sediment needed to create turbidity currents.

Large abyssal fans form only on passive continental margins. They are uncommon at active margins because in that environment, the sediment is swallowed by the trench. Furthermore, most of the world's largest rivers drain toward passive margins. The largest known fan is the Bengal fan, which covers about 4 million square kilometers beyond the mouth of the Ganges River in the Indian Ocean east of India. More than half of the sediment eroded from the rapidly rising Himalayas ends up in this fan. Interestingly, the Bengal fan has no associated submarine canyon, perhaps because the sediment supply is so great that the rapid accumulation of sediment prevents erosion of a canyon.

ACTIVE CONTINENTAL MARGINS

An active continental margin forms in a subduction zone, where an oceanic plate sinks beneath a continent. A long, narrow, steep-sided depression called a trench forms on the sea floor where the oceanic plate dives into the mantle (Fig. 11–19). Because an active margin has no gradual transition between continental and oceanic crust, it commonly has a narrower shelf than a passive margin.

The landward wall (the side toward the continent) of the trench is the continental slope of an active margin. It typically inclines at 4° or 5° in its upper part and steepens to 15° or more near the bottom of the trench. The continental rise is absent because sediment flows into the trench instead of accumulating on the ocean floor.

▶ II.7 ISLAND ARCS

In many parts of the Pacific Ocean and elsewhere, two oceanic plates converge. One dives beneath the other, forming a subduction zone and a trench. The deepest place on Earth is in the Mariana trench, north of New Guinea in the southwestern Pacific, where the ocean floor sinks to nearly 11 kilometers below sea level. Depths of 8 to 10 kilometers are common in other trenches.

Huge amounts of magma are generated in the subduction zone (Fig. 11–20). The magma rises and erupts on the sea floor to form submarine volcanoes next to the trench. The volcanoes eventually grow to become a chain of islands, called an **island arc.** The western Aleutian Islands are an example of an island arc. Many others occur at the numerous convergent plate boundaries in the southwestern Pacific (Fig. 11–21).

If subduction stops after an island arc forms, volcanic activity also ends. The island arc may then ride quietly on a tectonic plate until it arrives at another subduction zone at an active continental margin. However, the density of island arc rocks is relatively low, making them too buoyant to sink into the mantle. Instead, the island arc collides with the continent (Fig. 11–22). When this happens, the subducting plate commonly fractures on the seaward side of the island arc to form a new subduction zone. In this way, the island arc breaks away from the ocean plate and becomes part of the continent. The **accretion** of island arcs to continents in this manner played a major role in the geologic history of western North America and is explored more fully in Chapter 20.

Note the following points:

1. An island arc forms as magma rises from the mantle at an oceanic subduction zone.

2. The island arc eventually becomes part of a continent.

3. A continent cannot sink into the mantle at a subduction zone because of its buoyancy.

4. Thus, material is transferred from the mantle to a continent, but little or no material is transferred from continents to the mantle. This aspect of the plate tectonics model suggests that the amount of continental crust has increased throughout geologic time. However, some geologists feel that small

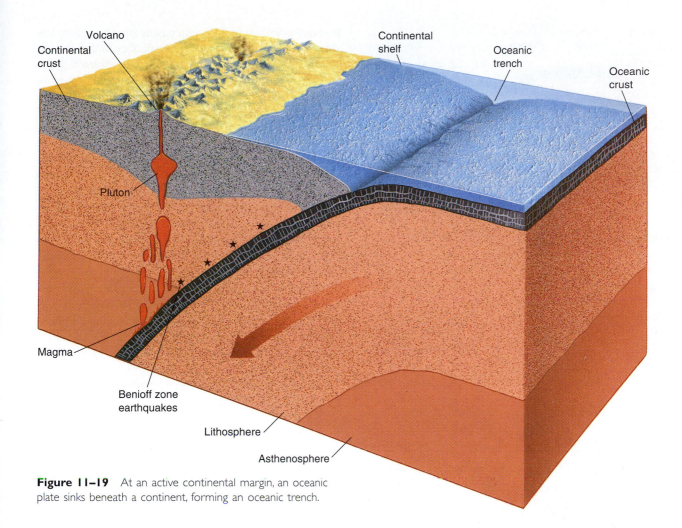

Figure 11–19   At an active continental margin, an oceanic plate sinks beneath a continent, forming an oceanic trench.

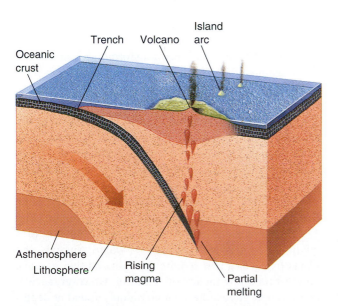

Figure 11–20   An island arc forms at a convergent boundary between two oceanic plates. One of the plates sinks, generating magma that rises to form a chain of volcanic islands.

Figure 11–21   Mataso is one of many volcanic islands in the Vanuatu island arc that formed along the Northern New Hebrides Trench in the South Pacific.

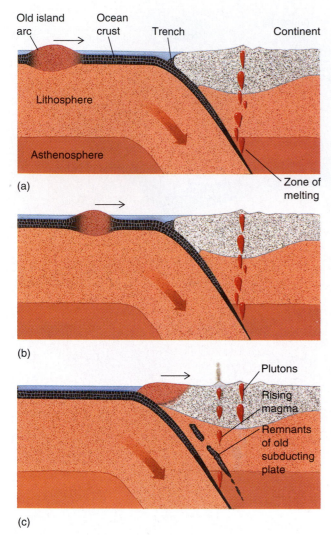

Old island arc   Ocean crust   Trench   Continent

Lithosphere

Asthenosphere

(a)

Zone of melting

(b)

Plutons

Rising magma

Remnants of old subducting plate

(c)

Figure 11–22 (a) An island arc is part of a lithospheric plate that is sinking into a subduction zone beneath a continent. (b) The island arc reaches the subduction zone but cannot sink into the mantle because of its low density. (c) The island arc is jammed onto the continental margin and becomes part of the continent. The subduction zone and trench step back to the seaward side of the island arc.

amounts of continental crust are returned to the mantle in subduction zones. This topic is discussed further in Chapter 12.

▶ **11.8   SEAMOUNTS AND OCEANIC ISLANDS**

A **seamount** is a submarine mountain that rises 1 kilometer or more above the surrounding sea floor. An **oceanic island** is a seamount that rises above sea level.

Both are common in all ocean basins but are particularly abundant in the southwestern Pacific Ocean. Seamounts and oceanic islands sometimes occur as isolated peaks on the sea floor, but they are more commonly found in chains. Dredge samples show that seamounts, like oceanic islands and the ocean floor itself, are made of basalt.

Seamounts and oceanic islands are submarine volcanoes that formed at a hot spot above a mantle plume. They form within a tectonic plate rather than at a plate boundary. An isolated seamount or short chain of small seamounts probably formed over a plume that lasted for only a short time. In contrast, a long chain of large islands, such as the Hawaiian Island–Emperor Seamount Chain, formed over a long-lasting plume. In this case the lithospheric plate migrated over the plume as the magma continued to rise. Each volcano formed directly over the plume and then became extinct as the moving plate carried it away from the plume. As a result, the seamounts and oceanic islands become progressively younger toward the end of the chain that is volcanically active today (Fig. 11–23).

After a volcanic island forms, it begins to sink. Three factors contribute to the sinking:

1. If the mantle plume stops rising, it stops producing magma. Then the lithosphere beneath the island cools and becomes denser, and the island sinks. Alternatively, the moving plate may carry the island away from the hot spot. This also results in cooling, contraction, and sinking of the island.

2. The weight of the newly formed volcano causes isostatic sinking.

3. Erosion lowers the top of the volcano.

These three factors gradually transform a volcanic island into a seamount (Fig. 11–24). If the Pacific Ocean plate continues to move at its present rate, the island of Hawaii may sink beneath the sea within 10 to 15 million years. A new submarine volcano, called Loihi, is currently forming off the southeast side of Hawaii. As the Pacific plate moves northwest and volcanism at Loihi increases, this seamount will become a new Hawaiian island.

Sea waves may erode a flat top on a sinking island, forming a flat-topped seamount called a **guyot** (Fig. 11–25). A reef commonly grows on the flat top of a guyot while it is still in shallow water. Animals and plants living in a reef require sunlight and thus can live only within a few meters of sea level. However, ancient reef-covered guyots are now commonly found at depths of more than 1 kilometer, showing that the guyots continued to sink after the reefs died.

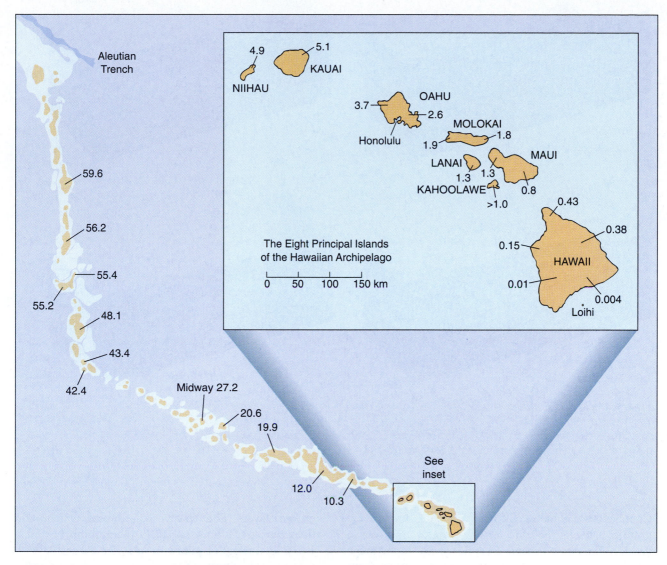

Figure 11–23 The Hawaiian Island–Emperor Seamount Chain becomes older in a direction going away from the island of Hawaii. The ages, in millions of years, are for the oldest volcanic rocks of each island or seamount.

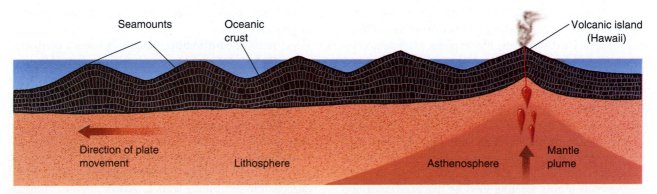

Figure 11–24 The Hawaiian Islands and Emperor Seamounts sink as they move away from the mantle plume.

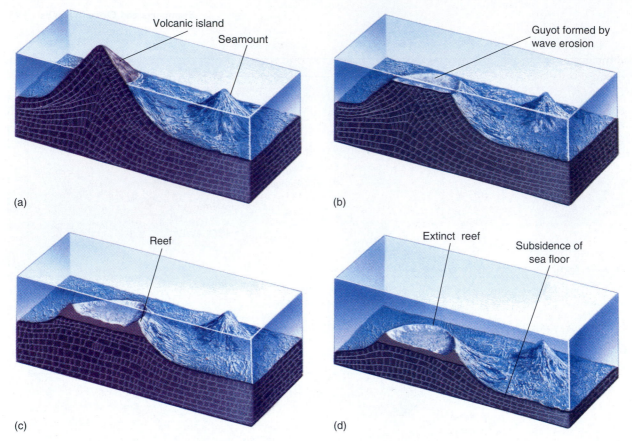

(a)

(b)

(c)

(d)

Figure 11–25 (a) A seamount is a volcanic mountain on the sea floor. Some rise above sea level to form volcanic islands. (b) Waves can erode a flat top on a sinking island to form a guyot. (c) A reef may grow on the guyot and (d) eventually become extinct if the guyot sinks below the sunlight zone or migrates into cooler latitudes.

SUMMARY

Continents are composed of relatively thick, low-density granite, whereas oceanic crust is mostly thin, dense basalt. Thin, dense oceanic crust lies at low topographic levels and forms ocean basins. Because of the great depth and remoteness of the ocean floor and oceanic crust, our knowledge of them comes mainly from **sampling** and **remote sensing**.

Stripes of normal and reversed **magnetic polarity** that are symmetrically distributed about the mid-oceanic ridge gave rise to the hypothesis of **sea-floor spreading**, which rapidly evolved into the modern plate tectonics theory.

The mid-oceanic ridge is a submarine mountain chain that extends through all of the Earth's major ocean basins. A **rift valley** runs down the center of many parts of the ridge, and the ridge and rift valley are both offset by numerous **transform faults**. The mid-oceanic ridge forms at the center of lithospheric spreading, where new oceanic crust is added to the sea floor.

Abyssal plains are flat areas of the deep sea floor where the rugged topography of the basaltic oceanic crust is covered by deep sea sediment. Oceanic crust varies from about 5 to 10 kilometers thick and consists

of three layers. The top layer is sediment, which varies from zero to 3 kilometers thick. Beneath this lies about 1 to 2 kilometers of **pillow basalt**. The deepest layer of oceanic crust is from 4 to 5 kilometers thick and consists of basalt dikes on top of gabbro. The base of this layer is the boundary between oceanic crust and mantle. The age of sea-floor rocks increases regularly away from the mid-oceanic ridge. No oceanic crust is older than about 200 million years because it recycles into the mantle at subduction zones.

A **passive continental margin** includes a **continental shelf**, a **slope**, and a **rise** formed by accumulation of **terrigenous** sediment. **Submarine canyons**, eroded by **turbidity currents**, notch continental margins and commonly lead into **abyssal fans**, where the turbidity currents deposit sediments on the continental rise. An **active continental margin**, where oceanic crust subducts beneath the margin of a continent, usually includes a narrow continental shelf and a continental slope that steepens rapidly into a **trench**. A trench is an elongate trough in the ocean floor formed where oceanic crust dives downward at a subduction zone. Trenches are the deepest parts of ocean basins.

Island arcs are common features of some ocean basins, particularly the southwestern Pacific. They are chains of volcanoes formed at subduction zones where two oceanic plates collide. **Seamounts** and **oceanic islands** form in oceanic crust as a result of volcanic activity over mantle plumes.

KEY WORDS

rock dredge   *180*
core   *180*
sea-floor drilling   *180*
echo sounder   *181*
seismic profiler   *181*
magnetometer   *181*
polarity   *182*
magnetic reversal   *182*
sea-floor spreading   *183*

rift valley   *185*
heat flow   *185*
transform fault   *185*
pillow basalt   *188*
terrigenous sediment   *188*
pelagic sediment   *188*
abyssal plains   *189*
passive continental
   margin   *189*

active continental
   margin   *189*
continental shelf   *189*
carbonate platform   *191*
continental slope   *191*
continental rise   *191*
submarine canyon   *191*
abyssal fan (submarine
   fan)   *191*

turbidity current   *192*
island arc   *192*
accretion   *192*
seamount   *194*
oceanic island   *194*
guyot   *194*

REVIEW QUESTIONS

1. Describe the main differences between oceans and continents.

2. Describe a magnetic reversal.

3. Explain how a rock preserves evidence of the orientation of the Earth's magnetic field at the time the rock formed.

4. Describe how the discovery of magnetic patterns on the sea floor confirmed the sea-floor spreading theory.

5. Sketch a cross section of the mid-oceanic ridge, including the rift valley.

6. Describe the dimensions of the mid-oceanic ridge.

7. Explain why the mid-oceanic ridge is topographically elevated above the surrounding ocean floor. Why does its elevation gradually decrease away from the ridge axis?

8. Explain the origin of the rift valley in the center of the mid-oceanic ridge.

9. Why is heat flow unusually high at the mid-oceanic ridge?

10. Why are the abyssal plains characterized by such low relief?

11. Sketch a cross section of oceanic crust from a deep sea basin. Label, describe, and indicate the approximate thickness of each layer.

12. Describe the two main types of sea-floor sediment. What is the origin of each type?

13. Compare the ages of oceanic crust with the ages of continental rocks. Why are they so different?

14. Sketch a cross section of both an active continental margin and a passive continental margin. Label the features of each. Give approximate depths below sea level of each of the features.

15. Explain why a continental shelf is made up of a foundation of granitic crust, whereas the deep ocean floor is composed of basalt.

16. Why does an active continental margin typically have a steeper continental slope than a passive margin? Why does an active margin typically have no continental rise?

17. Explain the relationships among submarine canyons, abyssal fans, and turbidity currents.

18. Why are turbidity currents often associated with earthquakes or with large floods in major rivers?

19. Explain the role played by an island arc in the growth of a continent.

20. Explain the origins of and differences between seamounts and island arcs.

21. Compare the ocean depths adjacent to an island arc and a seamount.

22. Why do oceanic islands sink after they form?

DISCUSSION QUESTIONS

1. How and why does an oceanic trench form?

2. The east coast of South America has a wide continental shelf, whereas the west coast has a very narrow shelf. Discuss and explain this contrast.

3. Seismic data indicate that continental crust thins where it joins oceanic crust at a passive continental margin, such as on the east coast of North America. Other than that, we know relatively little about the nature of the junction between the two types of crust. Speculate on the nature of that junction. Consider rock types, geologic structures, ages of rocks, and other features of the junction.

4. Discuss the topography of the Earth in an imaginary scenario in which all conditions are identical to present ones except that there is no water. In contrast, what would be the effect if there were enough water to cover all of the Earth's surface?

5. In Section 11.6 we stated that most of the world's largest rivers drain toward passive continental margins. Explain this observation.

Geologic Structures, Mountain Ranges, and Continents

Mountains form some of the most majestic landscapes on Earth. People find peace in the high mountain air and quiet valleys. But mountains also project nature's power. Storms swirl among the silent peaks that were lifted skyward millions of years ago by tectonic processes. Rocks have been folded as if squeezed by a giant's hand. In the first portion of this chapter, we will study how rocks behave when tectonic forces stress them. In the second portion, we will learn how tectonic forces raise mountains.

Tectonic forces contorted these once-horizontal sedimentary rocks near Carlin, Nevada, into tight folds. (David Matherly/Visuals Unlimited)

Geologic Structures

► 12.1   ROCK DEFORMATION

STRESS

Recall from Chapter 10 that stress is a force exerted against an object. Tectonic forces exert different types of stress on rocks in different geologic environments. The first, called **confining stress** or **confining pressure**, occurs when rock or sediment is buried (Fig. 12–1a). Confining pressure merely compresses rocks but does not distort them, because the compressive force acts equally in all directions, like water pressure on a fish. As you learned in Chapter 7, burial pressure compacts sediment and is one step in the lithification of sedimentary rocks. Confining pressure also contributes to metamorphism during deep burial in sedimentary basins.

In contrast, **directed stress** acts most strongly in one direction. Tectonic processes create three types of directed stress. Compression squeezes rocks together in one direction. It frequently acts horizontally, shortening the distance parallel to the squeezing direction (Fig. 12–1b). **Compressive stress** is common in convergent plate boundaries, where two plates converge and the rock crumples, just as car fenders crumple during a head-on collision. **Extensional stress** (often called **tensional stress**) pulls rock apart and is the opposite of tectonic compression (Fig. 12–1c). Rocks at a divergent plate boundary stretch and pull apart because they are subject to extensional stress. **Shear stress** acts in parallel but opposite directions (Fig. 12–1d). Shearing deforms rock by causing one part of a rock mass to slide past the other part, as in a transform fault or a transform plate boundary.

STRAIN

Strain is the deformation produced by stress. As explained in Chapter 10, a rock responds to tectonic stress by elastic deformation, plastic deformation, or brittle fracture. An elastically deformed rock springs back to its original size and shape when the stress is removed. During plastic deformation, a rock deforms like putty and retains its new shape. In some cases a rock will deform plastically and then fracture (Fig. 12–2).

Factors That Control Rock Behavior

Several factors control whether a rock responds to stress by elastic or plastic deformation or fails by brittle fracture:

1. *The nature of the material.* Think of a quartz crystal, a gold nugget, and a rubber ball. If you strike quartz with a hammer, it shatters. That is, it fails by

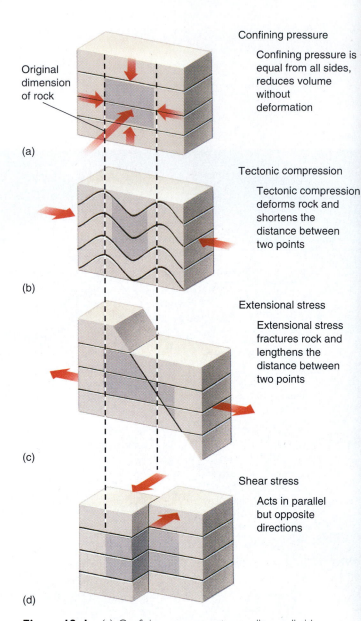

Figure 12–1   (a) Confining pressure acts equally on all sides of a rock. Thus, the rock is compressed much as a balloon is compressed if held under water. Rock volume decreases without deformation. (b) Tectonic compression shortens the distance parallel to the stress direction. Rocks fold or fracture to accommodate the shortening. (c) Extensional stress lengthens the distance parallel to the stress direction. Rocks commonly fracture to accommodate the stretching. (d) Shear stress deforms the rock parallel to the stress direction.

brittle fracture. In contrast, if you strike the gold nugget, it deforms in a plastic manner; it flattens and stays flat. If you hit the rubber ball, it deforms elastically and rebounds immediately, sending the hammer flying back at you. Initially, all rocks react to stress by deforming elastically. Near the Earth's

Figure 12–2 This rock (in the Nahanni River, Northwest Territories, Canada) folded plastically and then fractured.

Figure 12–3 A fold is a bend in rock. These are in quartzite in the Maria Mountains, California. (W. B. Hamilton, USGS)

surface, where temperature and pressure are low, different types of rocks behave differently with continuing stress. Granite and quartzite tend to behave in a brittle manner. Other rocks, such as shale, limestone, and marble, have greater tendencies to deform plastically.

2. *Temperature.* The higher the temperature, the greater the tendency of a rock to behave in a plastic manner. It is difficult to bend an iron bar at room temperature, but if the bar is heated in a forge, it becomes plastic and bends easily.

3. *Pressure.* High confining pressure also favors plastic behavior. During burial, both temperature and pressure increase. Both factors promote plastic deformation, so deeply buried rocks have a greater tendency to bend and flow than shallow rocks.

4. *Time.* Stress applied over a long time, rather than suddenly, also favors plastic behavior. Marble park benches in New York City have sagged plastically under their own weight within 100 years. In contrast, rapidly applied stress, such as the blow of a hammer, to a marble bench causes brittle fracture.

▶ **12.2 GEOLOGIC STRUCTURES**

Enormous compressive forces can develop at a convergent plate boundary, bending and fracturing rocks in the tectonically active region. In some cases the forces deform rocks tens or even hundreds of kilometers from the plate boundary. Because the same tectonic processes create great mountain chains, rocks in mountainous regions are commonly broken and bent. Tectonic forces also deform rocks at divergent and transform plate boundaries.

A **geologic structure** is any feature produced by rock deformation. Tectonic forces create three types of geologic structures: folds, faults, and joints.

FOLDS

A **fold** is a bend in rock (Fig. 12–3). Some folded rocks display little or no fracturing, indicating that the rocks deformed in a plastic manner. In other cases, folding occurs by a combination of plastic deformation and brittle fracture. Folds formed in this manner exhibit many tiny fractures.

If you hold a sheet of clay between your hands and exert compressive stress, the clay deforms into a sequence of folds (Fig. 12–4). This demonstration illustrates three characteristics of folds:

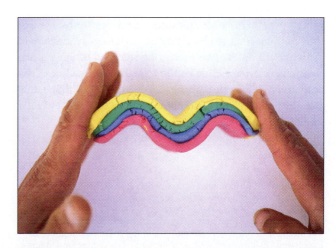

Figure 12–4 Clay deforms into a sequence of folds when compressed.

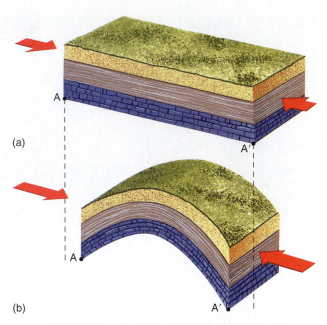

(a)

(b)

Figure 12–5 (a) Horizontally layered sedimentary rocks. (b) A fold in the same rocks. The forces that folded the rocks are shown by the arrows. Notice that points A and A′ are closer after folding.

1. Folding usually results from compressive stress. For example, tightly folded rocks in the Himalayas indicate that the region was subjected to compressive stress.

2. Folding always shortens the horizontal distances in rock. Notice in Figure 12–5 that the distance between two points, A and A′, is shorter in the folded rock than it was before folding.

3. Folds usually occur as a repeating pattern of many folds as in the illustration using clay.

Figure 12–6 shows that a fold arching upward is called an **anticline** and one arching downward is a **syncline**.[1] The sides of a fold are called the **limbs**. Notice that a single limb is shared by an anticline–syncline pair. A line dividing the two limbs of a fold and running along the crest of an anticline or the trough of a syncline is the fold **axis**. The **axial plane** is an imaginary plane that runs through the axis and divides a fold as symmetrically as possible into two halves.

In many folds, the axis is horizontal, as shown in Figure 12–6a. If you were to walk along the axis of a

---

[1]Properly, an upward-arched fold is called an anticline only if the oldest rocks are in the center and the youngest are on the outside of the fold. Similarly, a downward-arched fold is a syncline only if the youngest rocks are at the center and the oldest are on the outside. The age relationships become reversed if the rocks are turned completely upside down and folded. If the age relationships are unknown, as sometimes occurs, an upward-arched fold is called an *antiform*, and a downward-arched one is called a *synform*.

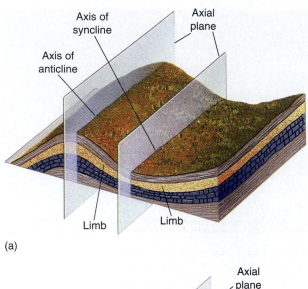

(a)

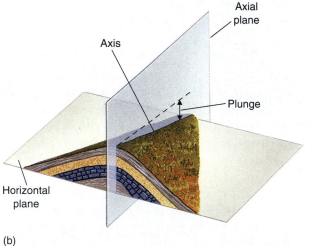

(b)

(c)

Figure 12–6 (a) An anticline, a syncline, and the parts of a fold. (b) A plunging anticline. (c) A syncline in southern Nevada.

Figure 12–7 A syncline lies beneath the mountain peak and an anticline forms the low point, or saddle, in the Canadian Rockies, Alberta.

horizontal anticline, you would be walking on a level ridge. In other folds, the axis is inclined or tipped at an angle called the **plunge**, as shown in Figure 12–6b. A fold with a plunging axis is called a **plunging fold**. If you were to walk along the axis of a plunging fold, you would be traveling uphill or downhill along the axis.

Even though an anticline is structurally a high point in a fold, anticlines do not always form topographic

ridges. Conversely, synclines do not always form valleys. Landforms are created by combinations of tectonic and surface processes. In Figure 12–7, the syncline lies beneath the peak and the anticline forms the saddle between two peaks.

Figure 12–8 summarizes the characteristics of five common types of folds. A special type of fold with only one limb is a **monocline**. Figure 12–9 shows a monocline

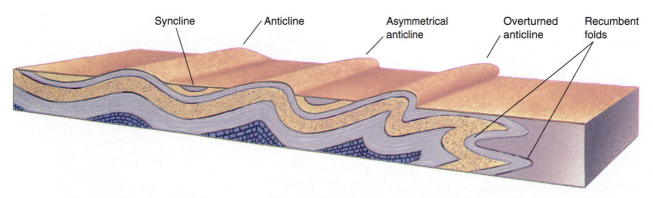

Figure 12–8 Cross-sectional view of five different kinds of folds. Folds can be symmetrical, as shown on the left, or asymmetrical, as shown in the center. If a fold has tilted beyond the perpendicular, it is overturned.

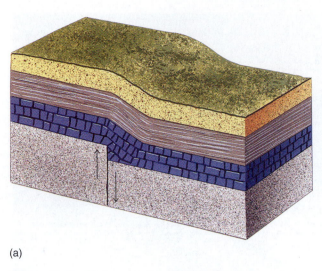

(a)

(b)

Figure 12–9 (a) A monocline formed where near-surface sedimentary rocks sag over a fault. (b) A monocline in southern Utah.

that developed where sedimentary rocks sag over an underlying fault.

A circular or elliptical anticlinal structure is called a **dome**. Domes resemble inverted bowls. Sedimentary layering dips away from the center of a dome in all directions (Fig. 12–10). A similarly shaped syncline is called a **basin**. Domes and basins can be small structures only a few kilometers in diameter or less. Frequently, however, they are very large and are caused by broad upward or downward movement of the continental crust. The Black Hills of South Dakota are a large structural dome. The Michigan basin covers much of the state of Michigan, and the Williston basin covers much of eastern Montana, northeastern Wyoming, the western Dakotas, and southern Alberta and Saskatchewan.

Although most folds form by compression, less commonly, crustal extension can also fold rocks. Figure 12–11 shows a block of rock that dropped down along a curved fault as the crust pulled apart. The block developed a syncline as it rotated and deformed while sliding downward. Folds formed by extension are usually broad, open folds in contrast to tight folds commonly formed by compression.

FAULTS

A **fault** is a fracture along which rock on one side has moved relative to rock on the other side (Fig. 12–12). **Slip** is the distance that rocks on opposite sides of a fault have moved. Movement along a fault may be gradual, or the rock may move suddenly, generating an earthquake. Some faults are a single fracture in rock; others consist of numerous closely spaced fractures called a **fault zone** (Fig. 12–13). Rock may slide hundreds of meters or many kilometers along a large fault zone.

Rock moves repeatedly along many faults and fault zones for two reasons: (1) Tectonic forces commonly persist in the same place over long periods of time (for example, at a tectonic plate boundary), and (2) once a fault forms, it is easier for movement to occur again

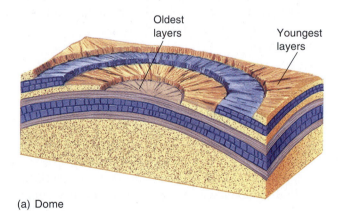

(a) Dome

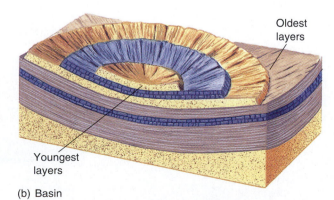

(b) Basin

Figure 12–10 (a) Sedimentary layering dips away from a dome in all directions, and the outcrop pattern is circular or elliptical. (b) Layers dip toward the center of a basin.

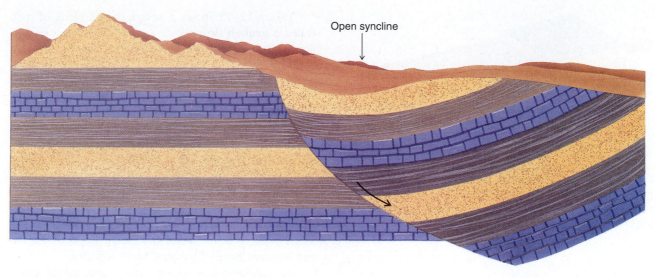

Open syncline

Figure 12–11 Folds can form by crustal extension. A syncline has developed in a down-dropped block of sedimentary rock as it slid down and rotated along a curved normal fault.

Figure 12–12 A small fault has dropped the right side of these volcanic ash layers downward about 60 centimeters relative to the left side. (Ward's Natural Science Establishment, Inc.)

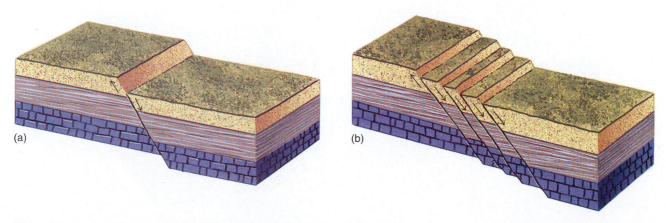

(a)

(b)

Figure 12–13 (a) Movement along a single fracture surface characterizes faults with relatively small slip. (b) Movement along numerous closely spaced faults in a fault zone is typical of faults with large slip.

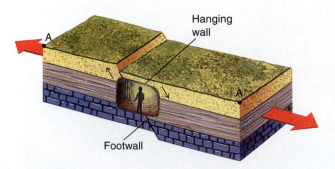

Figure 12–14 A normal fault accommodates extension of the Earth's crust. Large arrows show extensional stress direction. The overlying side of the fault is called the hanging wall, and the side beneath the fault is the footwall.

along the same fracture than for a new fracture to develop nearby.

Hydrothermal solutions often precipitate in faults to form rich ore veins. Miners then dig shafts and tunnels along veins to get the ore. Many faults are not vertical but dip into the Earth at an angle. Therefore, many veins have an upper side and a lower side. Miners referred to the side that hung over their heads as the **hanging wall** and the side they walked on as the **footwall**. These names are commonly used to describe both ore veins and faults (Fig. 12–14).

A fault in which the hanging wall has moved down relative to the footwall is called a **normal fault**. Notice that the horizontal distance between points on opposite sides of the fault, such as A and A′ in Figure 12–14, is greater after normal faulting occurs. Hence, a normal fault forms where tectonic tension stretches the Earth's crust, pulling it apart.

Figure 12–15 shows a wedge-shaped block of rock called a **graben** dropped downward between a pair of normal faults. The word *graben* comes from the German word for "grave" (think of a large block of rock settling downward into a grave). If tectonic forces stretch the crust over a large area, many normal faults may develop, allowing numerous grabens to settle downward between the faults. The blocks of rock between the downdropped grabens then appear to have moved upward relative to the grabens; they are called **horsts**.

Normal faults, grabens, and horsts are common where the crust is rifting at a spreading center, such as the mid-oceanic ridge and the East African rift zone. They are also common where tectonic forces stretch a single plate, as in the Basin and Range of Utah, Nevada, and adjacent parts of western North America.

In a region where tectonic forces squeeze the crust, geologic structures must accommodate crustal shortening. A fold accomplishes shortening. A **reverse fault** is another structure that accommodates shortening (Fig. 12–16). In a reverse fault, the hanging wall has moved up relative to the footwall. The distance between points A and A′ is shortened by the faulting.

A **thrust fault** is a special type of reverse fault that is nearly horizontal (Fig. 12–17). In some thrust faults, the rocks of the hanging wall have moved many kilometers over the footwall. For example, all of the rocks of Glacier National Park in northwestern Montana slid 50 to 100 kilometers eastward along a thrust fault to their present location. This thrust is one of many that formed from about 180 to 45 million years ago as compressive tectonic forces built the mountains of western North America. Most of those thrusts moved large slabs of rock, some even larger than that of Glacier Park, from west to east in a zone reaching from Alaska to Mexico.

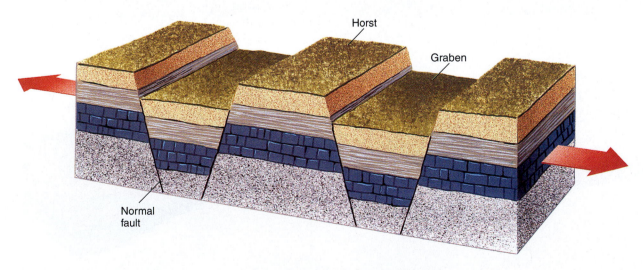

Figure 12–15 Horsts and grabens commonly form where tectonic forces stretch the Earth's crust.

(a)

(b)

Figure 12–16 (a) A reverse fault accommodates crustal shortening and reflects squeezing of the crust, shown by large arrows. (b) A small reverse fault in Zion National Park, Utah.

A **strike–slip fault** is one in which the fracture is vertical, or nearly so, and rocks on opposite sides of the fracture move horizontally past each other (Fig. 12–18). A transform plate boundary is a strike–slip fault. As explained previously, the famous San Andreas fault zone is a zone of strike–slip faults that form the border between the Pacific plate and the North American plate.

JOINTS

A **joint** is a fracture in rock and is therefore similar to a fault, except that in a joint rocks on either side of the fracture have not moved. We have already discussed columnar joints in basalt (Chapter 5) and jointing caused

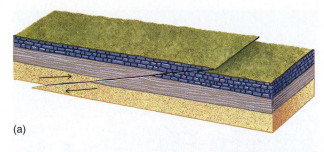

(a)

(b)

Figure 12–17 (a) A thrust fault is a low-angle reverse fault. (b) A small thrust fault near Flagstaff, Arizona. (Ward's Natural Science Establishment, Inc.)

by unloading and exfoliation (Chapter 6). Tectonic forces also fracture rock to form joints (Fig. 12–19). Most rocks near the Earth's surface are jointed, but joints become less abundant with depth because rocks become more plastic at deeper levels in the crust.

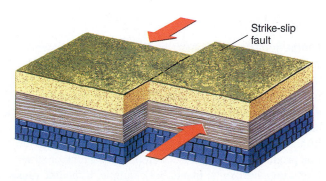

Figure 12–18 A strike–slip fault is nearly vertical, but movement along the fault is horizontal. The large arrows show direction of movement.

Figure 12–19 Joints, such as those in this sandstone along the Escalante River in Utah, are fractures along which the rock has not slipped.

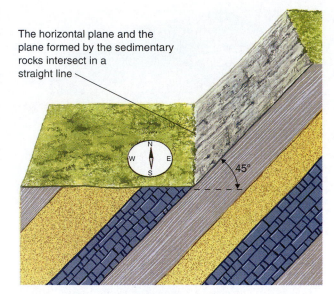

The horizontal plane and the plane formed by the sedimentary rocks intersect in a straight line

Figure 12–20 Strike is the compass direction of the intersection of a horizontal plane with a sedimentary bed or other planar feature in rock. Dip is the angle between a horizontal plane and the layering.

Joints and faults are important in engineering, mining, and quarrying because they are planes of weakness in otherwise strong rock. A dam constructed in jointed rock often leaks, not because the dam has a hole but because water follows the fractures and seeps around the dam. You can commonly see seepage caused by such leaks in the walls of a canyon downstream from a dam.

Strike and Dip

Faults, joints, sedimentary beds, slaty cleavage, and a wide range of other geologic features are planar surfaces in rock. Field geologists describe the orientations of sedimentary beds or other planes with two measurements called **strike** and **dip**. To understand these concepts, recall from elementary geometry that two planes intersect in a straight line. Strike is the compass direction of the line produced by the intersection of a tilted rock or structure with a horizontal plane. For example, if the line runs exactly north–south, the strike is 0° (you could also call it 180°). If the line points east, the strike is 90°. Dip is the angle of inclination of the tilted layer, also measured from the horizontal plane. In Figure 12–20 the dip is 45°.

GEOLOGIC STRUCTURES AND PLATE BOUNDARIES

Each of the three different types of plate boundaries produces different tectonic stresses and therefore different kinds of structures. Extensional stress at a divergent boundary (mid-oceanic ridges and continental rift boundaries) produces normal faults, and sometimes grabens, but little or no folding of rocks.

Where a transform boundary crosses continental crust, shear stress bends and fractures rock. Frictional drag between both sides of the fault may fold, fault, and uplift nearby rocks. Forces of this type have formed the San Gabriel Mountains along the San Andreas fault zone, as well as mountain ranges north of the Himalayas.

In contrast, compressive stress commonly dominates a convergent plate boundary. The compression produces folds, reverse faults, and thrust faults. These structures are common features of many mountain ranges formed at convergent plate boundaries. For example, subduction along the west coast of North America formed extensive regions of folded and thrust-faulted rocks in the western mountains. Similar structures are common in the Appalachian Mountains of eastern North America (Fig. 12–21), the Alps, and the Himalayas, all of which formed as the result of continent–continent collisions.

Although plate convergence commonly creates compressive stress, in some instances crustal extension and normal faulting are common. The Andes of western South America formed, and continue to grow today, by subduction of the Nazca plate beneath the western edge of

Figure 12–21   These sedimentary rocks in New Jersey were folded during the Appalachian orogeny.   (Breck P. Kent)

the South American plate. The two plates are converging, yet large grabens west of the mountains reflect crustal extension.

Mountain Ranges and Continents

▶ **12.3   MOUNTAINS AND MOUNTAIN RANGES**

MOUNTAIN-BUILDING PROCESSES

Mountains grow along each of the three types of tectonic plate boundaries. As you learned in Chapter 11, the world's largest mountain chain, the mid-oceanic ridge, formed at divergent plate boundaries beneath the ocean. Mountain ranges also originate at divergent plate boundaries on land. Mount Kilimanjaro and Mount Kenya, two volcanic peaks near the equator, lie along the East African rift. Other ranges, such as the San Gabriel Mountains of California, form at transform plate boundaries. However, the great continental mountain chains, including the Andes, Appalachians, Alps, Himalayas, and Rockies, all rose at convergent plate boundaries. Folding and faulting of rocks, earthquakes, volcanic eruptions, intrusion of plutons, and metamorphism all occur at a convergent plate boundary. The term **orogeny** refers to the process of mountain building and includes all of these activities.

Because plate boundaries are linear, mountains most commonly occur as long, linear, or slightly curved ranges and chains. For example, the Andes extend in a narrow band along the west coast of South America, and the Appalachians form a linear uplift along the east coast of North America.

Several processes thicken continental crust as tectonic forces build a mountain range. A subducting slab generates magma, which cools within the crust to form plutons or rises to the surface to form volcanic peaks. Both the plutons and volcanic rocks thicken the continental crust over a subduction zone by adding new material to it. In addition, the magmatic activity heats the lithosphere in the region above the subduction zone, causing it to become less dense and to rise isostatically. In a region where two continents collide, one continent may be forced beneath the other. This process, called **underthrusting**, can double the thickness of continental crust in the collision zone. Finally, compressive forces fold and crumple rock, squeezing the continent and increasing its thickness. Thus, addition of magma, heating, underthrusting, and folding all combine to thicken continental crust and lithosphere. As they thicken, the surface of the continent rises isostatically to form a mountain chain.

Opposing forces act on a rising mountain chain; the processes just described may continue to raise the mountains at the same time that other processes lower the peaks (Fig. 12–22). As a mountain chain grows higher and heavier, eventually the underlying rocks cannot support the weight of the mountains. The crustal rocks and

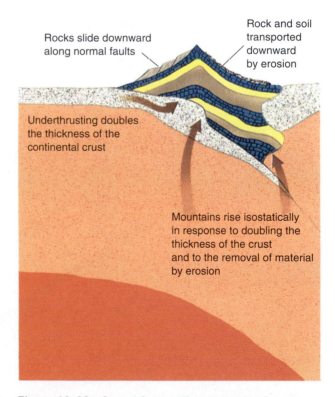

Rocks slide downward along normal faults

Rock and soil transported downward by erosion

Underthrusting doubles the thickness of the continental crust

Mountains rise isostatically in response to doubling the thickness of the crust and to the removal of material by erosion

Figure 12–22   Several factors affect the height of a mountain range. Today the Himalayas are being uplifted by continued underthrusting. At the same time, erosion removes the tops of the peaks, and the sides of the range slip downward along normal faults. As these processes remove weight from the range, the mountains rise by isostatic adjustment.

underlying lithosphere become so plastic that they flow outward from beneath the mountains. As an analogy, consider pouring cold honey onto a table top. At first, the honey piles up into a high, steep mound, but soon it begins to flow outward under its own weight, lowering the top of the mound.

Streams, glaciers, and landslides erode the peaks as they rise, carrying the sediment into adjacent valleys. When rock and sediment erode, the mountain becomes lighter and rises isostatically, just as a canoe rises when you step out of it. Eventually, however, the mountains erode away completely. The Appalachians are an old range where erosion is now wearing away the remains of peaks that may once have been the size of the Himalayas.

With this background, let us look at mountain building in three types of convergent plate boundaries.

▶ 12.4 ISLAND ARCS: MOUNTAIN BUILDING DURING CONVERGENCE BETWEEN TWO OCEANIC PLATES

As described in Chapter 11, an island arc is a volcanic mountain chain formed at an oceanic subduction zone. During subduction, one of the plates dives into the mantle, forming an oceanic trench and generating magma. This magma rises to the sea floor, where it erupts to build submarine volcanoes. These volcanoes may eventually grow above sea level, creating an arc-shaped volcanic island chain next to the trench.

A layer of sediment a half kilometer or more thick commonly covers the basaltic crust of the deep sea floor. Some of the sediment is scraped from the subducting slab and jammed against the inner wall (the wall toward the island arc) of the trench. Occasionally, slices of rock from the oceanic crust, and even pieces of the upper mantle, are scraped off and mixed in with the sea-floor sediment. The process is like a bulldozer scraping soil from bedrock and occasionally knocking off a chunk of bedrock along with the soil. The bulldozer process folds, shears, and faults sediment and rock. The rocks added to the island arc in this way are called a **subduction complex** (Fig. 12–23).

Growth of the subduction complex occurs by addition of the newest slices at the bottom of the complex. Consequently, this underthrusting forces the subduction complex upward, forming a sedimentary basin called a **forearc basin** between the subduction complex and the island arc. This process is similar to holding a flexible notebook horizontally between your two hands. If you move your hands closer together, the middle of the notebook bends downward to form a topographic depression analogous to a forearc basin. In addition, underthrusting thickens the crust, leading to isostatic uplift. The forearc

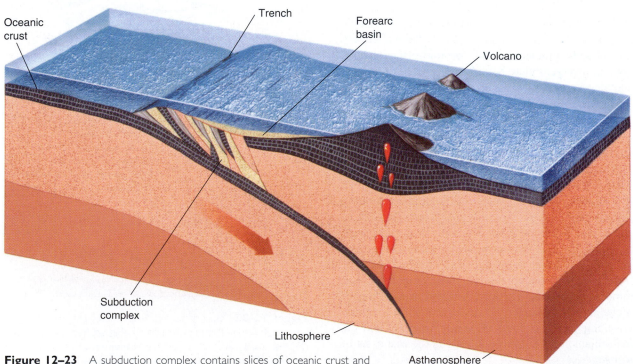

Figure 12–23 A subduction complex contains slices of oceanic crust and upper mantle scraped from the upper layers of a subducting plate.

Figure 12–24   The Cordillera Apolobamba in Bolivia rises over 6000 meters.

basin fills with sediment derived from erosion of the volcanic islands and also becomes a part of the island arc.

Island arcs are abundant in the Pacific Ocean, where convergence of oceanic plates is common. The western Aleutian Islands and most of the island chains of the southwestern Pacific are island arcs.

▶ 12.5   THE ANDES: SUBDUCTION AT A CONTINENTAL MARGIN

The Andes are the world's second highest mountain chain, with 49 peaks above 6000 meters (nearly 20,000 feet) (Fig. 12–24). The highest peak is Aconcagua, at 6962 meters. The Andes rise almost immediately from the Pacific coast of South America and thus start nearly at sea level. Igneous rocks make up most of the Andes, although the chain also contains folded sedimentary rocks, especially in the eastern foothills.

The supercontinent that Alfred Wegener called Pangea broke apart at the end of the Triassic Period. In the early Jurassic, the lithospheric plate that included South America started moving westward. To accommodate the westward motion, oceanic lithosphere began to dive into the mantle beneath the west coast of South America, forming a subduction zone by early Cretaceous time, 140 million years ago (Fig. 12–25a).

By 130 million years ago, vast amounts of basaltic magma were forming (Fig. 12–25b). Some of this magma rose to the surface to cause volcanic eruptions. Most of the remainder melted portions of the lower crust to form andesitic and granitic magma, as explained in Chapter 4.

This intrusive and volcanic activity occurred along the entire length of western South America, but in a band only a few tens of kilometers wide, directly over the zone of melting. As the oceanic plate sank beneath the continent, slices of sea-floor mud and rock were scraped from the subducting plate, forming a subduction complex similar to that of an island arc.

The rising magma heated and thickened the crust beneath the Andes, causing it to rise isostatically and form great peaks. When the peaks became sufficiently high and heavy, the weak, soft rock oozed outward under its own weight. This spreading formed a great belt of thrust faults and folds along the east side of the Andes (Fig. 12–25c).

The Andes, then, are a relatively narrow mountain chain consisting predominantly of igneous rocks formed by subduction at a continental margin. The chain also contains extensive sedimentary rocks on both sides of the mountains; these rocks formed from sediment eroded from the rising peaks. The Andes are a good general example of subduction at a continental margin, and this type of plate margin is called an **Andean margin**.

▶ 12.6   THE HIMALAYAN MOUNTAIN CHAIN: A COLLISION BETWEEN CONTINENTS

The world's highest mountain chain, the Himalayas, separates China from India and includes the world's highest peaks, Mount Everest and K2 (Fig. 12–26). If you were

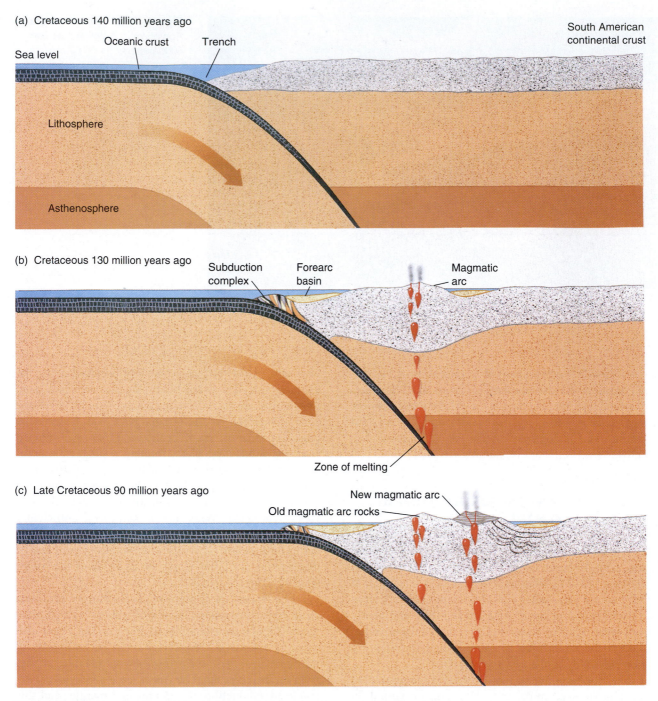

Figure 12–25 Development of the Andes, seen in cross section looking northward. (a) As the South American lithospheric plate moved westward in early Cretaceous time, about 140 million years ago, a subduction zone and a trench formed at the west coast of the continent. (b) By 130 million years ago, igneous activity began and a subduction complex and forearc basin formed. (c) In late Cretaceous time, the trench and region of igneous activity had both migrated eastward. Old volcanoes became dormant and new ones formed to the east.

to stand on the southern edge of the Tibetan Plateau and look southward, you would see the high peaks of the Himalayas. Beyond this great mountain chain lie the rainforests and hot, dry plains of the Indian subconti-nent. If you had been able to stand in the same place 100 million years ago and look southward, you would have seen only ocean. At that time, India was located south of the equator, separated from Tibet by thousands of kilo-

Figure 12–26 Machapuchare is a holy mountain in Nepal.

meters of open ocean. The Himalayas had not yet begun to rise.

FORMATION OF AN ANDEAN-TYPE MARGIN

About 120 million years ago, a triangular piece of lithosphere that included India split off from a large mass of continental crust near the South Pole. It began drifting northward toward Asia at a high speed, geologically speaking—perhaps as fast as 20 centimeters per year (Fig. 12–27a). As the Indian plate started to move, oceanic crust sank beneath Asia's southern margin, forming a subduction zone (Fig. 12–28b). As a result, volcanoes erupted, and granite plutons rose into southern Tibet. At this point, southern Tibet was an Andean-type continental margin, and it continued to be so from about 120 to 55 million years ago, while India drew closer to Asia.

CONTINENT–CONTINENT COLLISION

By about 55 million years ago, subduction had consumed all of the oceanic lithosphere between India and Asia (Fig. 12–28c). Then the two continents collided. Because both are continental crust, neither could sink deeply into the mantle. Igneous activity then ceased because subduction had stopped. The collision did not stop the northward movement of India, but it did slow it down to about 5 centimeters per year.

Continued northward movement of India was accommodated in two ways. The leading edge of India

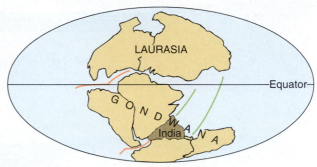

(a) 200 million years ago

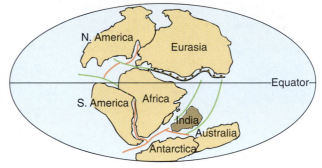

(b) 120 million years ago

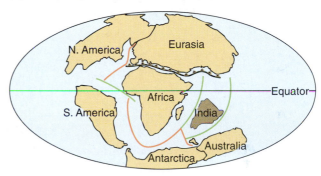

(c) 80 million years ago

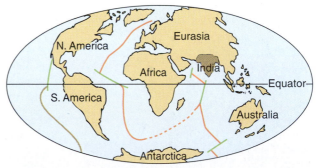

(d) 40 million years ago

Figure 12–27 (a) Gondwanaland and Laurasia formed shortly after 200 million years ago as a result of the early breakup of Pangea. Notice that India was initially part of Gondwanaland. (b) About 120 million years ago, India broke off from Gondwanaland and began drifting northward. (c) By 80 million years ago, India was isolated from other continents and was approaching the equator. (d) By 40 million years ago, it had moved 4000 to 5000 kilometers northward and collided with Asia.

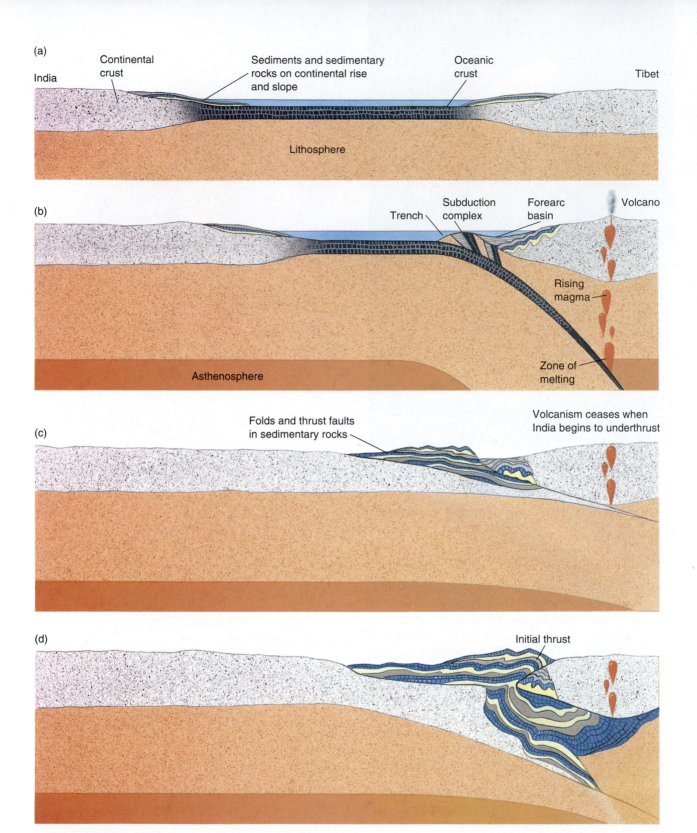

Figure 12–28 These cross-sectional views show the Indian and Asian plates before and during the collision between India and Asia. (a) Shortly before 120 million years ago, India, southern Asia, and the intervening ocean basin were parts of the same lithospheric plate. (In this figure, the amount of oceanic crust between Indian and Asian continental crust is abbreviated to fit the diagram on the page.) (b) When India began moving northward, the plate broke and subduction began at the southern margin of Asia. By 80 million years ago, an oceanic trench and subduction complex had formed. Volcanoes erupted, and granite plutons formed in the region now called Tibet. (c) By 40 million years ago, India had collided with Tibet. The leading edge of India was underthrust beneath southern Tibet. (d) Continued underthrusting and collision between the two continents has crushed Tibet and created the high Himalayas by folding and thrust faulting the sedimentary rocks. India continues to underthrust and crush Tibet today.

began to underthrust beneath Tibet. As a result, the thickness of continental crust in the region doubled. Thick piles of sediment that had accumulated on India's northern continental shelf were scraped from harder basement rock as India slid beneath Tibet. These sediments were pushed into folds and thrust faults (Fig. 12–28d). Some of the deeper thrusts extend downward into the basement rocks.

The second way in which India continued moving northward was by crushing Tibet and wedging China out of the way along huge strike–slip faults. India has pushed southern Tibet 1500 to 2000 kilometers northward since the beginning of the collision. These compressional forces have created major mountain ranges and basins north of the Himalayas.

THE HIMALAYAS TODAY

Today, the Himalayas contain igneous, sedimentary, and metamorphic rocks (Fig. 12–29). Many of the sedimentary rocks contain fossils of shallow-dwelling marine organisms that lived in the shallow sea of the Indian continental shelf. Plutonic and volcanic Himalayan rocks

formed when the range was an Andean margin. Rocks of all types were metamorphosed by the tremendous stresses and heat generated during the mountain building process.

The underthrusting of India beneath Tibet and the squashing of Tibet have greatly thickened continental crust and lithosphere under the Himalayas and the Tibetan Plateau to the north. Consequently, the region floats isostatically at high elevation. Even the valleys lie at elevations of 3000 to 4000 meters, and the Tibetan Plateau has an average elevation of 4000 to 5000 meters. One reason the Himalayas contain all of the Earth's highest peaks is simply that the entire plateau lies at such a high elevation. From the valley floor to the summit, Mount Everest is actually smaller than Alaska's Denali (Mount McKinley), North America's highest peak. Mount Everest rises about 3300 meters from base to summit, whereas Denali rises about 4200 meters. The difference in elevation of the two peaks lies in the fact that the base of Mount Everest is at about 5500 meters, but Denali's base is at 2000 meters.

Comparisons of older surveys with newer ones show that the tops of some Himalayan peaks are now rising rapidly—perhaps as fast as 1 centimeter per year. If this

Figure 12–29 Wildly folded sedimentary rocks on the Nuptse–Lhotse Wall from an elevation of 7600 meters on Mount Everest. (Galen Rowell)

rate were to continue, Mount Everest would double its height in about 1 million years, a short time compared with many other geologic events. However, normal faulting throughout the range is evidence that the mountains are oozing outward at the same time that they are rising. If the newly formed, steep mound of honey discussed earlier were covered with a layer of brittle chocolate frosting, the frosting would crack and slip apart in normal faults as the honey spread outward. The upper few kilometers of rocks of a rapidly rising mountain chain such as the Himalayas are like the frosting, and normal faulting is common in such regions. As blocks of rock slide off the mountains, they compress adjacent rock near the margins of the chain. In this way, normal faults in one region frequently form thrust faults and folds in a nearby region. But, at the same time, tectonic forces resulting from the continent–continent collision continue to push the mountains upward. No one knows when India will stop its northward movement or how high the mountains will become. However, we are certain that when the rapid uplift ends, the destructive forces—normal faulting and erosion—will lower the lofty peaks to form rolling hills.

THE TWO STEPS OF HIMALAYAN GROWTH

The Himalayan chain developed first as an Andean-type margin as oceanic crust sank beneath southern Asia. At that time, the geology of southern Asia was similar to the present geology of the Andes. Only later, after subduction had consumed all the oceanic crust between the two continents, did India and Asia collide. The two-step nature of the process is common to all continent–continent collisions because an ocean basin separating two continents must first be consumed by subduction before the continents can collide.

The Himalayan chain is only one example of a mountain chain built by a collision between two continents. The Appalachian Mountains formed when eastern North America collided with Europe, Africa, and South America between 470 and 250 million years ago. The European Alps formed during repeated collisions between northern Africa and southern Europe beginning about 30 million years ago. The Urals, which separate Europe from Asia, formed by a similar process about 250 million years ago.

▶ 12.7 THE ORIGIN OF CONTINENTS

Most geologists agree that the Earth formed by accretion of planetesimals, about 4.6 billion years ago. However, little evidence remains to trace our planet's earliest history. Some geologists argue that the entire Earth melted and was covered by an extensive magma ocean. Others contend that it was largely, but not completely, molten. In either scenario, the Earth was hot and active about 4.5 billion years ago. Magma rose to the surface and then cooled to form the earliest crust. From the evidence of a few traces of old ocean crust combined with calculations of the temperature and composition of the earliest upper mantle, geologists surmise that the first crust was composed of a type of ultramafic rock called **komatiite**. Komatiite is the volcanic equivalent of peridotite—the rock that now makes up the upper mantle. (Recall from Chapter 4 that ultramafic rocks have even higher magnesium and iron concentrations than basalt, which is a mafic rock.)

According to one hypothesis, heat-driven convection currents in a hot, active mantle initiated plate movement in this early crust. Dense komatiites dove into the mantle in subduction zones, where partial melting of the uppermost mantle created basaltic magma. As a result, the oceanic crust gradually became basaltic.

When did the earliest continental crust form? The 3.96-billion-year-old Acasta gneiss in Canada's Northwest Territories is the Earth's oldest known rock. It is metamorphosed granitic rock, similar to modern continental crust, and implies that at least some granitic crust had formed by that time.

Geologists have found grains of a mineral called zircon in a sandstone in western Australia. Although the sandstone is younger, the zircon gives radiometric dates of 4.2 billion years. Zircon commonly forms in granite. Geologists infer that the very old zircon initially formed in granite, which later weathered and released the zircon grains as sand. Eventually, the zircon became part of the younger sedimentary rock. Thus, these zircon grains suggest that granitic rocks existed 4.2 billion years ago. Geologists have also found granitic rocks nearly as old as the Acasta gneiss and the Australian zircon grains in Greenland and Labrador.

According to one model, the earliest continental crust formed by partial melting in oceanic subduction zones. Recall that island arcs form today by a similar mechanism. Thus, the first continents probably consisted of small granitic or andesitic blobs, like island arcs or microcontinents, sitting in a vast sea of basaltic crust.

Modeling suggests that about 40 percent of the present continental crust had formed by 3.8 billion years ago, and 50 percent had formed by 2.5 billion years ago. Thus, continental crust accumulated rapidly early in Earth history and more slowly after the end of Archean time. Geologists cannot calculate the rate of formation of continental crust precisely because they are not certain how much continental crust is recycled back into the mantle at subduction zones.

Most new continental crust now forms in subduction zones; a small amount forms over mantle plumes. Does

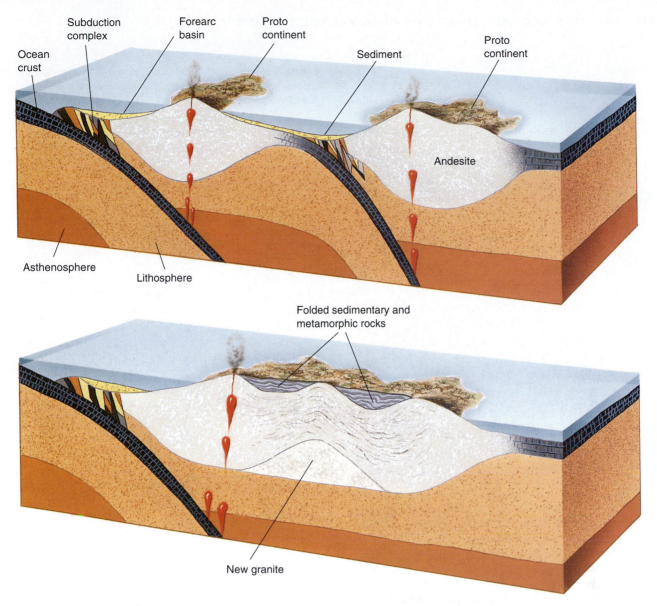

Figure 12–30 According to one model, the modern continents formed as island arcs sutured together. During the suturing, sediments eroded from the original islands were compressed, folded, and uplifted. Some were subjected to so much heat and pressure that they metamorphosed. New granite formed from partial melting of the crust at the subduction zone.

the evolution of continental crust early in Archean time suggest that modern-style plate tectonics had begun that early? Again, conflicting models have been proposed. Some geologists stress that the early Archean mantle was 200 to 400 degrees hotter than today's mantle. The high temperature should have caused rapid convection in the mantle and fast plate movements involving many small tectonic plates. In support of this model, some geologists point out that most Archean rocks are folded and sheared—a style of deformation that forms at modern convergent plate boundaries. They infer from this reasoning that horizontal plate movement has dominated tectonic activity from the beginning of Archean time to the present (Fig. 12–30).

However, other calculations and scant paleomagnetic evidence suggest that Archean plates moved at about the same speed as modern plates—between 1 and 16 centimeters per year. If Archean plates moved as slowly as modern plates do, how did the volume of Archean continental crust grow so rapidly? Another model suggests that early growth of continental crust, and perhaps even of oceanic crust, occurred mainly by "plume tectonics"—production of both basaltic and granitic magma over rising mantle plumes (Fig. 12–31). "Horizontal tectonics" became the dominant process only in late Archean time, after the mantle had cooled and convection slowed. Modern plate tectonics is dominated by horizontal plate movements.

217

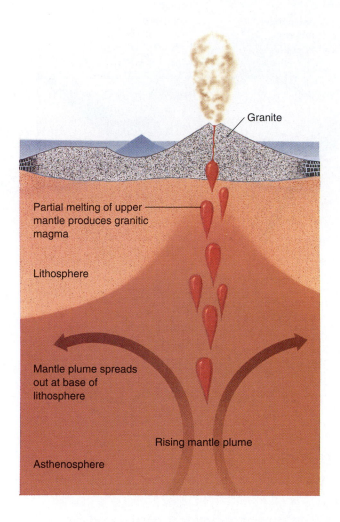

Granite

Partial melting of upper mantle produces granitic magma

Lithosphere

Mantle plume spreads out at base of lithosphere

Rising mantle plume

Asthenosphere

Figure 12–31 Another hypothesis contends that early continental crust formed over rising mantle plumes, in a process called "plume tectonics."

SUMMARY

Tectonic stress can be **confining stress, tectonic compression, extensional stress,** or **shear stress. Strain** is the distortion or deformation that results from stress.

When tectonic stress is applied to rocks, the rocks can deform in an **elastic** or **plastic** manner, or they may rupture by **brittle fracture.** The nature of the material, temperature, pressure, and rate at which the stress is applied all affect rock behavior under stress.

A **geologic structure** is any feature produced by deformation of rocks. Geologic structures consist of **folds,** which reflect predominant plastic rock behavior, and **faults** and **joints,** which form by rupture. Folds usually form when rocks are compressed.

Normal faults are usually caused by extensional stress, **reverse** and **thrust faults** are caused by compressional stress, and **strike–slip faults** form by shear stress, where blocks of crust slip horizontally past each other along vertical fractures. **Strike** is the direction in which rock layers are tilted, and **dip** is the angle of the bedding plane measured from the horizontal.

Mountains form when the crust thickens and rises isostatically. They become lower when crustal rocks flow outward or are worn away by erosion. If two converging plates carry oceanic crust, a volcanic **island arc** forms. If one plate carries oceanic crust and the other carries continental crust, an **Andean margin** develops. Andean margins are dominated by granitic plutons and andesitic volcanoes. They also contain rocks of a **subduction complex** and sedimentary rocks deposited in a **forearc basin.**

When two plates carrying continental crust converge, an Andean margin develops first as oceanic crust between the two continental masses is subducted. Later, when the two continents collide, one continent is underthrust beneath the other. The geology of mountain ranges formed by continent–continent collisions such as the **Himalayan chain** is dominated by vast regions of folded and thrust-faulted sedimentary and metamorphic rocks and by earlier formed plutonic and volcanic rocks.

The Earth's earliest crust was thin, ultramafic oceanic crust. According to one model, the first continental crust formed by partial melting in subduction zones or over mantle plumes.

KEY WORDS

REVIEW QUESTIONS

1. What is tectonic stress? Explain the main types of stress.

2. Explain the different ways in which rocks can respond to tectonic stress. What factors control the response of rocks to stress?

3. What is a geologic structure? What are the three main types of structures? What type(s) of rock behavior does each type of structure reflect?

4. At what type of tectonic plate boundary would you expect to find normal faults?

5. Explain why folds accommodate crustal shortening.

6. Draw a cross-sectional sketch of an anticline–syncline pair and label the parts of the folds. Include the axis and axial plane. Draw a sketch with a plunging fold.

7. Draw a cross-sectional sketch of a normal fault. Label the hanging wall and the footwall. Use your sketch to explain how a normal fault accommodates crustal extension. Sketch a reverse fault and show how it accommodates crustal shortening.

8. Explain the similarities and differences between a fault and a joint.

9. In what sort of a tectonic environment would you expect to find a strike–slip fault, a normal fault, and a thrust fault?

10. What mountain chain has formed at a divergent plate boundary? What are the main differences between this chain and those developed at convergent boundaries? Explain the differences.

11. Explain why erosion initially causes a mountain range to rise and then eventually causes the peak heights to decrease.

12. Describe the similarities and differences between an island arc and the Andes. Why do the differences exist?

13. Describe the similarities and differences between the Andes and the Himalayan chain. Why do the differences exist?

14. Draw a cross-sectional sketch of an Andean-type plate boundary to a depth of several hundred kilometers.

15. Draw a sequence of cross-sectional sketches showing the evolution of a Himalayan-type plate boundary. Why does this type of boundary start out as an Andean-type boundary?

16. What are the oldest Earth materials found to date? How old are they? What information do they provide us (what information can we infer from the data)?

17. Briefly outline one model for the formation of the continents.

DISCUSSION QUESTIONS

1. Discuss the relationships among types of lithospheric plate boundaries, predominant tectonic stress at each type of plate boundary, and the main types of geologic structures you might expect to find in each environment.

2. Why are thrust faults, reverse faults, and folds commonly found together?

3. Why do most major continental mountain chains form at convergent plate boundaries? What topographic and geologic features characterize divergent and transform plate boundaries in continental crust? Where do these types of boundaries exist in continental crust today?

4. Explain why extensional forces act on mountains rising in a tectonically compressional environment.

5. Explain why many mountains contain sedimentary rocks even though subduction leads to magma formation and the formation of igneous rocks.

6. Give a plausible explanation for the formation of the Ural Mountains, which lie in an inland portion of Asia.

7. Compare and explain the similarities and differences between the Andes and the Himalayan chain. How would the Himalayas, at their stage of development about 60 million years ago, have compared with the modern Andes?

8. Where would you be most likely to find large quantities of igneous rocks in the Himalayan chain: in the northern parts of the chain near Tibet or southward near India? Discuss why.

9. Where would you be most likely to find very old rocks: the sea floor, at the base of a growing mountain range, or within the central portion of the continent?

Mass Wasting

Every year, small landslides destroy homes and farmland. Occasionally, an enormous landslide buries a
town or city, killing thousands of people. Landslides cause
billions of dollars in damage every year, about equal to the
damage caused by earthquakes in 20 years. In many instances, losses occur because people do not recognize
dangers that are obvious to a geologist.

Consider three recent landslides that have affected
humans:

1. A movie star builds a mansion on the edge of a
   picturesque California cliff. After a few years, the
   cliff collapses and the house slides into the valley
   (Fig. 13–1a).

2. A ditch carrying irrigation water across a hillside in
   Montana leaks water into the ground. After years
   of seepage, the muddy soil slides downslope and
   piles against a house at the bottom of the hill
   (Fig. 13–1b).

3. Excavations for roads and high-rise buildings undercut the base of a steep hillside in Hong Kong.
   Suddenly, the slope slides, destroying everything in
   its path (Fig. 13–1c).

A landslide on steep, unstable slopes destroyed these expensive apartments and office buildings in Hong Kong. (Hong Kong Government Information Services)

(a)

(b)

(c)

Figure 13–1   Landslides cause billions of dollars in damage every year. (a) A few days after this photo was taken, the corner of the house hanging over the gully fell in.   (J. T. McGill, USGS) (b) A landslide, triggered by a leaking irrigation ditch, threatens a house in Darby, Montana. (c) An expensive landslide in Hong Kong.   (Hong Kong Government Information Services)

▶ **13.1   MASS WASTING**

Mass wasting is the downslope movement of Earth material, primarily under the influence of gravity. The word **landslide** is a general term for mass wasting and for the landforms created by such movements.

Think about the bedrock and soil on a hillside. Gravity constantly pulls them downward, but on any given day the rock and soil are not likely to slide down the slope. Their own strength and friction keep them in place. Eventually, however, natural processes or human activity may destabilize a slope to cause mass wasting. For example, a stream can erode the base of a rock cliff, undercutting it until it collapses. Rain, melting snow, or a leaking irrigation ditch can add weight and lubricate soil, causing it to slide downslope. Mass wasting occurs naturally in all hilly or mountainous terrain. Steep slopes are especially vulnerable, and landslide scars are common in mountainous country.

In recent years, the human population has increased dramatically. As the most desirable land has become overpopulated, large numbers of people have moved to more hostile and fragile terrain. In poor countries, people try to scratch out a living in mountains once considered too harsh for homes and farms. In wealthier nations, people have moved into the hills to escape congested cities. As a result, permanent settlements have grown in previously uninhabited steep terrain. Many of these slopes are naturally unstable. Construction and agriculture have destabilized others.

▶ **13.2   FACTORS THAT CONTROL MASS WASTING**

Imagine that you are a geological consultant on a construction project. The developers want to build a road at the base of a hill, and they wonder whether landslides will threaten the road. What factors should you consider?

STEEPNESS OF THE SLOPE

Obviously, the steepness of a slope is a factor in mass wasting. If frost wedging dislodges a rock from a steep cliff, the rock tumbles to the valley below. However, a similar rock is less likely to roll down a gentle hillside.

TYPE OF ROCK AND ORIENTATION OF ROCK LAYERS

If sedimentary rock layers dip in the same direction as a slope, the upper layers may slide over the lower ones. Imagine a hill underlain by shale, sandstone, and limestone oriented so that their bedding lies parallel to the

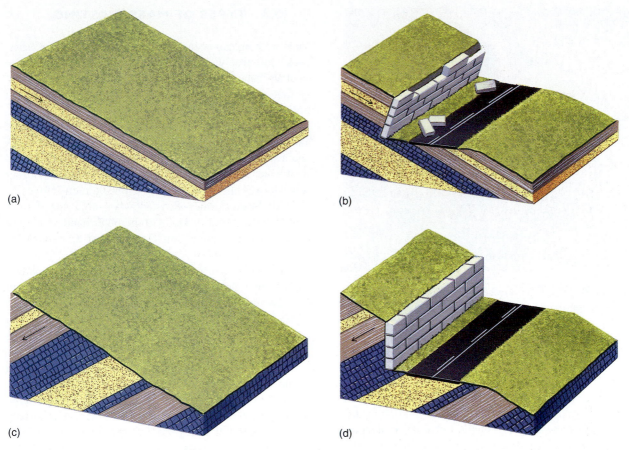

(a)

(b)

(c)

(d)

Figure 13–2   (a) Sedimentary rock layers dip parallel to this slope. (b) If a road cut undermines the slope, the dipping rock provides a good sliding surface, and the slope may fail. (c) Sedimentary rock layers dip at an angle to this slope. (d) The slope may remain stable even if it is undermined.

slope, as shown in Figure 13–2a. If the base of the hill is undercut (Fig. 13–2b), the upper layers may slide over the weak shale. In contrast, if the rock layers dip at an angle to the hillside, the slope may be stable even if it is undercut (Figs. 13–2c and 13–2d).

Several processes can undercut a slope. A stream or ocean waves can erode its base. Road cuts and other types of excavation can also destabilize it. Therefore, a geologist or engineer must consider not only a slope's stability before construction, but how the project might alter its stability.

THE NATURE OF UNCONSOLIDATED MATERIALS

The **angle of repose** is the maximum slope or steepness at which loose material remains stable. If the slope becomes steeper than the angle of repose, the material slides. The angle of repose varies for different types of material. Rocks commonly tumble from a cliff to collect at the base as angular blocks of talus. The angular blocks

interlock and jam together. As a result, talus typically has a steep angle of repose, up to 45°. In contrast, rounded sand grains do not interlock and therefore have a lower angle of repose (Fig. 13–3).

WATER AND VEGETATION

To understand how water affects slope stability, think of a sand castle. Even a novice sand-castle builder knows that the sand must be moistened to build steep walls and

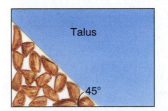
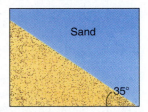

Figure 13–3   The angle of repose is the maximum slope that can be maintained by a specific material.

Figure 13–4 The angle of repose depends on both the type of material and its water content. Dry sand forms low mounds, but if you moisten the sand, you can build steep, delicate towers with it.

towers (Fig. 13–4). But too much water causes the walls to collapse. Small amounts of water bind sand grains together because the electrical charges of water molecules attract the grains. However, excess water lubricates the sand and adds weight to a slope. When some soils become water saturated, they flow downslope, just as the sand castle collapses. In addition, if water collects on impermeable clay or shale, it may provide a weak, slippery layer so that overlying rock or soil can move easily.

Roots hold soil together and plants absorb water; therefore, a highly vegetated slope is more stable than a similar bare one. Many forested slopes that were stable for centuries slid when the trees were removed during logging, agriculture, or construction.

Mass wasting is common in deserts and regions with intermittent rainfall. For example, southern California has dry summers and occasional heavy winter rain. Vegetation is sparse because of summer drought and wildfires. When winter rains fall, bare hillsides often become saturated and slide. Mass wasting occurs for similar reasons during infrequent but intense storms in deserts.

EARTHQUAKES AND VOLCANOES

An earthquake may cause mass wasting by shaking an unstable slope, causing it to slide. A volcanic eruption may melt snow and ice near the top of a volcano. The water then soaks into the slope to release a landslide.

▶ 13.3 TYPES OF MASS WASTING

Mass wasting can occur slowly or rapidly. In some cases, rocks fall freely down the face of a steep mountain. In other instances, rock or soil creeps downslope so slowly that the movement may be unnoticed by a casual observer.

Mass wasting falls into three categories: flow, slide, and fall (Fig. 13–5). To understand these categories, think again of building a sand castle. Sand that is saturated with water flows down the face of the structure. During **flow**, loose, unconsolidated soil or sediment moves as a fluid. Some slopes flow slowly—at a speed of 1 centimeter per year or less. On the other hand, mud with a high water content can flow almost as rapidly as water.

If you undermine the base of a sand castle, part of the wall may break away and slip downward. Movement of coherent blocks of material along fractures is called **slide**. Slide is usually faster than flow, but it still may take several seconds for the block to slide down the face of the castle.

If you take a huge handful of sand out of the bottom of the castle, the whole tower topples. This rapid, free-falling motion is called **fall**. Fall is the most rapid type of mass wasting. In extreme cases like the face of a steep cliff, rock can fall at a speed dictated solely by the force of gravity and air resistance.

Table 13–1 outlines the characteristics of flow, slide, and fall. Details of these three types of mass wasting are explained in the following sections.

FLOW

Types of flow include creep, debris flow, earthflow, mudflow, and solifluction.

Creep

As the name implies, **creep** is the slow, downhill movement of rock or soil under the influence of gravity. Individual particles move independently of one another, and the slope does not move as a consolidated mass. A creeping slope typically moves at a rate of about 1 centimeter per year, although wet soil can creep more rapidly. During creep, the shallow soil layers move more rapidly than deeper material (Fig. 13–6). As a result, anything with roots or a foundation tilts downhill. In many hillside cemeteries, older headstones are tilted, whereas newer ones are vertical (Fig. 13–7). Over the years, soil creep has tipped the older monuments, but the newer ones have not yet had time to tilt.

(Continued on p. 226)

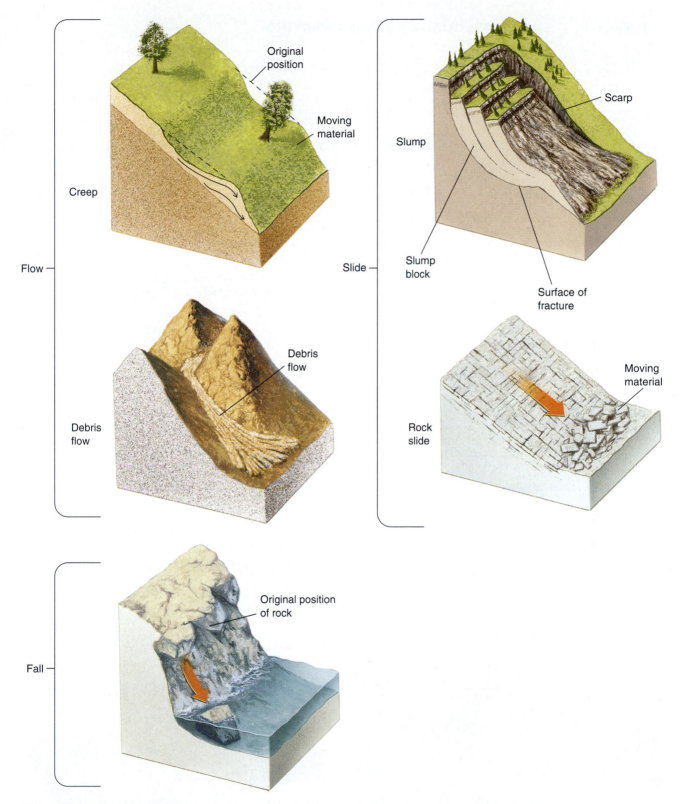

Figure 13–5 Flow, slide, and fall are the three categories of mass wasting.

Table 13–1 • SOME CATEGORIES OF MASS WASTING

TYPE OF MOVEMENT	DESCRIPTION	SUBCATEGORY	DESCRIPTION	COMMENTS
Flow	Individual particles move downslope independently of one another, not as a consolidated mass. Typically occurs in loose, unconsolidated regolith.	Creep	Slow, visually imperceptible movement	Trees on creep slopes develop pistol-butt shape
		Debris flow	More than half the particles larger than sand size; rate of movement varies from less than 1 m/year to 100 km/hr or more.	Common in arid regions with intermittent heavy rainfall, or can be triggered by volcanic eruption
		Earthflow and mudflow	Movement of fine-grained particles with large amounts of water	
		Solifluction	Movement of waterlogged soil generally over permafrost	Can occur on very gradual slopes
Slide	Material moves as discrete blocks; can occur in regolith or bedrock	Slump	Downward slipping of a block of Earth material, usually with a backward rotation on a concave surface	Trees on slump blocks remain rooted
		Rockslide	Usually rapid movement of a newly detached segment of bedrock	
Fall	Material falls freely in air; typically occurs in bedrock.	—	—	Occurs only on steep cliffs

Trees have a natural tendency to grow straight upward. As a result, when soil creep tilts a growing tree, the tree develops a J-shaped curve in its trunk called pistol butt (Fig. 13–8). If you ever contemplate buying hillside land for a home site, examine the trees. If they have pistol-butt bases, the slope is probably creeping, and creeping soil may tear a building apart.

Figure 13–6 Creep has bent layering in sedimentary rocks in a downslope direction. (Ward's Natural Science Establishment, Inc.)

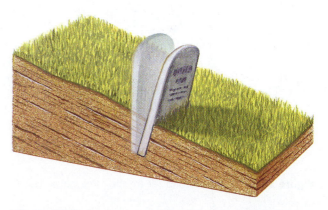

Figure 13–7 During creep, the soil surface moves more rapidly than deeper layers, so the tombstone embedded in the soil tilts downhill.

Figure 13–8   If a hillside creeps as a tree grows, the tree develops pistol butt.

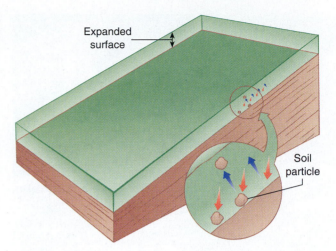

Figure 13–9   When soil expands due to freezing or absorption of water by clays, soil particles move outward, perpendicular to the slope. But when the soil shrinks again, particles sink vertically downward. The net result is a small downhill movement with each expansion–contraction cycle.

Creep can also result from freeze–thaw cycles in the spring and fall in temperate regions. Recall that water expands when it freezes. When damp soil freezes, expansion pushes it outward at a right angle to the slope. However, when the Sun melts the frost, the particles fall vertically downward, as shown in Figure 13–9. This movement creates a net downslope displacement. The displacement in a single cycle is small, but the soil may freeze and thaw once a day for a few months, leading to a total movement of a centimeter or more every year.

Other factors that cause creep include expansion and shrinking of clay-rich soils during alternating wet and dry seasons, and activities of burrowing animals. Both of these processes move soil downslope in a manner similar to that of freeze–thaw cycles.

Debris Flows, Mudflows, and Earthflows

In a debris flow, mudflow, or earthflow, wet soil flows downslope as a plastic or semifluid mass. If heavy rain falls on unvegetated soil, the water can saturate the soil to form a slurry of mud and rocks. A slurry is a mixture of water and solid particles that flows as a liquid. Wet concrete is a familiar example of a slurry. It flows easily and is routinely poured or pumped from a truck.

The advancing front of a flow often forms a tongue-shaped lobe (Fig. 13–10). A slow-moving flow travels at a rate of about 1 meter per year, but others can move as fast as a car speeding along an interstate highway. Flows can pick up boulders and automobiles and smash houses, filling them with mud or even dislodging them from their foundations.

Different types of flows are characterized by the sizes of the solid particles. A **debris flow** consists of a mixture of clay, silt, sand, and rock fragments in which more than half of the particles are larger than sand. In contrast, mudflows and earthflows are predominantly sand and mud. Some **mudflows** have the consistency of wet concrete, and others are more fluid. Because of its high water content, a mudflow may race down a stream channel at speeds up to 100 kilometers per hour. An **earthflow** contains less water than a mudflow and is therefore less fluid.

Solifluction

In temperate regions, soil moisture freezes in winter and thaws in summer. However, in very cold regions such as the Arctic and high mountain ranges, a layer of permanently frozen soil or subsoil, called permafrost, lies about a half meter to a few meters beneath the surface. Because ice is impermeable, summer meltwater cannot percolate

Figure 13–10 The 1980 eruption of Mount St. Helens melted large quantities of ice. The meltwater triggered characteristic lobe-shaped debris flows. (M. Freidman, USGS)

Figure 13–11 Arctic solifluction is characterized by lobes and a hummocky surface in Greenland. (R. B. Colton, USGS)

Figure 13–12 (a) In slump, blocks of soil or rock remain intact as they move downslope. (b) Trees tilt back into the hillside on this slump along the Quesnell River, British Columbia.

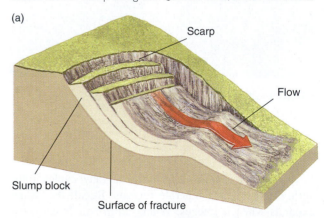

downward, and it collects on the ice layer. This leads to two characteristics of these soils:

1. Water cannot penetrate the ice layer, so it collects near the surface. As a result, even though many Arctic regions receive little annual precipitation, bogs and marshes are common.

2. Ice, especially ice with a thin film of water on top, is slippery. Therefore, permafrost soils are particularly susceptible to mass wasting.

Solifluction is a type of mass wasting that occurs when water-saturated soil flows downslope. It is most common in permafrost regions, where the permanent ice layer causes overlying soil to become waterlogged, although it can also occur in the absence of permafrost (Fig. 13–11). Solifluction can occur on a very gentle slope, and the soil typically flows at a rate of 0.5 to 5 centimeters per year.

SLIDE

In some cases, a large block of rock or soil, or sometimes an entire mountainside, breaks away and **slides** downslope as a coherent mass or as a few intact blocks. Two types of slides occur: slump and rockslide.

A **slump** occurs when blocks of material slide downhill over a gently curved fracture in rock or regolith (Fig. 13–12). Trees remain rooted in the moving blocks. However, because the blocks rotate on the concave fracture, trees on the slumping blocks are tilted backward. Thus, you can distinguish slump from creep be-

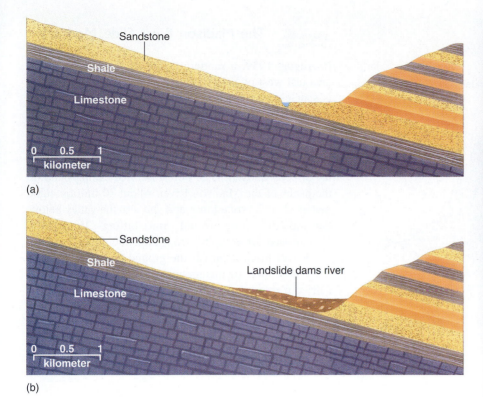

Figure 13–13   A profile of the Gros Ventre hillside (a) before and (b) after the slide. (c) About 38 million cubic meters of rock and soil broke loose and slid downhill during the Gros Ventre slide.

(c)

cause slump tilts trees uphill, whereas creep tilts them downhill. At the lower end of a large slump, the blocks often pile up to form a broken, jumbled, hummocky topography.

It is useful to identify slump because it often recurs in the same place or on nearby slopes. Thus, a slope that shows evidence of past slump is not a good place to build a house.

During a **rockslide**, or **rock avalanche**, bedrock slides downslope over a fracture plane. Characteristically, the rock breaks up as it moves and a turbulent mass of rubble tumbles down the hillside. In a large avalanche, the falling debris traps and compresses air beneath and within the tumbling blocks. The compressed air reduces friction and allows some avalanches to attain speeds of 500 kilometers per hour. The same mechanism allows a snow or ice avalanche to cover a great distance at a high speed.

Rock Avalanche near Kelly, Wyoming

A mountainside above the Gros Ventre River near Kelly, Wyoming, was composed of a layer of sandstone resting on shale, which in turn was supported by a thick bed of limestone (Fig. 13–13). The rocks dipped 15° to 20° toward the river and parallel to the slope. Over time, the Gros Ventre River had undercut the sandstone, leaving the slope above the river unsupported. In the spring of 1925, snowmelt and heavy rains seeped into the ground, saturating the soil and bedrock and increasing their weight. The water collected on the shale, forming a slippery surface. Finally, the sandstone layer broke loose and slid over the shale. In a few moments, approximately 38 million cubic meters of rock tumbled into the valley. The sandstone crumbled into blocks that formed a 70-meter-high natural dam across the Gros Ventre River. Two years later, the lake overflowed the dam, washing it out and creating a flood downstream that killed several people.

FALL

If a rock dislodges from a steep cliff, it falls rapidly under the influence of gravity. Several processes commonly detach rocks from cliffs. Recall from our discussion of weathering that when water freezes and thaws, the alternate expansion and contraction can dislodge rocks from cliffs and cause rockfall. Rockfall also occurs when a cliff is undercut. For example, if ocean waves or

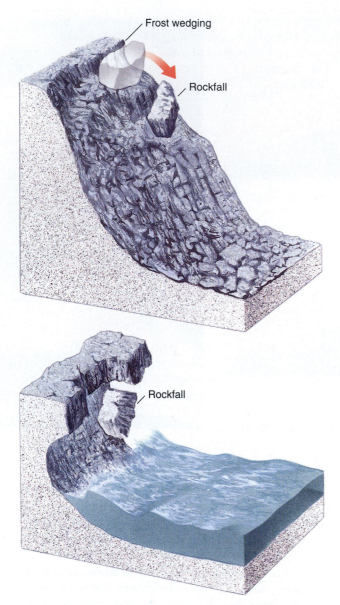

Frost wedging

Rockfall

Rockfall

Figure 13–14 Rockfall commonly occurs in spring or fall when freezing water dislodges rocks from cliffs. Undercutting of cliffs by waves, streams, or construction can also cause rockfall.

a stream undercuts a cliff, rock above the waterline may tumble (Fig. 13–14).

▶ 13.4 THREE CASE STUDIES: MASS WASTING TRIGGERED BY EARTHQUAKES AND VOLCANIC ERUPTIONS

In many cases, an earthquake or volcanic eruption causes comparatively little damage, but it triggers a devastating landslide. Consider the following case studies.

The Madison River Slide, Montana

In August 1959, a moderate-size earthquake jolted the area just west of Yellowstone National Park. This region is sparsely populated, and most of the buildings in the area are wood-frame structures that can withstand quakes. As a result, the earthquake itself caused little property damage and no loss of life. However, the quake triggered a massive rockslide from the top of Red Mountain, which lay directly above a U.S. Forest Service campground on the banks of the Madison River. About 30 million cubic meters of rock broke loose and slid into the valley below, burying the campground and killing 26 people. Compressed air escaping from the slide created intense winds that lifted a car off the ground and carried it into trees more than 10 meters away. The slide's momentum carried it more than 100 meters up the mountain on the opposite side of the valley. The debris dammed the Madison River, forming a lake that was later named Quake Lake. Figure 13–15 shows the debris and some of the damage caused by this slide.

Nevado del Ruiz, Colombia

Recall the 1985 eruption of Nevado del Ruiz volcano in central Colombia that was briefly described in Chapter 5. The eruption itself caused only minor damage, but heat from the ash and lava melted large quantities of ice and snow that lay on the mountain. The rushing water mixed with ash, rock, and soil on the mountainside, forming a

Figure 13–15 This landslide near Yellowstone Park buried a campground, killing 26 people. (Donald Hyndman)

mudflow that raced down gullies and stream valleys to the town of Armero, 48 kilometers from the mountain. The mudflow buried and killed 22,000 people in Armero and caused additional loss of life and property damage in a dozen other villages in nearby valleys.

Mass Wasting in Washington

Several volcanoes in western Washington State have been active in recent geologic history. The 1980 eruption of Mount St. Helens blew away the entire north side of the mountain. The heat of the eruption melted glaciers and snowfields near the summit, and the water mixed with volcanic ash and soil to create mammoth mudflows. Although the eruption and mudflows killed 63 people, the total loss of life was small compared with the annihilation of Armero. Why was the death toll so much lower in the Mount St. Helens eruption? Is a catastrophic mass wasting event possible or likely elsewhere in Washington State?

The answer to the first question is twofold. First, geologists predicted the Mount St. Helens eruption. As a result, the United States Forest Service evacuated many residents and withdrew water from reservoirs so that the dams would partially contain the anticipated mudflows. Second, and perhaps more important, the region around the mountain is a forested park and there are no cities in the immediate vicinity.

Could other eruptions in Washington and nearby locations lead to much greater disasters? Unfortunately, the answer is yes. Mount Baker is an active, glacier-covered volcano that lies north of Seattle. Steam and gases still escape periodically from its crater (Fig. 13–16). A large eruption could melt the glaciers on the mountain to create mudflows similar to those that devastated the valleys below Mount St. Helens in 1980. Mount Baker lies 20 kilometers upriver from the town of Glacier, Washington, and less than 50 kilometers upriver from the city of Bellingham, which has a population of over 50,000. Recall that the city of Armero was 48 kilometers from Nevado del Ruiz. We can imagine an eruption initiating a mudflow that follows the valley leading to Glacier or even to Bellingham and buries the towns.

We are not predicting an eruption of Mount Baker or a disaster in Glacier or Bellingham; we are simply stating that both are geologically plausible. So the question remains: What should be done in response to the hazard? It is impractical to move an entire city. Therefore, the only alternative is to monitor the mountain continuously and hope that it will not erupt or, if it does, that it will provide enough warning that urban areas can be evacuated in time.

▶ 13.5 PREDICTING AND AVOIDING LANDSLIDES

Landslides commonly occur on the same slopes as earlier landslides because the geologic conditions that cause mass wasting tend to be constant over a large area and remain constant for long periods of time. Thus, if a hillside has slumped, nearby hills may also be vulnerable to mass wasting. In addition, landslides and mudflows commonly follow the paths of previous slides and flows. If

Figure 13–16 A wisp of steam rises near the summit of Mount Baker, in the lower center of the photograph.

an old mudflow lies in a stream valley, future flows may follow the same valley.

Many towns were founded decades or centuries ago, before geologic disasters were understood. Often the choice of a town site was not dictated by geologic considerations but by factors related to agriculture, commerce, or industry, such as proximity to rivers and ocean harbors and the quality of the farmland. Once a city is established, it is virtually impossible to move it. Furthermore, geologists' warnings that a disaster might occur are often ignored. After all, predictions of earthquakes and volcanic eruptions are sometimes incorrect. Even in areas known to be active, a quake or eruption may not occur for decades or even centuries.

Awareness and avoidance are the most effective defenses against mass wasting. Geologists construct maps of slope and soil stability by combining data on soil and bedrock stability, slope angle, and history of slope failure in the area. They include evaluations of the probability of a triggering event, such as a volcanic eruption or earthquake. Building codes then regulate or prohibit construction in unstable areas. For example, according to the *United States Uniform Building Code*, a building cannot be constructed on a sandy slope steeper than 27°, even though the angle of repose of sand is 35°. Thus, the law leaves a safety margin of 8°. Architects can obtain permission to build on more precipitous slopes if they anchor the foundation to stable rock.

SUMMARY

Mass wasting is the downhill movement of rock and soil under the influence of gravity. The stability of a slope and the severity of mass wasting depend on (1) steepness of the slope, (2) orientation and type of rock layers, (3) nature of unconsolidated materials, (4) climate and vegetation, and (5) earthquakes or volcanic eruptions.

Mass wasting falls into three categories: flow, slide, and fall. During **flow**, a mixture of rock, soil, and water moves as a viscous fluid. **Creep** is a slow type of flow that occurs at a rate of about 1 centimeter per year. A **debris flow** consists of a mixture in which more than half the particles are larger than sand. **Earthflows** and **mudflows** are mass movements of predominantly fine-grained

particles mixed with water. Earthflows have less water than mudflows and are therefore less fluid. **Solifluction** is a type of flow that occurs when water-saturated soil moves downslope, usually over permafrost.

Slide is the movement of a coherent mass of material. **Slump** is a type of slide in which the moving mass travels on a concave surface. In a **rockslide**, a newly detached segment of bedrock slides along a tilted bedding plane or fracture. **Fall** occurs when particles fall or tumble down a steep cliff.

Earthquakes and volcanic eruptions trigger devastating mass wasting. Damage to human habitation can be averted by proper planning and engineering.

KEY WORDS

mass wasting *222*	**slide** *224*	**debris flow** *227*	**solifluction** *228*
angle of repose *223*	**fall** *224*	**mudflow** *227*	**slump** *228*
flow *224*	**creep** *224*	**earthflow** *227*	**rockslide** *229*
			rock avalanche *229*

REVIEW QUESTIONS

1. List and describe each of the factors that control slope stability.

2. What is the angle of repose? Why is the angle of repose different for different types of materials?

3. Explain how a small amount of water might increase slope stability, whereas a landslide might occur on the same slope during heavy rainfall or rapid snowmelt.

4. How does vegetation affect slope stability?

5. Why is mass wasting common in deserts and semiarid lands?

6. How do volcanic eruptions cause landslides?

7. How do earthquakes cause landslides?

8. Discuss the differences among flow, slide, and fall. Give examples of each.

9. Compare and contrast creep, debris flow, and mudflow.

10. What does a pistol-butt tree trunk tell you about slope stability?

11. Why is solifluction more likely to occur in the Arctic than in temperate or tropical regions?

12. Compare and contrast slump and rockslide.

13. Explain how trees are bent but not killed by slump. How are trees affected by rockslide?

14. How do landslides reach and destroy towns and villages many kilometers from the steep slopes where the slides originate?

DISCUSSION QUESTIONS

1. The Moon is considerably less massive than the Earth, and therefore its gravitational force is less. It has no atmosphere and therefore no rainfall. The interior of the Moon is cool, and thus it is geologically inactive. Would you expect mass wasting to be a common or an uncommon event in mountainous areas of the Moon? Defend your answer.

2. Explain how wildfires affect slope stability and mass wasting.

3. What types of mass wasting (if any) would be likely to occur in each of the following environments?   a. A very gradual (2 percent) slope in a heavily vegetated tropical rainforest.   b. A steep hillside composed of alternating layers of conglomerate, shale, and sandstone, in a region that experiences distinct dry and rainy seasons. The dip of the rock layers is parallel to the slope.   c. A hillside similar to that of b, in which the rock layers are oriented perpendicular to the slope.   d. A steep hillside composed of clay in a rainy environment in an active earthquake zone.

4. Identify a hillside in your city or town that might be unstable. Using as much data as you can collect, discuss the magnitude of the potential danger. Would the landslide be likely to affect human habitation?

5. Explain how the mass wasting triggered by earthquakes and volcanoes can have more serious effects than the earthquake or volcano itself. Is this always the case?

6. How do mudflows and debris flows transport automobile-sized boulders?

7. Develop a strategy for minimizing loss of life from mass wasting if Mount Baker should show signs of an impending eruption similar to those shown by Mount St. Helens in the spring of 1980. How would your strategy apply to towns such as Armero?

CHAPTER 14

Streams and Lakes

About 1.3 billion cubic kilometers of water exist at the Earth's surface. If the surface were perfectly level, water would form a layer 2 kilometers thick surrounding the entire planet. Of this huge quantity, 97.5 percent is salty seawater, and another 1.8 percent is frozen into the great ice caps of Antarctica and Greenland. Only about 0.65 percent is fresh water in streams, underground reservoirs, lakes, and wetlands. Thus, although the hydrosphere contains a great amount of water, only a tiny fraction is fresh and liquid.

The clear water of the Big Sandy River flows from the Wind River Mountains of Wyoming.

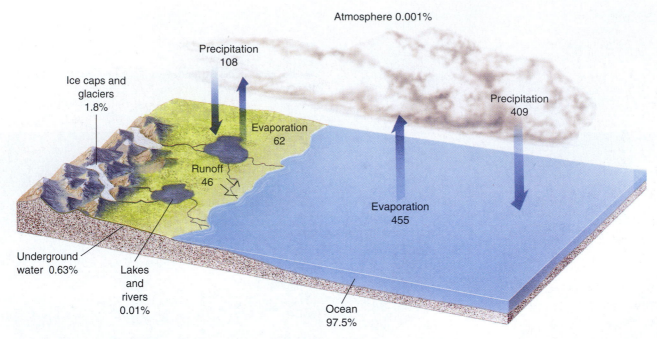

Figure 14–1 The hydrologic cycle shows that water circulates constantly among the sea, the atmosphere, and the land. Numbers indicate thousands of cubic kilometers of water transferred each year. Percentages show proportions of total global water in different portions of the Earth's surface.

▶ 14.1 THE WATER CYCLE

Water evaporates from the sea, falls as rain, and flows from land back to the sea. The circulation of water among sea, land, and the atmosphere is called the **hydrologic cycle**, or the water cycle (Fig. 14–1).

Water evaporates from sea and land to form clouds and invisible water vapor in the atmosphere. Water also evaporates directly from plants as they breathe, a process called **transpiration**. Atmospheric moisture then returns to the Earth's surface as **precipitation**: rain, snow, hail, and sleet.

Water that falls onto land can follow three different paths:

1. **Surface water** flowing to the sea in streams and rivers is called **runoff**. Surface water may stop temporarily in a lake or wetland, but eventually it flows to the oceans.

2. Some water seeps into the ground to become part of a vast subterranean reservoir known as **ground water**. Although surface water is more conspicuous, 60 times more water is stored as ground water than in all streams, lakes, and wetlands combined. Ground water also seeps through bedrock and soil toward the sea, although it flows much more slowly than surface water.

3. The remainder of water that falls onto land evaporates or transpires back into the atmosphere.

▶ 14.2 STREAMS

Geologists use the term **stream** for all water flowing in a channel, regardless of the stream's size. The term **river** is commonly used for any large stream fed by smaller ones called **tributaries**. Most streams run year round, even during times of drought, because they are fed by ground water that seeps into the stream bed.

Normally a stream flows in its **channel**. The floor of the channel is called the **bed**, and the sides of the channel are the **banks**. When rainfall is heavy or when snow melts rapidly, a **flood** may occur. During a flood, a stream overflows its banks and spreads over adjacent land called a **flood plain**.

STREAM FLOW

A slow stream flows at 0.25 to 0.5 meter per second (1 to 2 kilometers per hour), whereas a steep, flooding stream may race along at about 7 meters per second (25 kilometers per hour). Three factors control current velocity: (1) the gradient of the stream; (2) the discharge; and (3) the shape and roughness of the channel.

Gradient

Gradient is the steepness of a stream. The lower Mississippi River has a shallow gradient and drops only 10 centimeters per kilometer of stream length. In contrast, a tumbling mountain stream may drop 40 meters or more per kilometer. Obviously, if all other factors are equal, a stream flows more rapidly down a steep channel than a gradual one.

Discharge

Discharge is the amount of water flowing down a stream. It is expressed as the volume of water flowing past a point per unit time, usually in cubic meters per second (m^3/sec). The largest river in the world is the Amazon, with a discharge of 150,000 m^3/sec. In contrast, the Mississippi River, the largest in North America, has a discharge of about 17,500 m^3/sec, approximately one ninth that of the Amazon.

A stream's discharge can change dramatically from month to month or even during a single day. For example, the Selway River, a mountain stream in Idaho, has a discharge of 100 to 130 m^3/sec during early summer, when mountain snow is melting rapidly. During the dry season in late summer, the discharge drops to about 10 to 15 m^3/sec (Fig. 14–2). A desert stream may dry up completely during summer but become the site of a flash flood during a sudden thunderstorm.

Stream velocity increases when discharge increases. Thus, a stream flows faster during flood, even though its gradient is unchanged. The velocity of a stream also generally increases in a downstream direction because tributaries add to the discharge.

Channel Shape and Roughness

Friction between flowing water and the stream channel slows current velocity. Consequently, water flows more slowly near the banks than near the center of a stream. If you paddle a canoe down a straight stream channel, you move faster when you stay away from the banks.

The total friction depends on both the shape of a stream channel and its roughness. If streams of equal cross-sectional area are compared, a semicircular channel has the least surface in contact with the water and therefore imposes the least friction. If other factors are equal, a stream with this shape will flow more rapidly than one that is either wide and shallow or narrow and deep.

A rough channel creates more friction than a smooth one. Boulders in the stream bed increase turbulence and resistance, so a stream flows more slowly through a rough channel than a smooth one (Fig. 14–3).

▶ 14.3 STREAM EROSION

A stream may erode sediment and bedrock from its channel. When it does so, it carries the sediment and deposits it in its bed or flood plain farther downstream, or on a delta where it enters the sea or a lake.

STREAM ENERGY: THE ABILITY OF A STREAM TO ERODE AND CARRY SEDIMENT

The ability of a stream to erode and carry sediment depends on its energy. The energy of a stream is proportional to both velocity and discharge. A rapid, high-volume stream is a high-energy stream. It can move

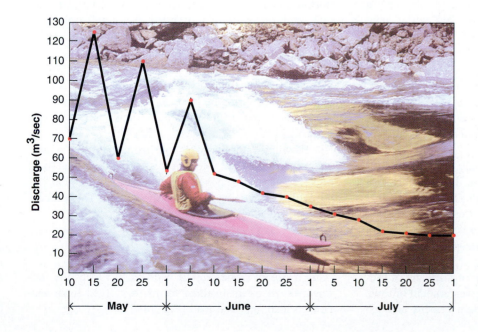

Figure 14–2 The hydrograph for the Selway River in the spring and summer of 1988 shows that the discharge varied from 125 to 15 m^3/sec.

Figure 14–3 A boulder-choked stream bed in the North Fork of Trapper Creek, Bitterroot Mountains, Montana, creates turbulence and resistance to flow.

boulders as well as smaller particles and can carry a large load of sediment. In contrast, a slow, low-volume stream flows with much less energy; it moves only fine sediment and carries a much smaller sediment load.

The **competence** of a stream is a measure of the largest particle it can carry. It depends mostly on current velocity. Thus, a swiftly flowing stream can carry cobbles, boulders, and even automobiles during a big flood; but a slow stream with the same volume carries only silt and clay.

Velocity controls competence only when a stream is deep enough to cover the particles, but in a shallow stream, discharge becomes critical. To illustrate this point, think of a tiny but very steep stream tumbling over boulders. Although it may flow at great speed, it cannot move the boulders because it is not deep enough to cover them completely.

The **capacity** of a stream is the total amount of sediment it can carry past a point in a given amount of time. Capacity is proportional to both current velocity and dis-

charge. Thus, a fast, large stream can carry more sediment than a slow, small one.

Because the ability of a stream to erode and carry sediment is proportional to its velocity and discharge, most erosion and sediment transport occur during the few days each year when the stream is flooding. Relatively little erosion and sediment transport occur during the remainder of the year. To see this effect for yourself, look at any stream during low water. It will most likely be clear, indicating little erosion or sediment transport. Look at the same stream when it is flooding. It will probably be muddy and dark, indicating that the stream is eroding its bed and banks and carrying a large load of sediment.

STREAM EROSION

A stream weathers and erodes its bed and banks by three processes: hydraulic action, abrasion, and solution (Fig. 14–4).

Hydraulic action is the process in which flowing water erodes sediment directly. To demonstrate hydraulic action, point a garden hose at bare dirt. In a short time the water will erode a small hole, displacing soil and small pebbles. Similarly, a stream can erode its bed and banks, especially when the current is moving swiftly.

Although it can erode loose soil, water by itself is not abrasive and is ineffective at wearing away solid rock. However, when a stream carries sand and other sediment, the grains grind against each other and against rocks in the channel in a process called **abrasion**. Thus, a sediment-laden stream is like flowing sandpaper.

Abrasion rounds sediment of all sizes, from sand to boulders (Fig. 14–5). It also erodes bedrock and forms **potholes** in a stream bed. A pothole forms where the current recirculates cobbles trapped in a small hollow in bedrock (Fig. 14–6). Over time, the cobbles can abrade a deep circular hole in the bedrock.

In cold climates, ice is an abrasive agent. In winter, ice on a frozen stream expands and gouges the stream banks. During spring breakup, the flooding stream drives great sheets of ice into the stream banks to erode rock and soil.

Flowing water dissolves ions from rocks and minerals in the stream bed. Most of a stream's dissolved load, however, comes from weathering of soils by ground water, which eventually seeps into the stream. This process is described in Chapter 6.

SEDIMENT TRANSPORT

After a stream erodes soil or bedrock, it carries the sediment downstream in three forms: dissolved load, suspended load, and bed load (Fig. 14–7).

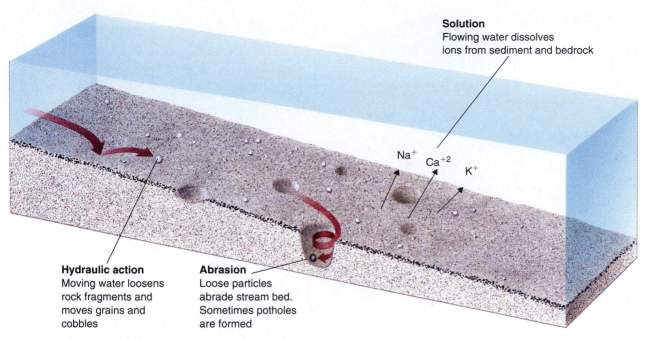

Solution
Flowing water dissolves
ions from sediment and bedrock

Na^+ Ca^{+2} K^+

Hydraulic action
Moving water loosens
rock fragments and
moves grains and
cobbles

Abrasion
Loose particles
abrade stream bed.
Sometimes potholes
are formed

Figure 14–4   A stream weathers and erodes its channel by hydraulic action, abrasion, and solution.

Figure 14–5   Stream abrasion has rounded these rocks in the Bitterroot River, Montana.

Ions dissolved in water are called **dissolved load**. A stream's ability to carry dissolved ions depends mostly on its discharge and its chemistry, not its velocity. Thus, even the still waters of a lake or ocean contain dissolved substances; that is why the sea and some lakes are salty.

Figure 14–6   Potholes form in bedrock when a stream recirculates cobbles.   (Hubbard Scientific)

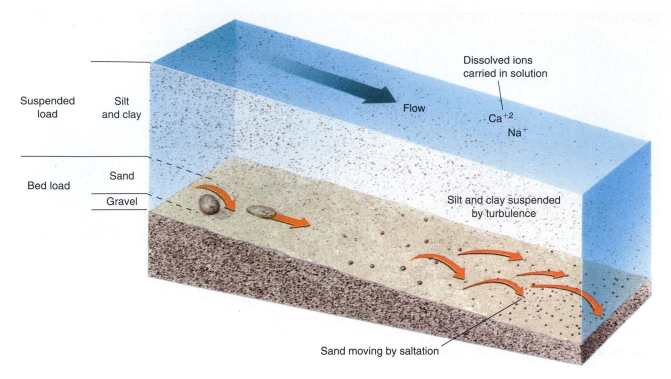

Figure 14–7 A stream carries dissolved ions in solution, silt and clay in suspension, and larger particles as bed load.

Even a slow stream can carry fine sediment. If you place loamy soil in a jar of water and shake it up, the sand grains settle quickly. But the smaller silt and clay particles remain suspended in the water as **suspended load**, giving it a cloudy appearance. Clay and silt are small enough that even the slight turbulence of a slow stream keeps them in suspension. A rapidly flowing stream can carry sand in suspension.

During a flood, when stream energy is highest, the rushing water can roll boulders and cobbles along the bottom as **bed load**. Sand also moves in this way, but if the stream velocity is sufficient, sand grains bounce over the stream bed in a process called **saltation**. Saltation occurs because a turbulent stream flows with many small chaotic currents. When one of these currents scours the stream bed, it lifts sand and carries it a short distance before dropping it back to the bed. The falling grains strike other grains and bounce them up into the current. The overall effect is one of millions of sand grains hopping and bouncing downstream over the stream bed.

The world's two muddiest rivers—the Yellow River in China and the Ganges River in India—each carry more than 1.5 *billion* tons of sediment to the ocean every year. The sediment load of the Mississippi River is about 450 million tons per year. Most streams carry the greatest proportion of sediment in suspension, less in solution, and the smallest proportion as bed load (Fig. 14–8).

▶ 14.4 STREAM DEPOSITS

A large, swift stream can carry all sizes of particles from clay to boulders. When the current slows down, its competence decreases and the stream deposits the largest particles in the stream bed. If current velocity continues to decrease—as a flood wanes, for example—finer particles settle out on top of the large ones. Thus, a stream **sorts** its sediment according to size. A waning flood might deposit a layer of gravel, overlain by sand and finally topped by silt and clay (Fig 14–9).

Streams also sort sediment in the downstream direction. Many mountain streams are choked with boulders and cobbles, but far downstream, their deltas are composed mainly of fine silt and clay. This downstream sorting is curious because stream velocity generally increases in the downstream direction. Competence increases with velocity, so a river should be able to transport larger particles than its tributaries. One explanation for downstream sorting is that abrasion wears away the boulders and cobbles to sand and silt as the sediment moves downstream over the years. Thus, only the fine sediment reaches the lower parts of most rivers.

A stream deposits its sediment in three environments: (1) **Channel deposits** form in the stream channel itself; (2) **alluvial fans** and **deltas** form where stream

Bed load
30 million
metric tons

Suspension
300 million
metric tons

Solution
120 million
metric tons

Figure 14–8 The Mississippi River carries the greatest proportion of its sediment as suspended load. Numbers indicate sediment load per year.

gradient suddenly decreases as a stream enters a flat plain, a lake, or the sea; and (3) flood plain deposits accumulate on a flood plain adjacent to the stream channel.

Figure 14–9 The lower portion of this photograph shows coarse gravel overlain by finer sediment. The coarse gravel was deposited during a flood; the finer sediment accumulated as the flood waned. The upper gravel accumulated during a second flood. (G. R. Roberts)

CHANNEL DEPOSITS

A **bar** is an elongate mound of sediment. Bars are transient features that form in the stream channel and on the banks. They commonly form in one year and erode the next. Rivers used for commercial navigation must be recharted frequently because bars shift from year to year.

Imagine a winding stream such as that in Figure 14–10. The water on the outside of the curve A-A′ moves faster than the water on the inside. The stream erodes its outside bank because the current's inertia drives it into the outside bank. At the same time, the slower water on the inside point of the bend deposits sediment, forming a **point bar**. A **mid-channel bar** is a sandy and gravelly deposit that forms in the middle of a stream channel.

Most streams flow in a single channel. In contrast, a **braided stream** flows in many shallow, interconnecting channels (Fig. 14–11). A braided stream forms where more sediment is supplied to a stream than it can carry. The stream dumps the excess sediment, forming mid-channel bars. The bars gradually fill a channel, forcing the stream to overflow its banks and erode new channels. As a result, a braided stream flows simultaneously in several channels and shifts back and forth across its flood plain.

Braided streams are common in both deserts and glacial environments because both produce abundant sediment. A desert yields large amounts of sediment because it has little or no vegetation to prevent erosion. Glaciers grind bedrock into fine sediment, which is carried by streams flowing from the melting ice.

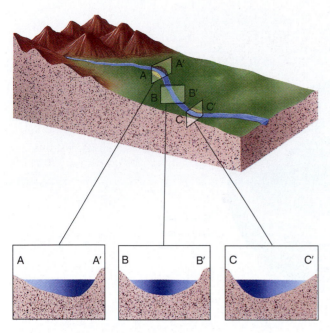

Figure 14–10 In a winding stream, the current flows most rapidly on the outside of a bend, and slowest on the inside bend. In a straight section, the current is fastest in the center of the channel. The dark-shaded zone in each cross section shows the area with fastest flow.

ALLUVIAL FANS AND DELTAS

If a steep mountain stream flows onto a flat plain, its gradient and velocity decrease abruptly. As a result, it deposits most of its sediment in a fan-shaped mound called an alluvial fan. Alluvial fans are common in many arid and semiarid mountainous regions (Fig. 14–12).

A stream also slows abruptly where it enters the still water of a lake or ocean. The sediment settles out to form a nearly flat landform called a delta. Part of the delta lies above water level, and the remainder lies slightly below water level. Deltas are commonly fan-shaped, resembling the Greek letter "delta" (Δ).

Both deltas and alluvial fans change rapidly. Sediment fills channels, which are then abandoned while new channels develop, as in a braided stream. As a result, a stream feeding a delta or fan splits into many channels called **distributaries**. A large delta may spread out in this manner until it covers thousands of square kilometers (Fig. 14–13). Most fans, however, are much smaller, covering a fraction of a square kilometer to a few square kilometers.

Figure 14–14 shows that the Mississippi River has flowed through seven different delta channels during the past 5000 to 6000 years. But, in recent years, engineers have built great systems of levees in attempts to stabilize the channels. If the Mississippi River were left alone, it would probably abandon the lower 500 kilometers of its present path and cut into the channel of the Atchafalaya River to the west. However, this part of the delta is heavily industrialized, and it is impractical to allow the river to change its course, to flood towns in some areas and leave shipping lanes and wharves high and dry in others.

► 14.5 DOWNCUTTING AND BASE LEVEL

A stream erodes downward into its bed and laterally against its banks. Downward erosion is called **downcutting** (Fig. 14–15). The **base level** of a stream is the deep-

Figure 14–11 The Chaba River in the Canadian Rockies is braided because glaciers pour more sediment into the stream than the water can carry.

Figure 14–12 This alluvial fan in the Canadian Rockies formed where a steep mountain stream deposits most of its sediment as it enters a flat valley.

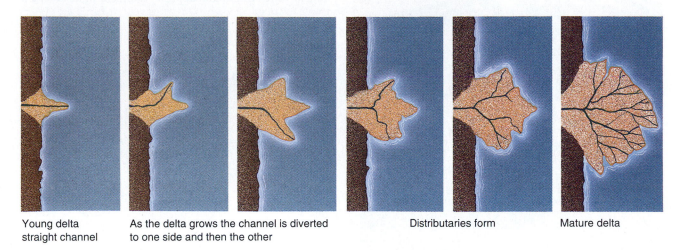

Young delta
straight channel

As the delta grows the channel is diverted
to one side and then the other

Distributaries form

Mature delta

Figure 14–13 A delta forms and grows with time where a stream deposits its sediment as it flows into a lake or the sea.

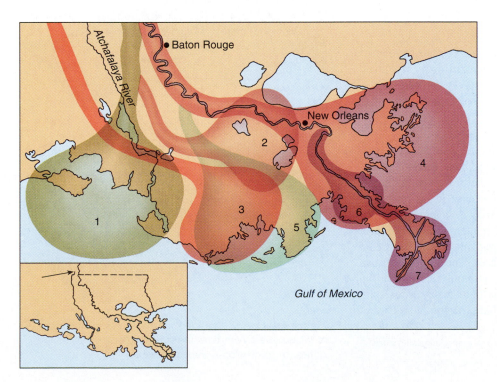

Figure 14–14 The Mississippi River has flowed into the sea by seven different channels during the past 6000 years. As a result, the modern delta is composed of seven smaller deltas formed at different times. The oldest delta is numbered 1, and the current delta is 7.

Figure 14–15 Deer Creek, a tributary of Grand Canyon, has downcut its channel into solid sandstone.

est level to which it can erode its bed. For most streams,[1] the lowest possible level of downcutting is sea level, which is called the **ultimate base level**. This concept is straightforward. Water can only flow downhill. If a stream were to cut its way down to sea level, it would stop flowing and hence would no longer erode its bed.

In addition to ultimate base level, a stream may have a number of **local**, or **temporary, base levels**. For example, a stream stops flowing where it enters a lake. It then stops eroding its channel because it has reached a temporary base level (Fig. 14–16a). A layer of rock that resists erosion may also establish a temporary base level because it flattens the stream gradient. Thus, the stream slows down and erosion decreases. The top of a waterfall is a temporary base level commonly established by resistant rock. Niagara Falls is held up by a resistant layer of dolomite over softer shale. As the falling water

[1] We say "for most streams" because a few empty into valleys that lie below sea level.

erodes the shale, it undermines the dolomite cap, which periodically collapses. As a result, Niagara Falls has retreated 11 kilometers upstream since its formation about 9000 years ago (Fig. 14–17).

If a stream has numerous temporary base levels, it erodes its bed in the steep places where it flows rapidly, and it deposits sediment in the low-gradient stretches where it flows more slowly (Fig. 14–16b). Over time, erosion and deposition smooth out the irregularities in the gradient. An idealized **graded stream** has a smooth, concave profile (Fig. 14–16c). Once a stream becomes graded, the rate of channel erosion becomes equal to the rate at which the stream deposits sediment in its channel.

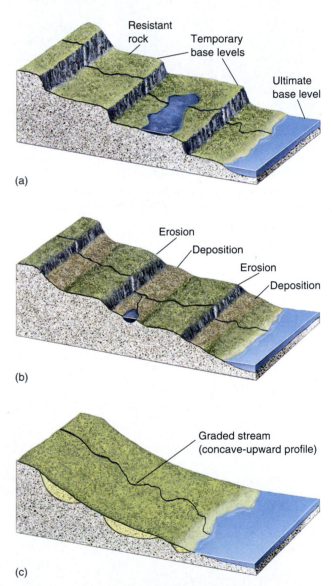

Figure 14–16 An ungraded stream (a) has many temporary base levels. With time, the stream smooths out the irregularities (b) to develop a graded profile (c).

Figure 14–17   Niagara Falls has eroded 11 kilometers upstream in the last 9000 years and continues to erode today. (Hubertus Kanus/Photo Researchers, Inc.)

Figure 14–18   A steep mountain stream eroded a V-shaped valley into soft shale in the Canadian Rockies.

Thus, there is no net erosion or deposition, and the stream profile no longer changes. An idealized graded stream such as this does not actually exist in nature, but many streams come close.

SINUOSITY OF A STREAM CHANNEL

A steep mountain stream usually downcuts rapidly into its bed. As a result, it cuts a relatively straight channel with a steep-sided, V-shaped valley (Fig. 14–18). The stream maintains its relatively straight path because it flows with enough energy to erode and carry off any material that slumps into its channel.

In contrast, a low-gradient stream is less able to erode downward into its bed. Much of the stream energy is directed against the banks, causing **lateral erosion**. Lateral erosion undercuts the valley sides and widens a stream valley.

Most low-gradient streams flow in a series of bends called **meanders** (Fig. 14–19). A random event may initiate the development of meanders. For example, if the right bank of a slow stream collapses into the channel, the current cannot quickly erode the material and carry it away. Instead, the obstruction deflects the current toward the left bank, where it undercuts the bank and causes another slump. The new slump then deflects the current back to the right bank. Thus, a single, random cave-in creates a sinuous current that erodes the stream's

outside bends. Over time, the meanders propagate downstream.

A meandering stream wanders back and forth across its entire flood plain, forming a wide valley with a flat bottom. A meandering channel seems to be the natural pattern for a low-gradient stream, but geologists do not fully understand why. Most agree that meanders mini-

Figure 14–19   A low-gradient stream commonly flows in a series of looping bends called meanders. This one is in Baffin Island, Canada.

mize flow resistance of the channel and allow the stream to expend its energy uniformly throughout its channel.

In many valleys, meanders become so pronounced that the outside of one meander approaches that of another. Given enough time, the stream erodes the narrow neck of land separating the two meanders and creates a new channel, abandoning the old meander loop. Because the current no longer flows through the entrance and exit of the abandoned meander, sediment accumulates at those points, isolating the old meander from the stream to form an **oxbow lake** (Fig. 14–20).

Most streams do not maintain the same sinuosity throughout their entire length. For example, a meandering stream may develop a straight channel if it flows into an area of steeper gradient, or it may become braided if it encounters a supply of excess sediment.

LANDFORM EVOLUTION AND TECTONIC REJUVENATION

According to a model popular in the first half of this century, streams erode mountain ranges and create landforms in a particular sequence. At first, the streams cut steep, V-shaped valleys. Over time, erosion decreases the gradient, and the valleys widen into broad flood plains. Eventually, the entire landscape flattens, forming a large, low, featureless plain called a **peneplain**.

This model of continuous leveling of the Earth's surface tells only half the story because over geologic time, tectonic forces uplift the land and interrupt the simple, idealized sequence. A stream is **rejuvenated** when tectonic activity raises the land. As the land rises, the stream becomes steeper. As a result, its energy increases and it erodes downward into its bed.

If tectonic uplift steepens the gradient of a meandering stream, the stream may downcut rapidly, preserving the winding channel by cutting it deeply into bedrock, forming **incised meanders** (Fig. 14–21). If the stream occupied a broad flood plain before tectonic rejuvenation, floodwaters of the newly incised channel would no longer reach the old flood plain. The old, abandoned flood plain at a higher elevation is then called a **stream terrace** (Fig. 14–22).

Incised meanders and stream terraces can form without tectonic uplift. If the climate becomes wetter, the discharge increases and a graded stream then cuts downward into its bed, abandoning the old flood plain. The rate of downcutting also increases when temporary base level is lowered. For example, a stream may stop

Figure 14–20 An oxbow lake forms where a stream erodes through a meander neck. This one is in the Flathead River, Montana.

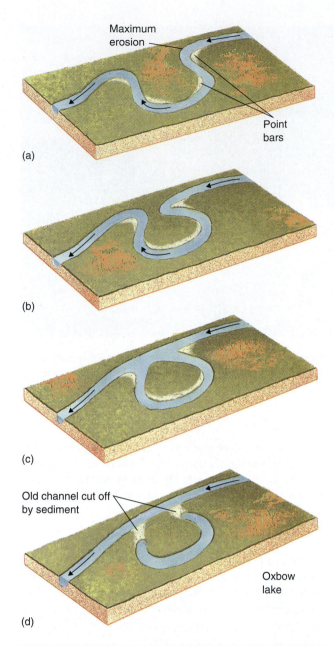

(a) Maximum erosion — Point bars

(b)

(c)

(d) Old channel cut off by sediment — Oxbow lake

(e)

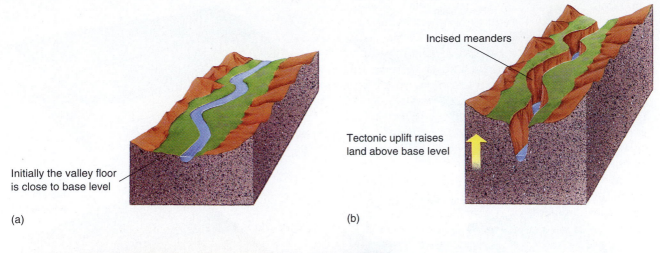

Incised meanders

Tectonic uplift raises
land above base level

Initially the valley floor
is close to base level

(a)   (b)

(c)

Figure 14–21 (a) If tectonic uplift steepens the gradient of a meandering stream, (b) the stream may downcut into bedrock to form incised meanders. (c) The Escalante River cut an incised meander into sandstone.

downcutting at a temporary base level that is supported by a layer of resistant rock. When the stream finally erodes through this layer, it may start downcutting through softer rock below, quickly deepening its channel and abandoning its flood plain.

STREAMS THAT FLOW THROUGH MOUNTAIN RANGES

Some streams flow through a mountain range, plateau, or ridge of resistant rock. Why don't they flow around the mountains rather than cutting directly through them?

Rejuvenation often causes such odd behavior. Imagine a stream flowing across a plain. If tectonic forces uplift the center of the plain, and if the uplift occurs slowly, the stream may cut through the rising bedrock to keep its original course (Fig. 14–23). A stream of this type is said to be **antecedent** because it existed before

the uplift rose. The Grand Canyon is an antecedent stream channel; it formed as the Colorado River cut downward through more than 1600 meters of sedimentary rock, as the Colorado Plateau rose.

A stream may also cut through mountains by the process of **superposition**. If an old mountain range is covered with younger sedimentary rocks, a stream cutting its channel into the sedimentary rocks is unaffected by the buried mountains. Eventually, the stream may downcut until it reaches the buried mountains. At this point the channel may be too deep to shift laterally, and the stream may cut through the range rather than flow around it (Fig. 14–24).

Stream Piracy

Consider two streams flowing in opposite directions from a mountain range. One of the streams may cut downward

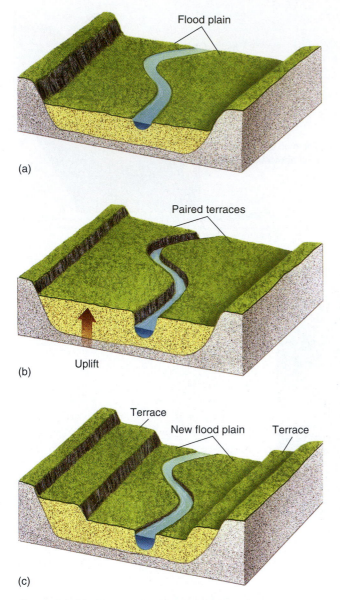

Figure 14-22   Formation of terraces. (a) A stream has formed a broad flood plain. (b) Tectonic uplift or climatic change causes the stream to downcut into its bed. As the stream cuts downward, the old flood plain becomes a terrace above the new stream level. (c) A new flood plain forms at the lower level.

faster than the other because one side of the range is steeper than the other, because it receives more rainfall, or because the rock on one side is softer than rock on the other (Fig. 14–25). The stream that is downcutting more rapidly will also cut its way backward into the mountains. This process is called **headward erosion**. If headward erosion continues, the more deeply eroded stream may intercept the higher stream on the opposite side of the range. The stream at higher elevation then reverses di-

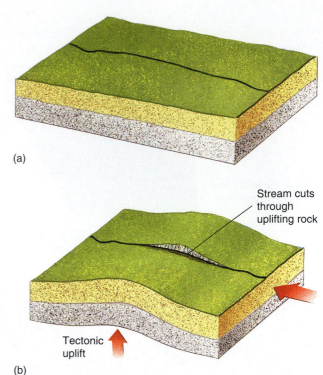

Figure 14–23   An antecedent stream forms where tectonic forces form a ridge, but a stream erodes its bed as rapidly as the land rises. Thus, the stream is able to maintain its original path by cutting through the rising ridge.

rection and flows into the lower one. This sequence of events is called **stream piracy**.

▶ 14.6   DRAINAGE BASINS

The region drained by a stream and its tributaries is called a **drainage basin**. Mountain ranges or other raised areas called **drainage divides** separate adjacent drainage basins. For example, streams on the western slope of the Rocky Mountains are parts of the Colorado and Columbia River drainage basins, which ultimately empty into the Pacific Ocean. Tributaries on the eastern slope of the Rockies are parts of the Mississippi and Rio Grande basins, which flow into the Gulf of Mexico. A drainage basin can be large, like the Mississippi basin, or as small as a single mountain valley.

In most drainage basins, the pattern of tributaries resembles the veins in a leaf. Each tributary forms a V pointing downstream where it joins the main stream. This type of system is called a **dendritic drainage pattern** (Fig. 14–26a). Dendritic drainages develop where streams flow over uniform bedrock. Because they are not

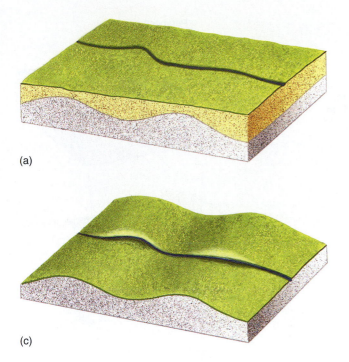

(a)

(b)

(c)

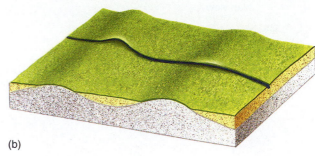

Figure 14–24 A superposed stream forms where a stream flows over young sedimentary rock that has buried an ancient mountain range. As the stream erodes downward, it cuts into the old mountains, maintaining its course.

deflected by resistant layers of bedrock, the streams take the shortest route downslope.

In some regions, bedrock is not uniform. For example, Figure 14–26b shows a layer of sandstone lying over granite. The rocks were faulted and tilted after the sandstone formed. As a result, parallel outcrops of easily eroded sandstone alternate with bands of hard granite. Streams followed the softer sandstone, forming a series of long, straight, parallel channels intersected at right angles by short tributaries. This type of drainage pattern, called a **trellis pattern**, is common in the tilted rocks of the Appalachian Mountains. A **rectangular pattern** can

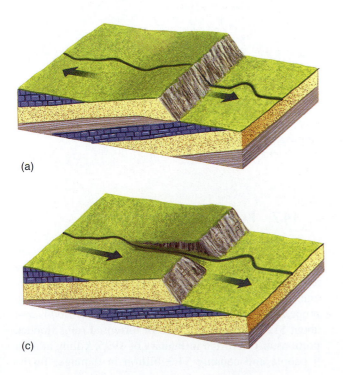

(a)

(c)

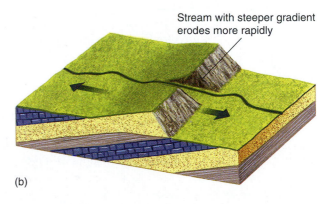

Stream with steeper gradient erodes more rapidly

(b)

Figure 14–25 Stream piracy occurs when a stream on the steeper side of a ridge erodes downward more rapidly than the stream on the opposite slope. Eventually, the steeper stream cuts through the ridge to intersect the higher stream. The higher stream then reverses direction to flow into the lower one.

Map view

Perspective view

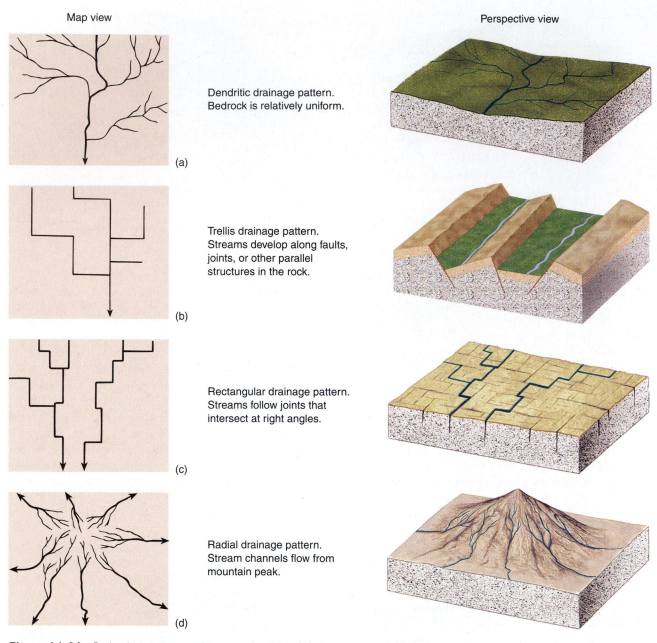

Dendritic drainage pattern. Bedrock is relatively uniform.

(a)

Trellis drainage pattern. Streams develop along faults, joints, or other parallel structures in the rock.

(b)

Rectangular drainage pattern. Streams follow joints that intersect at right angles.

(c)

Radial drainage pattern. Stream channels flow from mountain peak.

(d)

Figure 14–26 Bedrock structures and topography control drainage patterns. (a) A dendritic pattern reflects homogeneous bedrock. (b) Parallel faults and contrasting rock types may form a trellis drainage pattern. (c) A rectangular pattern reflects bedrock fractures that intersect at right angles. (d) A radial pattern forms on a peak.

develop if streams follow faults or joints that intersect at right angles. In this case, the main stream and its tributaries are of approximately the same length (Fig. 14–26c). A **radial drainage pattern** develops where a number of streams originate on a mountain and radiate outward from the peak (Fig. 14–26d).

The next time you fly in an airplane, look out the window and try to determine the type of drainage pattern below and what it tells you about the geology of the area.

▶ 14.7 FLOODS

When stream discharge exceeds the volume of a stream channel, water overflows onto the flood plain, creating a flood. Floods in the United States cause an annual average of 85 human deaths and more than $1 billion in property damage. The 1993 Mississippi River floods cost about $12 billion. Two weeks of torrential rains flooded portions of California in January of 1995, killing at least 9 people and causing $1.3 billion in damage. In the summer of 1993, raging rivers killed more than 150 people in southern China, flash floods and related landslides killed 1800 people in Nepal, and 1350 people died in monsoon floods in northern India and Bangladesh.

Although flooding is a natural event, human activities can increase flood frequency and severity. Stream discharge increases where forests are logged, prairies plowed, or cities paved. Without trees, shrubs, or grass to absorb rain, water runs over the land and seeps through the ground to increase stream flow. In cities, nearly all rainwater runs off pavement directly into nearby streams.

Floods are costly in part because people choose to live in flood plains. Many riverbank cities originally grew as ports, to take advantage of the easy transportation afforded by rivers. Rivers are no longer the important transportation arteries that they were 100 years ago, but the flood plain cities continue to thrive and grow.

Although we normally think of floods as destructive events, river and flood plain ecosystems depend on floods. For example, cottonwood tree seeds germinate only after a flood. Many species of fish gradually lose out to stronger competitors during normal flows, but have adapted better to floods so that their populations increase as a result of flooding.

In some cases, flooding even benefits humans. Frequent small floods dredge bigger channels, which reduce the severity of large floods. Flooding streams carry large sediment loads and deposit them on flood plains to form fertile soil. So, paradoxically, the same floods that cause death and disaster create the rich soils that make flood plains so attractive for farming.

FLOODS AND STREAM SIZE

Rapid snowmelt or a single intense thunderstorm can flood a small stream. In 1976, a series of summer storms saturated soil and bedrock near Rocky Mountain National Park, northwest of Denver. Then a large thunderhead dropped 19 centimeters of rain in 1 hour in the headwaters of Big Thompson Canyon. The Big Thompson River flooded, filling its narrow valley with a deadly, turbulent wall of water. Some people in the valley tried to escape by driving toward the mouth of the canyon, but traffic clogged the two-lane road and trapped motorists in their cars, where they drowned. A few residents tried to escape by scaling the steep canyon walls, but the rising waters caught some of them. Within a few hours, 139 people had died and five were missing. By the next day, the flood was over (Fig. 14–27).

Big Thompson Canyon is a relatively small drainage basin that flooded rapidly during a locally heavy rain. In contrast, the Mississippi River basin covers 3.2 million square kilometers, and the river itself has a discharge of 17,500 m³/sec. A sudden downpour in any of its small tributaries would have no effect on the Mississippi. A

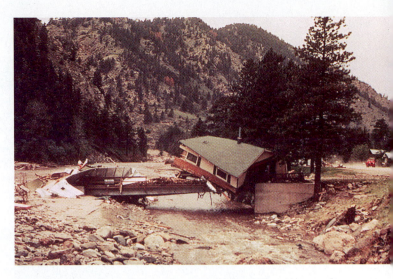

Figure 14–27   The 1976 Big Thompson Canyon flood in Colorado lifted this house from its foundation and carried it downstream.   (USGS)

flood on the Mississippi results from large amounts of rainfall over a broad area. The Mississippi River flood of 1993 plagued parts of the Mississippi flood plain for two months.

FLOOD FREQUENCY

Many streams flood regularly, some every year. In any stream, small floods are more common than large ones, and the size of floods can vary greatly from year to year. A ten-year flood is the largest flood that occurs in a given stream on an average of once every ten years. A 100-year flood is the largest that occurs on an average of once every 100 years. For example, a stream may rise 2 meters above its banks during a ten-year flood, but 7 meters during a 100-year flood. Thus, a 100-year flood is higher and larger, but less frequent, than a ten-year flood.

NATURAL LEVEES

As a stream rises to flood stage, both its discharge and velocity increase. It expends much of this increased energy eroding its bed and banks, thus increasing its sediment load.

The current of a flooding stream slows abruptly where water leaves the channel to flow onto the flood plain. The sudden decrease in current velocity causes the stream to deposit sand on the stream banks. The sand forms ridges called **natural levees** at the margins of the channel (Fig. 14–28). Farther out on the flood plain, the flood water carries finer particles, mostly clay and silt. As a flood wanes, this fine sediment settles onto the flood plain, renewing and enriching the soil.

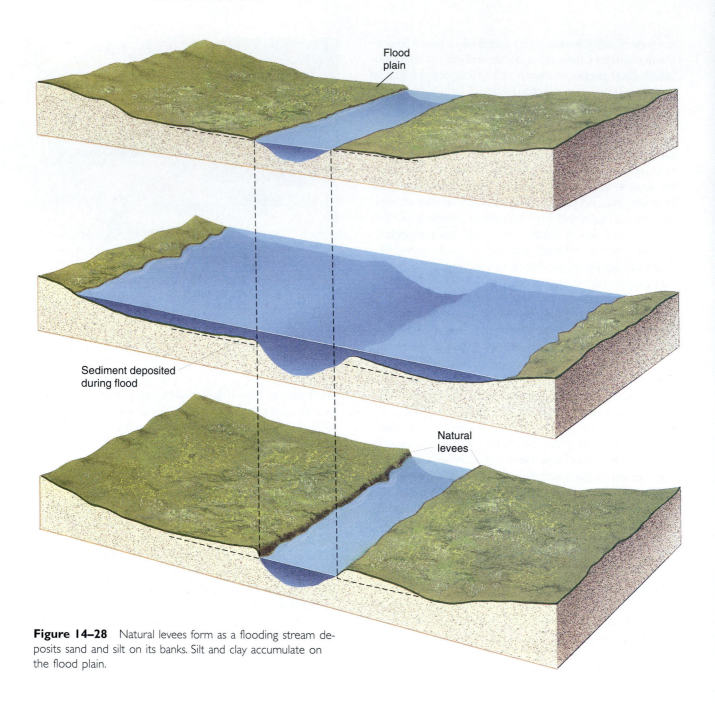

Figure 14–28 Natural levees form as a flooding stream deposits sand and silt on its banks. Silt and clay accumulate on the flood plain.

Flood Control and the 1993 Mississippi River Floods

During the late spring and summer of 1993, heavy rain soaked the upper Midwest. Thirteen centimeters fell in already saturated central Iowa in a single day. In mid-July, 2.5 centimeters of rain fell in 6 minutes in Papillion, Nebraska. As a result of the intense rainfall over such a large area, the Mississippi River and its tributaries flooded. In Fargo, North Dakota, the Red River, fed by a day-long downpour, rose 1.2 meters in 6 hours, flood-

ing the town and backing up sewage into homes and the Dakota Hospital. In St. Louis, Missouri, the Mississippi crested 14 meters above normal and 1 meter above the highest previously recorded flood level. At its peak, the flood inundated nearly 44,000 square kilometers in a dozen states. Damage to homes and businesses on the flood plain reached $12 billion. Forty-five people died.

During the 1993 Mississippi River flood, control projects saved entire towns and prevented millions of dollars in damage. However, in some cases, described in the following section, control measures increased damage. Geologists, engineers, and city planners are studying these conflicting results to plan for the next flood, which may be years or decades away.

ARTIFICIAL LEVEES AND CHANNELS

An artificial levee is a wall built of earth, rocks, or concrete along the banks of a stream to prevent rising water from spilling out of the stream channel onto the flood plain. In the past 70 years, the U.S. Army Corps of Engineers has spent billions of dollars building flood control structures, including 11,000 kilometers of levees along the banks of the Mississippi and its tributaries (Fig. 14–29).

As the Mississippi River crested in July 1993, flood waters surged through low areas of Davenport, Iowa, built on the flood plain. However, the business district of nearby Rock Island, Illinois, remained mostly dry. In 1971, Rock Island had built levees to protect low-lying areas of the town, whereas Davenport had not built levees. Hannibal, Missouri, had just completed levee construction to protect the town when the flood struck. The $8 million project in Hannibal saved the Mark Twain home and museum and protected the town and surrounding land from flooding.

Unfortunately, two major problems plague flood control projects that rely on artificial levees: Levees are temporary solutions to flooding, and in some cases they cause higher floods along nearby reaches of the river.

In the absence of levees, when a stream floods, it deposits mud and sand on the flood plain. When artificial levees are built, the stream cannot overflow during small floods, so it deposits the sediment in its channel, raising the level of the stream bed. After several small floods, the entire stream may rise *above* its flood plain, contained only by the levees (Fig. 14–30). This configuration creates the potential for a truly disastrous flood because if the levee should be breached during a large

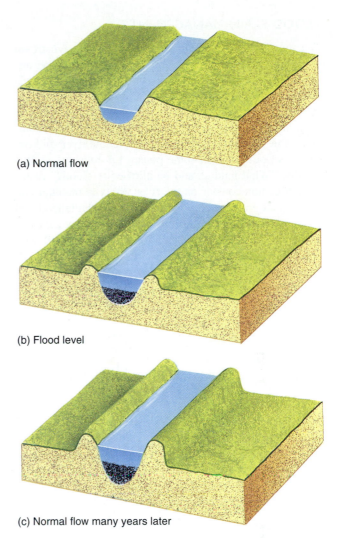

(a) Normal flow

(b) Flood level

(c) Normal flow many years later

Figure 14–30 Artificial levees cause sediment to accumulate in a stream channel, eventually raising the channel above the level of the flood plain and creating the potential for a disastrous flood.

Figure 14–29 Flood waters pouring from the channel of the Missouri River through a broken levee onto the flood plain in Boone County, Missouri. (Stephen Levin)

flood, the entire stream then flows out of its channel and onto the flood plain. As a result of levee building and channel sedimentation, portions of the Yellow River in China now lie 10 meters above its flood plain. Thus, levees may solve flooding problems in the short term, but in a longer time frame they may cause even larger and more destructive floods.

Engineers have tried to solve the problem of channel sedimentation by dredging artificial channels across meanders. When a stream is straightened, its velocity increases and it scours more sediment from its channel. This solution, however, also has its drawbacks. A straightened stream is shorter than a meandering one, and consequently the total volume of its channel is reduced. Therefore, the channel cannot contain as much excess water, and flooding is likely to increase downstream.

FLOOD PLAIN MANAGEMENT

Because levees can worsen upstream and downstream flooding, the levees that save one city may endanger another. In many cases, attempts at controlling floods either do not work, or they shift the problem to a different place.

An alternative approach to flood control is to abandon some flood control projects and let the river spill out onto its flood plain in some places. Of course the question is: "What land should be allowed to flood?" Every farmer or homeowner on the river wants to maintain the levees that protect his or her land. Currently, federal and state governments are establishing wildlife reserves in some flood plain areas. Since no development is allowed in these reserves, they will flood during the next high water. However, a complete river management plan involves complex political and economic considerations.

Figure 14–31 Lago Nube in Bolivia's Cordillera Apolobamba.

▶ 14.8 LAKES

Lakes and lake shores are some of the most attractive recreational and living environments on Earth. Clean, sparkling water, abundant wildlife, beautiful scenery, aquatic recreation, and fresh breezes all come to mind when we think of going to the lake. Despite the great value that we place on them, lakes are among the most fragile and ephemeral landforms. Modern, post–ice age humans live in a special time in Earth history when the Earth's surface is dotted with numerous beautiful lakes.

THE LIFE CYCLE OF A LAKE

A **lake** is a large, inland body of standing water that occupies a depression in the land surface (Fig. 14–31). Streams flowing into the lake carry sediment, which fills the depression in a relatively short time, geologically speaking. Soon the lake becomes a swamp, and with time the swamp fills with more sediment and vegetation to become a meadow or forest with a stream flowing through it.

If most lakes fill quickly with sediment, why are they so abundant today? Most lakes exist in places that were covered by glaciers during the latest ice age. About 18,000 years ago, great continental ice sheets extended well south of the Canadian border, and mountain glaciers scoured their alpine valleys as far south as New Mexico and Arizona. Similar ice sheets and alpine glaciers existed in higher latitudes of the Southern Hemisphere. We are just now emerging from that glacial episode.

The glaciers created lakes in several different ways. Flowing ice eroded numerous depressions in the land

surface, which then filled with water. The Finger Lakes of upper New York State and the Great Lakes are examples of large lakes occupying glacially scoured depressions.

The glaciers also deposited huge amounts of sediment as they melted and retreated. Because mountain glaciers flow down stream valleys, some of these great piles of glacial debris formed dams across the valleys. When the glaciers melted, streams flowed down the valleys but were blocked by the dams. Many modern lakes occupy glacially dammed valleys (Fig. 14–32).

Figure 14–32 A mountain lake dammed by glacial debris in the Sierra Nevada.

Figure 14–33 Kettle lakes in Montana's Flathead Valley formed when blocks of glacial ice melted.

In addition, the melting glaciers left huge blocks of ice buried in the glacial sediment. As the ice blocks melted, they left depressions that filled with water. Many thousands of small lakes and ponds, called **kettles** or **pothole lakes**, formed in this way. Kettles are common in the northern United States and the southern Canadian prairie (Fig. 14–33).

Most of these glacial lakes formed within the past 10,000 to 20,000 years, and sediment is rapidly filling them. Many smaller lakes have already become swamps. In the next few hundred to few thousand years, many of the remaining lakes will fill with mud. The largest, such as the Great Lakes, may continue to exist for tens of thousands of years. But the life spans of lakes such as these are limited, and it will take another glacial episode to replace them.

Lakes also form by nonglacial means. A volcanic eruption can create a crater that fills with water to form a lake, such as Crater Lake, Oregon. Other lakes form in abandoned river channels, such as the oxbow lakes on the Mississippi River flood plain, or in flat lands with shallow ground water, such as Lake Okeechobee of the Florida Everglades. These types of lakes, too, fill with sediment and, as a result, have limited lives.

A few lakes, however, form in ways that extend their lives far beyond that of a normal lake. For example, Russia's Lake Baikal is a large, deep lake lying in a depression created by an active fault. Although rivers pour sediment into the lake, movement of the fault repeatedly deepens the basin. As a result, the lake has existed for more than a million years, so long that indigenous species of seals and other animals and fish have evolved in its ecosystem.

FRESH-WATER AND SALTY LAKES

Most lakes contain fresh water because the constant flow of streams both into and out of them keeps salt from accumulating. A few lakes are salty; some, such as Utah's Great Salt Lake, are saltier than the oceans. A salty lake forms when streams flow into the lake but no streams flow out. Streams carry salts into the lake, but water leaves the lake only by evaporation and a small amount of seepage into the ground. Evaporation removes pure water, but no salts. Thus, over time the small amounts of dissolved salts carried in by the streams concentrate in the lake water. Salty lakes usually occur in desert and semiarid basins, where dry air and sunshine evaporate water rapidly.

SUMMARY

Only about 0.65 percent of the Earth's water is fresh. The rest is salty seawater and glacial ice. **Evaporation, transpiration, precipitation,** and **runoff** continuously recycle water among land, sea, and the atmosphere in the **hydrologic cycle**. About 60 times more fresh water is stored as **ground water** than as **surface water**.

A **stream** is any body of water flowing in a **channel**. A **flood** occurs when a stream overflows its banks and flows over its **flood plain**. The velocity of a stream is determined by its **gradient, discharge,** and channel shape and roughness.

The ability of a stream to erode and carry sediment depends on its velocity and its discharge. Stream **competence** is a measure of the largest particle it can carry.

Capacity is the total amount of sediment a stream can carry past a point in a given amount of time. Most erosion and sediment transport occur when a stream is flooding. A stream weathers and erodes its channel and flood plain by **hydraulic action, abrasion,** and **solution**. A stream transports sediment as **dissolved load, suspended load,** and **bed load**. Most sediment is carried as suspended load. Streams deposit sediment in **channel deposits, alluvial fans, deltas,** and as **flood plain deposits**. A **braided stream** flows in many shallow, interconnecting channels.

Ultimate base level is the lowest elevation to which a stream can erode its bed. It is usually sea level. A lake or resistant rock can form a **local,** or **temporary, base**

level. A **graded stream** has a smooth, concave profile. Steep mountain streams form straight channels and V-shaped valleys, whereas lower-gradient streams form **meanders** and wide valleys. Tectonic uplift, increased rainfall, and lowering of base level all can **rejuvenate** a stream, causing it to cut down into its bed to form **incised meanders** and abandon an old flood plain to form a **stream terrace**. **Headward erosion** can cause **stream piracy**. A **drainage basin** can be characterized by den-

dritic, **trellis**, **rectangular**, or **radial** patterns, depending on bedrock geology.

Floods occur when a stream flows out of its channel and onto the **flood plain**. Floods occur periodically in all streams. Artificial levees and other flood control measures reduce flood severity in some areas but may increase flood severity at other times and places.

Many modern **lakes** were created by recent glaciers; as a result, we live in an unusual time of abundant lakes.

KEY WORDS

REVIEW QUESTIONS

1. What proportion of the Earth's free water is useful for drinking and irrigation? Why is the proportion so small?

2. In which physical state (solid, liquid, or vapor) does most of the Earth's free water exist? Which physical state accounts for the least?

3. Describe the movement of water through the hydrologic cycle.

4. Describe the factors that determine the velocity of stream flow and describe how those factors interact.

5. For each of the following pairs of streams (or segments of streams), which would move faster?   a. Two streams have equal gradients and discharges, but one is narrow and deep while the other has a semicircular cross section. b. Two streams have equal gradients and channel shapes, but one has a greater discharge.   c. Two streams have equal channel shapes and discharges, but one has a steeper gradient.   d. Two streams have equal gradients, channel shapes, and discharges, but one is choked with boulders and the other is lined by smooth rock surfaces.

6. Describe the factors that control the competence of a stream.

7. Describe the factors that affect stream capacity.

8. Distinguish among the three types of stream erosion: hydraulic action, solution, and abrasion.

9. List and explain three ways in which sediment can be transported by a stream. Which type of transport is independent of stream velocity? Explain.

10. In what transport mode is most sediment carried by a stream?

11. Why do braided streams often develop in glacial and desert environments?

12. How is an alluvial fan similar to a delta? How do they differ?

13. Give two examples of natural features that create temporary base levels. Why are they temporary?

14. Draw a profile of a graded stream and an ungraded one.

15. Explain how a stream forms and shapes a valley.

16. In what type of terrain would you be likely to find a V-shaped valley? Where would you be likely to find a meandering stream?

17. What is a meander and how may it become an oxbow lake?

18. How can a stream become rejuvenated? Give an example of a landform created by a rejuvenated stream.

19. Explain the difference between an antecedent and a superposed stream.

20. Why are most lakes short-lived landforms?

21. What geologic conditions create a long-lived lake?

DISCUSSION QUESTIONS

1. In certain regions, stream discharge rises rapidly and dramatically during and after a rainfall. In other regions, stream discharge increases slowly and less dramatically. Draw a graph of these two different types of behavior, with time on the X axis and discharge on the Y axis. How do rock and soil type and vegetation affect the relationship between rainfall and discharge?

2. Gold dust settles out in regions where stream velocity slows down. If you were panning for gold, would you look for   a. a graded stream or one with many temporary base levels?   b. the inside or the outside of a stream bend?   c. a rocky stream bed or a sandy stream bed?   d. a steep gradient or a shallow gradient portion of a stream? Give reasons for each of your choices.

3. Stream flow is often reported as stream depth at a certain point. Discuss the relationship between depth and discharge. If the depth doubled, how would the discharge be affected?

4. If you were buying a house located in a flood plain, what evidence would you look for to determine past flood activity?

5. Examine Figure 14–34. If you were building a house near this river, would you choose site A or B? Defend your choice.

6. Defend the statement that most stream erosion occurs in a relatively short time when the stream is in flood.

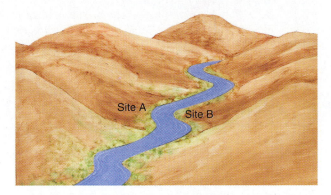

Figure 14–34   Two possible house sites along a river.

7. Imagine that a 100-year flood has just occurred on a river near your home. You want to open a small business in the area. Your accountant advises that your business and building have an economic life expectancy of 50 years. Would it be safe to build on the flood plain? Why or why not?

8. What type of drainage pattern would you expect in the following geologic environments?   a. Platform sedimentary rocks.   b. A batholith fractured by numerous faults.   c. A flat plain with a composite volcano in the center. Defend your answers.

Ground Water

If you drill a hole into the ground in most places, its bottom fills with water after a few days. The water appears even if no rain falls and no streams flow nearby. The water that seeps into the hole is part of the vast reservoir of subterranean **ground water** that saturates the Earth's crust in a zone between a few meters and a few kilometers below the surface.

Ground water is exploited by digging wells and pumping the water to the surface. It provides drinking water for more than half of the population of North America and is a major source of water for irrigation and industry. However, deep wells and high-speed pumps now extract ground water more rapidly than natural processes replace it in many parts of the central and western United States. In addition, industrial, agricultural, and domestic contaminants seep into ground water in many parts of the world. Such pollution is often difficult to detect and expensive to clean up.

Over geologic time, seeping ground water dissolves bedrock to create caverns such as this one. (William Palmer/Visuals Unlimited)

Well-sorted sediment

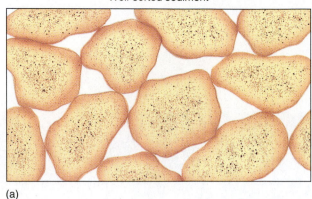

(a)

Poorly sorted sediment

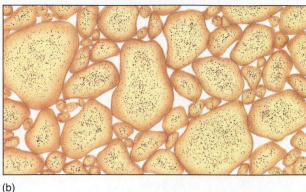

(b)

Sedimentary rock with cementing material between grains

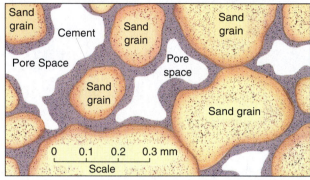

(c)

Figure 15–1 Different materials have different amounts of open pore space between grains. (a) Well-sorted sediment consists of equal-size grains and has a high porosity, about 30 percent in this case. (b) In poorly sorted sediment, small grains fill the spaces among the large ones, and porosity is lower. In this drawing it is about 15 percent. (c) Cement partly fills pore space in sedimentary rock, lowering the porosity.

▶ 15.1 CHARACTERISTICS OF GROUND WATER

POROSITY AND PERMEABILITY

In the upper few kilometers of the Earth, bedrock and soil contain small cracks and voids that are filled with air or ground water. The proportional volume of these open spaces is called the **porosity** of rock or soil. The porosity of sand and gravel is typically high—40 percent or higher. Mud can have a porosity of 90 percent or more because the tiny clay particles are electrically attracted to water. Most rocks have lower porosities than loose sediment. Sandstone and conglomerate can have 5 to 30 percent porosity (Fig. 15–1). Shale typically has a porosity less than 10 percent. Igneous and metamorphic rocks have very low porosities unless they are fractured.

Porosity indicates the amount of water that rock or soil can hold; in contrast, **permeability** is the ability of rock or soil to transmit water (or any other fluid). Water can flow rapidly through material with high permeability. Most materials with high porosity also have high permeability, but permeability also depends on how well the pores are connected and on pore size.

The connections between pores affect permeability because the pores, no matter how large, must be connected for water to flow through rock or soil. Uncemented sand and gravel are both porous and permeable because their pores are large and well connected. Thus, water flows easily through them. Sedimentary rocks such as sandstone and conglomerate can have high permeabilities if cement has not filled the pores and channels. The permeability of many other rocks depends on the density of fractures in the rock.

If the pores are very small, electrical attractions between water and soil particles slow the passage of water. Clay typically has a high porosity, but because its pores are so small it commonly has a very low permeability and transmits water slowly.

THE WATER TABLE

When rain falls, it usually soaks into the ground. Water does not descend into the crust indefinitely, however. Below a depth of a few kilometers, the pressure from overlying rock closes the pores, making bedrock both nonporous and impermeable. Water accumulates on this

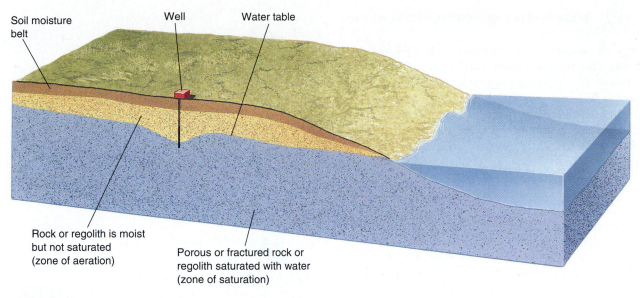

Soil moisture belt

Well

Water table

Rock or regolith is moist but not saturated (zone of aeration)

Porous or fractured rock or regolith saturated with water (zone of saturation)

Figure 15–2   The water table is the top of the zone of saturation near the Earth's surface. It intersects the land surface at lakes and streams and is the level of standing water in a well.

impermeable barrier, filling pores in the rock and soil above it. This completely wet layer of soil and bedrock above the barrier is called the **zone of saturation**. The **water table** is the top of the zone of saturation (Fig. 15–2). Above the water table lies the **unsaturated zone**, or **zone of aeration**. In this layer, the rock or soil may be moist but not saturated.

Gravity pulls ground water downward. However, electrical forces can pull water upward through small channels, just as water rises in a paper towel dipped in water (Fig. 15–3). This upward movement of water is called **capillary action**. Thus, a **capillary fringe** 30 to 60 centimeters thick rises from the water table.

Topsoil usually contains abundant litter and humus, which retain moisture. Thus, in most humid environments, topsoil is wetter than the unsaturated zone beneath it. This moist surface layer is called the **soil moisture belt**, and it supplies much of the water needed by plants.

If you dig into the unsaturated zone, the hole does not fill with water. However, if you dig below the water table into the zone of saturation, you have dug a **well**, and the water level in a well is at the level of the water table. During a wet season, rain seeps into the ground to **recharge** the ground water, and the water table rises. During a dry season, the water table falls. Thus, the water level in most wells fluctuates with the seasons.

An **aquifer** is any body of rock or soil that can yield economically significant quantities of water. An aquifer must be both porous and permeable so that water flows into a well to replenish water that is pumped out. Sand and gravel, sandstone, limestone, and highly fractured bedrock of any kind make excellent aquifers. Shale, clay, and unfractured igneous and metamorphic rocks are poor aquifers.

Figure 15–3   Capillary action pulls water upward from the water table just as it does in this paper towel dipped into dyed water.

▶ 15.2 MOVEMENT OF GROUND WATER

Nearly all ground water seeps slowly through bedrock and soil. Ground water flows at about 4 centimeters per day (about 15 meters per year), although flow rates may be much faster or slower depending on permeability. Most aquifers are like sponges through which water seeps, rather than underground pools or streams. However, ground water can flow very rapidly through large fractures in bedrock, and in a few regions underground rivers flow through caverns.

Ground water flows from zones where the water table is highest toward areas where it is lowest. In general, the water table is higher beneath a hill than it is beneath an adjacent valley (Fig. 15–4). The arrows in Figure 15–5a show that some ground water flows from a hill to a valley along the sloping surface of the water table. Much of it, however, flows in curved paths, descending below the valley floor and then rising beneath the lowest part of the valley.

Why does ground water flow upward against the force of gravity? The key to this phenomenon lies in water pressure. Ground water flows from areas of high water pressure toward areas of low pressure. The water pressure at any point is proportional to the weight of water above that point. Ground water beneath a hill is under greater pressure than water beneath the valley because the water table is higher beneath the hill. Thus, the water pressure beneath the hill forces the water upward beneath the valley. Because ground water flows from high places to low ones, the water table becomes flatter during a dry season.

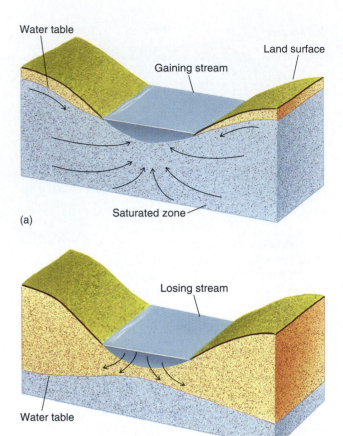

(a)

(b)

Figure 15–5 (a) In a moist climate the water table lies above the stream and ground water seeps *into* the stream. (b) A desert stream lies above the water table. Water seeps *from* the stream bed to recharge the ground-water reservoir beneath the desert.

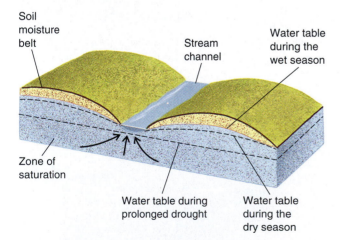

Figure 15–4 The water table follows topography, rising beneath a hill and sinking beneath a valley. Dashed lines show that it also sinks during drought and rises during the rainy season. The arrows show ground-water flow.

Streams flow through most valleys. Because ground water rises beneath the valley floor, it continually feeds the stream. That is why streams continue to flow even when rain has not fallen for weeks or months. A stream that is recharged by ground water is called an **effluent** (or **gaining**) **stream** (Fig. 15–5a). Ground water also seeps into most lakes because lakes occupy low parts of the land.

In a desert, however, the water table commonly lies below a stream bed, and water seeps downward from the stream to the water table (Fig. 15–5b). Such a stream is an **influent**, or **losing**, **stream**. Desert stream channels are dry most of the time. When they do run, the water often flows from nearby mountains where precipitation is greater, although a desert storm can also fill the channel briefly. Thus, a desert stream feeds the ground-water reservoir, but in temperate climates, ground water feeds the stream.

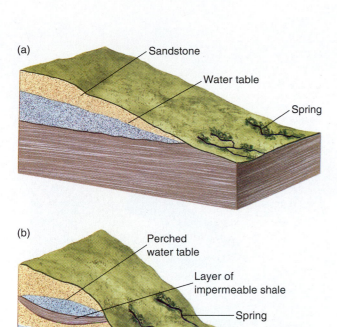

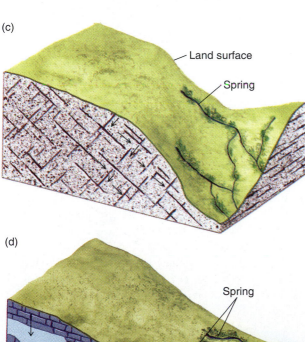

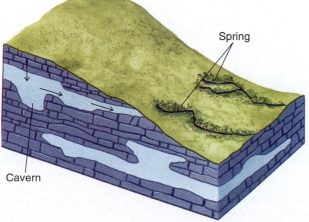

SPRINGS

A spring occurs where the water table intersects the surface and water flows or seeps onto the land (Fig. 15–6). In some places, a layer of impermeable rock or clay lies above the main water table, creating a locally saturated zone, the top of which is called a **perched water table** (Fig. 15–6b). Hillside springs often flow from a perched water table.

ARTESIAN WELLS

Figure 15–7 shows a tilted sandstone aquifer sandwiched between two layers of shale. An inclined aquifer bounded top and bottom by impermeable rock is an **artesian aquifer**. Water in the lower part of the aquifer is under pressure from the weight of water above. Therefore, if a well is drilled through the shale and into the sandstone, water rises in the well without being pumped. A well of this kind is called an **artesian well**. If pressure is sufficient, the water spurts out onto the land surface.

▶ 15.3   USE OF GROUND WATER

Ground water is a particularly valuable resource because

1. It is abundant. Sixty times more fresh water exists underground than in surface reservoirs.
2. It is stored below the Earth's surface and remains available for use during dry periods.
3. In some regions, ground water flows from wet environments to arid ones, making water available in dry areas.

GROUND-WATER DEPLETION

If ground water is pumped from a well faster than it can flow through the aquifer, a **cone of depression** forms around the well (Fig. 15–8). If the aquifer has good permeability, water flows back toward the well in a few days or weeks after the pump is turned off, and the cone of depression disappears. Near the desert town of Cave Creek, Arizona, ground water is pumped onto a golf

Figure 15–6   Springs form where the water table intersects the land surface. This situation can occur where (a) the land surface intersects a contact between permeable and impermeable rock layers; (b) a layer of impermeable rock or clay lies "perched" above the main water table; (c) water flows from fractures in otherwise impermeable bedrock; and (d) water flows from caverns onto the surface.

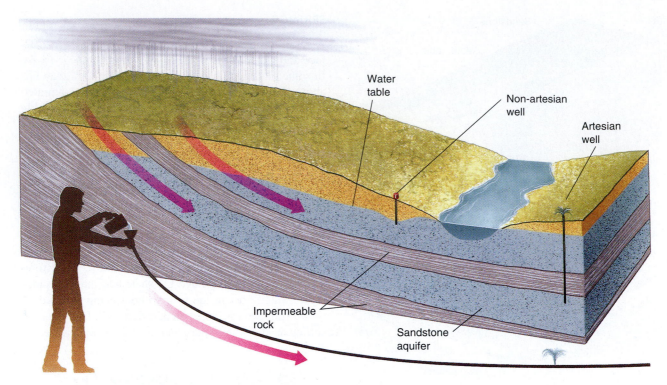

Figure 15–7 An artesian aquifer forms where a tilted layer of permeable rock, such as sandstone, lies sandwiched between layers of impermeable rock, such as shale. Water rises in an artesian well without being pumped. A hose with a hole shows why an artesian well flows spontaneously.

course so fast that the cone of depression sometimes leaves the wells of nearby homeowners dry.

If water is continuously pumped more rapidly than it can flow through the aquifer, or if many wells extract water from the same aquifer, the water table drops. Before the development of advanced drilling and pumping technologies, human impact on ground water was minimal. Today, however, deep wells and high-speed pumps can extract ground water more rapidly than the hydrologic cycle recharges it. In that case, an aquifer becomes de-

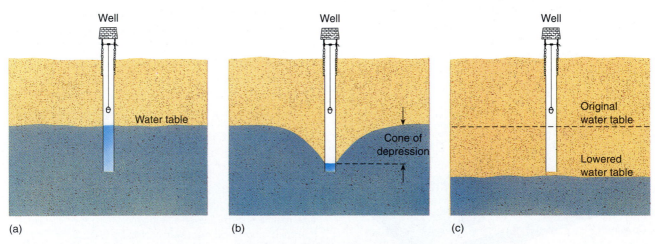

Figure 15–8 (a) A well is drilled into an aquifer. (b) A cone of depression forms because a pump draws water faster than the aquifer can recharge the well. (c) If the pump continues to extract water at the same rate, the water table falls.

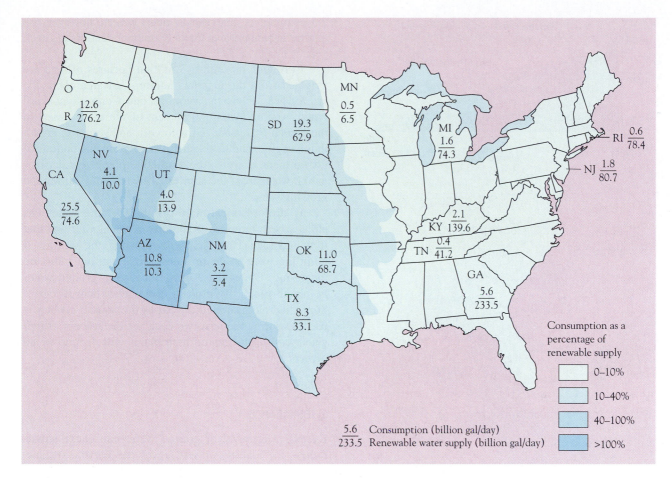

Figure 15–9   Ground-water consumption compared to recharge rates. Arizona consumes ground water faster than the aquifers are recharged. In Georgia, recharge is much faster than consumption.

pleted and is no longer able to supply enough water to support the farms or cities that have overexploited it. This situation is common in the arid and semiarid western United States (Fig. 15–9).

The High Plains and the Ogallala Aquifer

Most of the high plains in western and midwestern North America receive scant rainfall, yet the soil is fertile. Early farmers prospered in rainy years and suffered during drought. In the 1930s, two events combined to change agriculture in this region. One was a widespread drought that destroyed crops and exposed the soil to erosion. Dry winds blew across the land, eroding the parched soil and carrying it for hundreds and even thousands of kilometers. Thousands of families lost their farms, and the region was dubbed the Dust Bowl. The second event was

the arrival of irrigation technology. Electric power lines were built to service rural regions, and affordable pumps and irrigation systems were developed. With the specter of drought fresh in people's memories and the tools to avert future calamities available, the age of modern irrigation began.

Figure 15–10 shows a map and cross section of the Ogallala aquifer beneath the central high plains. The aquifer extends almost 900 kilometers from the Rocky Mountains eastward across the prairie and from Texas into South Dakota. It consists of a layer of permeable sand and gravel 50 to 100 meters thick. The top of the aquifer lies 15 to 100 meters below the surface. Before intensive pumping began, the aquifer contained more than 3 billion acre-feet of water (an acre-foot is an acre of water 1 foot deep—325,851 gallons).

The Ogallala aquifer filled when the last Pleistocene glaciers melted in the Rocky Mountains hundreds of kilometers to the west. Water now flows southeastward

Figure 15–10 The Ogallala aquifer supplies water to much of the High Plains. (a) A cross-sectional view of the aquifer shows that much of its water originates in the Rocky Mountains and flows slowly as ground water beneath the High Plains. (b) A map showing the extent of the aquifer.

through the aquifer at a rate of about a meter per day, and the recharge rate (the rate at which natural processes would raise the water level in the aquifer if no water were being extracted) varies from 0.4 centimeter to about 2.5 centimeters per year. In many places, however, the aquifer is being drawn down by tens of centimeters each year.

Currently, more than 3 million hectares of land are irrigated from the Ogallala aquifer—an area about the size of Massachusetts, Vermont, and Connecticut combined. Approximately 40 percent of the cattle raised in the United States are fed with corn and sorghum grown in this region, and large quantities of grain and cotton are grown there as well. About 150,000 pumps extract irrigation water from the aquifer 24 hours a day during the growing season.

Hydrologists estimate that half of the water has already been removed from parts of the aquifer and that it would take 1000 years to recharge the aquifer if pumping were to cease today. But pumping rates are increasing. Under these conditions, deep ground water is, for all practical purposes, a nonrenewable resource. Some hydrologists predict that, as the aquifer's water is used up during the next two decades, the amount of irrigated land in the central high plains will decline by 80 percent.

SUBSIDENCE

Excessive removal of ground water can cause **subsidence**, the sinking or settling of the Earth's surface. When water is withdrawn from an aquifer, rock or soil particles may shift to fill space left by the lost water. As a result, the volume of the aquifer decreases and the overlying ground subsides (Fig. 15–11). Removal of oil from petroleum reservoirs has the same effect.

Subsidence rates can reach 5 to 10 centimeters per year, depending on the rate of ground-water removal and the nature of the aquifer. Some areas in the San Joaquin Valley of California, one of the most productive agricultural regions in the world, have sunk nearly 10 meters. Subsidence creates particularly severe problems when it affects a city. For example, Mexico City is built on an old marsh. Over the years, as the weight of buildings and roadways has increased and ground water has been extracted, parts of the city have settled as much as 8.5 meters. Many millions of dollars have been spent to maintain this complex city on its unstable base. Similar problems are found in Phoenix, Arizona, in the Houston–Galveston area of Texas, and in other U.S. cities.

Unfortunately, subsidence is not a reversible process. When rock and soil contract, their porosity is permanently reduced so that ground-water reserves cannot be completely recharged even if water becomes abundant again.

Figure 15–11 A dropping water table caused subsidence and structural damage in Jacksonville, Florida. (Wendell Metzen/Bruce Coleman, Inc.)

SALT-WATER INTRUSION

Two types of ground water are found in coastal areas: fresh water and salty water that seeps in from the sea. Fresh water floats on top of salty water because it is less dense. If too much fresh water is pumped to the surface, the salty water rises into the aquifer and contaminates wells (Fig. 15–12). **Salt-water intrusion** has affected many of south Florida's coastal ground-water reservoirs.

Figure 15–12 Salt-water intrusion can pollute coastal aquifers. (a) Fresh water lies above salt water, and the water in the well is fit to drink. (b) If too much fresh water is removed, the water table falls. The level of salt water rises and contaminates the well.

▶ 15.4 GROUND-WATER POLLUTION

Love Canal

Love Canal in Niagara Falls, New York, was excavated to provide water to an industrial park that was never built. After it lay abandoned for several years, the Hooker Chemical Company purchased part of the old canal early in the 1940s. During the following years, the company disposed of approximately 19,000 tons of chemical wastes by loading them into 55-gallon steel drums and dumping the drums in the canal. In 1953, the company covered one of the sites with dirt and sold the land to the Board of Education of Niagara Falls for $1. The city then built a school and playground on the site.

During the following decades, the buried drums rusted through and the chemical wastes seeped into the ground water. In the spring of 1977, heavy rains raised the water table to the surface, and the area around Love Canal became a muddy swamp. But it was no ordinary swamp; the leaking drums had contaminated the ground water with toxic and carcinogenic compounds. The poi-

sonous fluids soaked the playground, seeped into basements of nearby homes, and saturated gardens and lawns. Children who attended the school and adults who lived nearby developed epilepsy, liver malfunctions, skin sores, rectal bleeding, and severe headaches. In the years that followed, an abnormal number of pregnant women suffered miscarriages, and large numbers of babies were born with birth defects.

The Love Canal incident is not unique. In December 1979, the U.S. Congress passed the Comprehensive Environmental Response, Compensation, and Liability Act (CERCLA), commonly known as the Superfund. This law provides an emergency fund to clean up chemical hazards and imposes fines for maintaining a dump site that pollutes the environment. After the Superfund was established, the Environmental Protection Agency (EPA) identified 20,766 hazardous waste sites in the United States. By 1989, the General Accounting Office estimated that there were 400,000 hazardous waste sites. Many are small, involving a few rusting drums in a backlot, but others contaminate large aquifers.

More than 50 percent of the people in the United States drink ground water. The EPA has established maximum tolerance levels for a variety of chemicals that may be present in water drawn from wells. Recent studies show that 45 percent of municipal ground-water supplies in the United States are contaminated with synthetic organic chemicals. According to the Association of Ground Water Scientists and Engineers, approximately 10 million Americans drink water that does not meet EPA standards.

Wells in 38 states contain pesticide levels high enough to threaten human health. Every major aquifer in New Jersey is contaminated. In Florida, where 92 percent of the population drinks ground water, more than 1000 wells have been closed because of contamination, and over 90 percent of the remaining wells have detectable levels of industrial or agricultural chemicals. It is common to read about cities and towns in the United States where a certain type of cancer or other disease afflicts a much greater percentage of the population than the national average. In many cases, contaminated drinking water is suspected to be the cause of the disease.

AQUIFER CONTAMINATION

There are many sources of ground-water pollution (Fig. 15–13). **Point source pollution** arises from a specific site such as a septic tank, a gasoline spill, or a factory.

In contrast, **non-point source pollution** is generated over a broad area. Fertilizers and pesticides spread over fields fall into this latter category.

Once a pollutant enters an aquifer, the natural flow of ground water disperses it as a growing **plume** of contamination. Because ground water flows slowly, usually at a few centimeters per day, the plume also spreads slowly.

Some contaminants, such as gasoline and diesel fuel, are lighter than water and float on top of the water table as they spread (Fig. 15–14a). Others are water soluble and mix with ground water. Mixing dilutes many contaminants, diminishing their toxic effects. However, because ground water moves so slowly, dilution occurs slowly. Still other contaminants are nonsoluble and denser than water. These chemicals sink to an impermeable layer and then flow slowly downslope (Fig. 15–14b).

Many contaminants persist in a polluted aquifer for much longer times than they do in a stream or lake. The rapid flow of water through streams and lakes replenishes their water quickly, but ground water flushes much more slowly. In addition, oxygen, which decomposes many contaminants, is less abundant in ground water than in surface water.

TREATING A CONTAMINATED AQUIFER

The treatment, or **remediation**, of a contaminated aquifer commonly occurs in a series of steps.

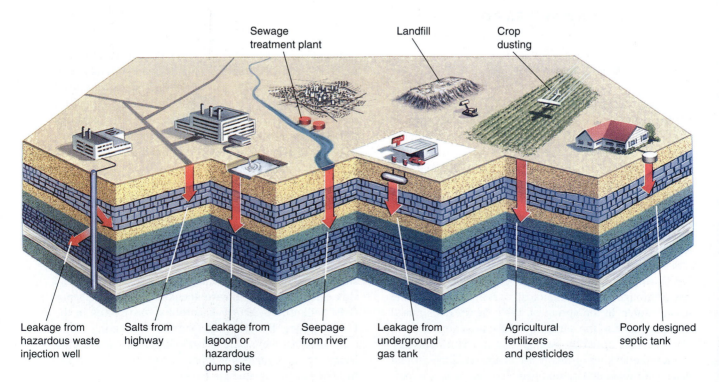

Sewage treatment plant   Landfill   Crop dusting

Leakage from hazardous waste injection well   Salts from highway   Leakage from lagoon or hazardous dump site   Seepage from river   Leakage from underground gas tank   Agricultural fertilizers and pesticides   Poorly designed septic tank

Figure 15–13 Point and non-point sources pollute ground water.

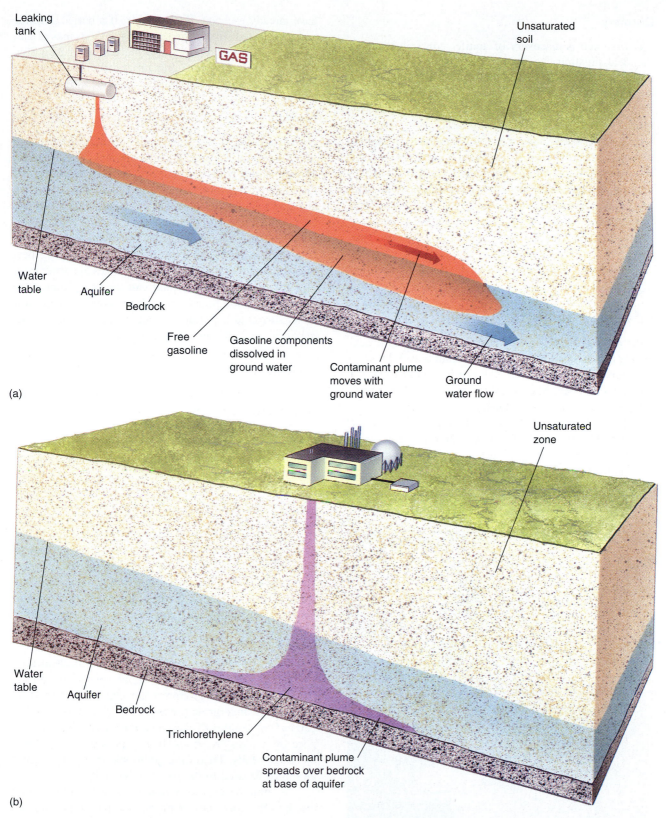

Figure 15–14 (a) Gasoline and many other contaminants are lighter than water. As a result, they float and spread on top of the water table. Soluble components may dissolve and migrate with the ground water. (b) Other pollutants, such as trichlorethylene, are heavier than water and may sink to the base of an aquifer.

Discovery

The first step is discovery of aquifer contamination. In the ideal situation, the contaminant source is discovered before a pollutant can enter the aquifer. For example, if a train derailment dumps pesticide or fuel from ruptured tank cars, the pollutant can usually be contained and removed before it reaches the water table.

In other cases, however, pollutants enter an aquifer before they are detected. This situation commonly occurs when the pollutant comes from hidden sources, such as leaking underground gasoline tanks, unlined landfills and waste dumps such as Love Canal, and agricultural fertilizer and pesticides. In such cases, the first report of polluted ground water often comes from a homeowner whose well water suddenly tastes like gasoline or, worse, from epidemics such as those in the area surrounding Love Canal.

Elimination of the Source

After a contaminant source has been discovered, the next step is to eliminate it. If an underground tank is leaking, the remaining liquid in the tank can be pumped out and the tank dug from the ground. If a factory is discharging toxic chemicals into an unlined dump, courts may issue an injunction ordering the factory to stop using the dump (Fig. 15–15). Elimination of the source prevents additional material from entering the ground water, but it does not solve the problem posed by the pollutants that

Figure 15–15 An oil refinery in New Jersey. New Jersey suffers from some of the worst ground-water pollution in North America as a result of heavy industry.

have already escaped. For example, if a buried gasoline tank has leaked slowly for years, many thousands of gallons of gas may have entered the underlying aquifer. Once the tank has been dug up and the source eliminated, people must deal with the gasoline in the aquifer.

Monitoring

A **hydrogeologist** is a scientist who studies ground water and related aspects of surface water. When aquifer contamination is discovered, a hydrogeologist monitors the contaminants to determine how far, in what direction, and how rapidly the plume is moving and whether the contaminant is becoming diluted. In an area where many houses obtain water from wells, the hydrogeologist may take samples from tens or hundreds of wells and repeat the sampling monthly. He or she analyzes the water samples for the contaminant and can thereby monitor the movement of the plume through the aquifer. If too few wells surround a pollution source, the hydrogeologist may drill wells to monitor the plume.

Modeling

After measuring the rate at which the contaminant plume is spreading, the hydrogeologist develops a computer model to predict future spread of the contaminant through the aquifer. The model considers the permeability of the aquifer, directions of ground-water flow, and mixing rates of ground water (to predict dilution effects).

Remediation

Remediation is so difficult and expensive that of 400,000 hazardous waste sites identified by the General Accounting Office, only 1300 of the most dangerous sites had been placed on a national priority list for cleanup by 1993 (Fig. 15–16). At present, approximately 217 of these projects have been completed, at an average cost of $27 million per site. Future cleanup rates depend in part on government enforcement and budget allocations.

Several processes are currently used to clean up a contaminated aquifer and the source of its contamination. Contaminated ground water can be contained by building an underground barrier to isolate it from other parts of the aquifer. If the contaminant does not decompose by natural processes, it may pollute the aquifer if the barrier fails. Therefore, additional treatment eventually must be used to destroy or remove it.

In some cases, hydrogeologists drill wells around and into the contaminant plume and pump the polluted ground water to the surface. The contaminated water is then collected in tanks, where it is treated to destroy the pollutant. Containment and pumping are often used simultaneously.

Figure 15–16 A hazardous waste dump site photographed in 1989 shows violations of environmental protection laws. (Jeff Amberg/Gamma Liaison)

Bioremediation uses microorganisms to decompose a contaminant. Specialized microorganisms can be fine-tuned by genetic engineers to destroy a particular contaminant without damaging the ecosystem. Once a specialized microorganism is developed, it is relatively inexpensive to breed it in large quantities. The microorganisms are then pumped into the contaminant plume, where they attack the pollutant. When the contaminant is destroyed, the microorganisms run out of food and die, leaving a clean aquifer. Bioremediation can be among the cheapest of all cleanup procedures.

Chemical remediation is similar to bioremediation. If a chemical compound reacts with a pollutant to produce harmless products, the compound can be injected into an aquifer to destroy contaminants. Common reagents used in chemical remediation include oxygen and dilute acids and bases. Oxygen may react with a pollutant directly or provide an environment favorable for microorganisms, which then degrade the pollutant. Thus, contamination can sometimes be reduced simply by pumping air into the ground. Acids or bases neutralize certain contaminants or precipitate dissolved pollutants.

In some extreme cases, reclamation teams dig up the entire contaminated portion of an aquifer. The contaminated soil is treated by incineration or with chemical processes to destroy the pollutant. The treated soil is then used to fill the hole.

▶ **15.5   GROUND WATER AND NUCLEAR WASTE DISPOSAL**

In a nuclear reactor, radioactive uranium nuclei split into smaller nuclei, many of which are also radioactive. Most of these radioactive waste products are useless and must be disposed of without exposing people to the radioactivity. In the United States, military processing plants, 111 commercial nuclear reactors, and numerous laboratories and hospitals generate approximately 3000 tons of radioactive wastes every year.

Chemical reactions cannot destroy radioactive waste because radioactivity is a nuclear process, and atomic nuclei are unaffected by chemical reactions. Therefore, the only feasible method for disposing of radioactive wastes is to store them in a place safe from geologic hazards and human intervention and to allow them to decay naturally. The U.S. Department of Energy defines a permanent repository as one that will isolate radioactive wastes for 10,000 years.[1] For a repository to keep radioactive waste safely isolated for such a long time, it must meet at least three geologic criteria:

1. It must be safe from disruption by earthquakes and volcanic eruptions.

2. It must be safe from landslides, soil creep, and other forms of mass wasting.

3. It must be free from floods and seeping ground water that might corrode containers and carry wastes into aquifers.

The Yucca Mountain Repository

In December 1987, the U.S. Congress chose a site near Yucca Mountain, Nevada, about 175 kilometers from Las Vegas, as the national burial ground for all spent reactor fuel unless sound environmental objections were found. Since that time, numerous studies of the geology, hydrology, and other aspects of the area have been conducted.

The Yucca Mountain site is located in the Basin and Range province, a region noted for faulting and volcanism related to ongoing tectonic extension of the Earth's crust. Bedrock at the Yucca Mountain site is welded tuff, a hard volcanic rock. The tuffs erupted from several large volcanoes that were active from 16 to 6 million years ago. Later volcanism created the Lathrop Wells cinder cone 24 kilometers from the proposed repository. The last eruption near Lathrop Wells occurred 15,000 to 25,000 years ago. In addition, geologists have mapped 32 faults that have moved during the past 2 million years adjacent to the Yucca Mountain site. The site itself is lo-

[1]This number is derived from human and political considerations more than scientific ones. The National Academy of Sciences issued a report stating that the radioactive wastes will remain harmful for one million years.

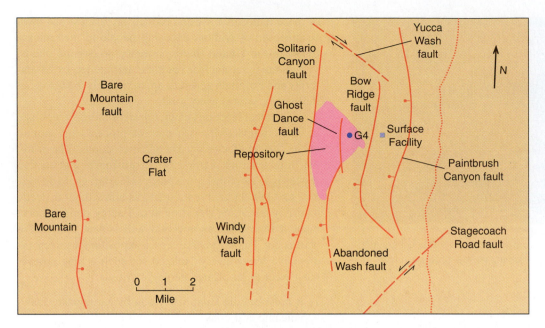

Figure 15–17 A map of the proposed Yucca Mountain repository shows numerous faults where rock has fractured and moved during the past 2 million years. (Redrawn from *Geotimes*, January 1989)

cated within a structural block bounded by parallel faults (Fig. 15–17). Critics of the Yucca Mountain site argue that recent earthquakes and volcanoes prove that the area is geologically active.

The environment is desert dry, and the water table lies 550 meters beneath the surface. The repository will consist of a series of tunnels and caverns dug into the tuff 300 meters beneath the surface and 250 meters above the water table. Thus, it is designed to isolate the waste from ground water. However, geologists have suggested that an earthquake could drive deep ground water upward, where it would become contaminated by the radioactive wastes. In addition, because radioactive decay produces heat, the wastes may be hot enough to convert the water to steam. Steam trapped underground could build up enough pressure to rupture containment vessels and cavern walls.

Other scientists are concerned that slow seepage of water from the surface will percolate through the repository site to the water table sometime between 9000 and 80,000 years from now. The lower end of this estimate is within the 10,000-year mandate for isolation. If the climate becomes appreciably wetter, which is possible over thousands of years, ground-water flow may accelerate and the water table may rise. If rocks beneath the site were fractured by an earthquake, then contaminated

ground water might disperse more rapidly than predicted. Furthermore, critics point out that construction of the repository will involve blasting and drilling, and these activities could fracture underlying rock, opening conduits for flowing water.

To stop development of the Yucca Mountain site, the state of Nevada refused to issue air quality permits to operate drilling rigs at the repository. In December 1995, U.S. Energy Secretary Hazel O'Leary announced that permanent storage for spent nuclear fuel cannot begin at Yucca Mountain before the year 2015. Supporters of the repository argue that we need nuclear power, and therefore as a society we must accept a certain level of risk. Furthermore, the Yucca Repository is safer than the temporary storage sites now being used. At present, 29,000 tons of radioactive waste lie in unstable temporary storage at nuclear power plants across the United States.

▶ 15.6 CAVERNS AND KARST TOPOGRAPHY

Just as streams erode valleys and form flood plains, ground water also creates landforms. Rainwater reacts with atmospheric carbon dioxide to produce a slightly acidic solution that is capable of dissolving limestone. This reaction is reversible: The dissolved ions can precipitate to form calcite again. (Recall that limestone is composed of calcite.)

CAVERNS

Most **caverns**, also called caves, form where slightly acidic water seeps through a crack in limestone, dissolving the rock and enlarging the crack. Most caverns form at or below the water table. If the water table drops, the chambers are opened to air. Caverns can be huge. The largest chamber in Carlsbad Caverns in New Mexico is taller than the U.S. Capitol building and is broad enough to accommodate 14 football fields. While caves form when limestone dissolves, most caves also contain features formed by deposition of calcite. Collectively, all mineral deposits formed by water in caves are called **speleothems**. Some are long, pointed structures hanging from the ceilings; others rise from the floors.

When a solution of water, dissolved calcite, and carbon dioxide percolates through the ground, it is under pressure from water in the cracks above it. If a drop of this solution seeps into the ceiling of a cavern, the pressure decreases suddenly because the drop comes in contact with the air. The high humidity of the cave prevents the water from evaporating rapidly, but the lowered pressure allows some of the carbon dioxide to escape as a gas. When the carbon dioxide escapes, the drop becomes less acidic. This decrease in acidity causes some of the dissolved calcite to precipitate as the water drips from the ceiling. Over time, a beautiful and intricate **stalactite** grows to hang icicle-like from the ceiling of the cave (Fig. 15–18).

Only a portion of the dissolved calcite precipitates as the drop seeps from the ceiling. When the drop falls to the floor, it spatters and releases more carbon dioxide. The acidity of the drop decreases further, and another minute amount of calcite precipitates. Thus, a **stalagmite** builds from the floor upward to complement the stalactite. Because stalagmites are formed by splashing water, they tend to be broader than stalactites. As the two features continue to grow, they may eventually join to form a **column**.

SINKHOLES

If the roof of a cavern collapses, a **sinkhole** forms on the Earth's surface. A sinkhole can also form as limestone dissolves from the surface downward (Fig. 15–19). A well-documented sinkhole formed in May 1981 in Winter Park, Florida. During the initial collapse, a three-bedroom house, half a swimming pool, and six Porsches in

Figure 15–18 Stalactites, stalagmites, and columns form as calcite precipitates in a limestone cavern. Luray Caverns, Virginia. (Breck P. Kent)

a dealer's lot fell into the underground cavern. Within a few days, the sinkhole had grown to about 200 meters wide and 50 meters deep and had devoured additional buildings and roads (Fig. 15–20).

Although sinkhole formation is a natural event, the problem can be intensified by human activities. The Winter Park sinkhole formed when the water table dropped, removing support for the ceiling of the cavern. The water table fell as a result of a severe drought augmented by excessive removal of ground water by humans.

KARST TOPOGRAPHY

Karst topography forms in broad regions underlain by limestone and other readily soluble rocks. Caverns and

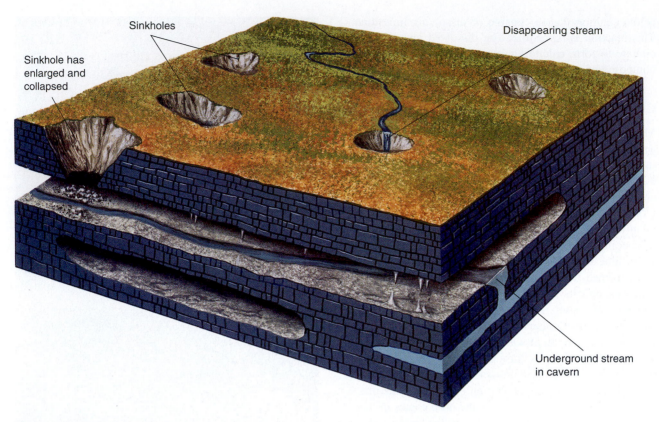

Figure 15–19 Sinkholes and caverns form in limestone. Streams commonly disappear into sinkholes and flow through the caverns to emerge elsewhere.

sinkholes are common features of karst topography. Surface streams often pour into sinkholes and disappear into caverns. In the area around Mammoth Caves in Kentucky, streams are given names such as Sinking Creek, an indication of their fate. The word **karst** is derived from a region in Croatia where this type of topog-

Figure 15–20 This sinkhole in Winter Park, Florida, collapsed suddenly in May 1981, swallowing several houses and a Porsche agency. (Wide World Photos/Associated Press)

raphy is well developed. Karst landscapes are found in many parts of the world.

▶ 15.7 HOT SPRINGS AND GEYSERS

At numerous locations throughout the world, hot water naturally flows to the surface to produce **hot springs**. Ground water can be heated in three different ways:

1. The Earth's temperature increases by about 30°C per kilometer in the upper portion of the crust. Therefore, if ground water descends through cracks to depths of 2 to 3 kilometers, it is heated by 60° to 90°. The hot water then rises because it is less dense than cold water. However, it is unusual for fissures to de-

scend so deep into the Earth, and this type of hot spring is uncommon.

2. In regions of recent volcanism, magma or hot igneous rock may remain near the surface and can heat ground water at relatively shallow depths. Hot springs heated in this way are common throughout western North and South America because these regions have been magmatically active in the recent past and remain so today. Shallow magma heats the hot springs and geysers of Yellowstone National Park.

3. Many hot springs have the odor of rotten eggs from small amounts of hydrogen sulfide (H_2S) dissolved in the hot water. The water in these springs is heated by chemical reactions. Sulfide minerals, such

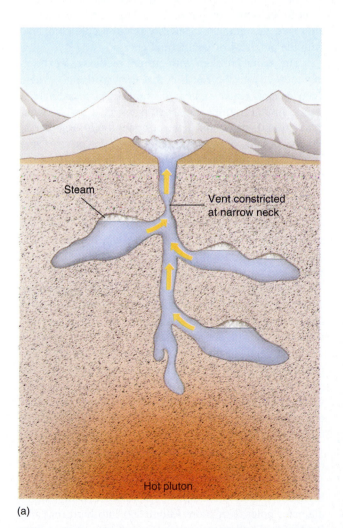

(a)

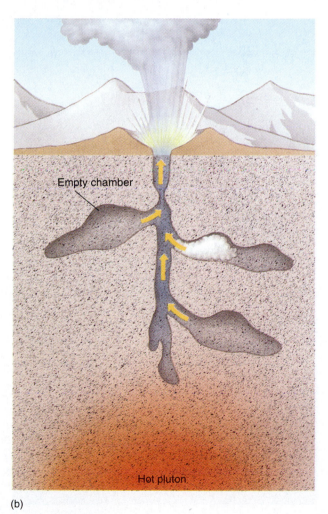

(b)

Figure 15–21 (a) Before a geyser erupts, ground water seeps into underground chambers and is heated by hot igneous rock. Foam constricts the geyser's neck, trapping steam and raising pressure. (b) When the pressure exceeds the strength of the blockage, the constriction blows out. Then the hot ground water flashes into vapor and the geyser erupts.

as pyrite (FeS_2), react chemically with water to produce hydrogen sulfide and heat. The hydrogen sulfide rises with the heated ground water and gives it the strong odor.

Most hot springs bubble gently to the surface from cracks in bedrock. However, **geysers** violently erupt hot water and steam. Geysers generally form over open cracks and channels in hot underground rock. Before a geyser erupts, ground water seeps into the cracks and is heated by the rock (Fig. 15–21). Gradually, steam bubbles form and start to rise, just as they do in a heated teakettle. If part of the channel is constricted, the bubbles accumulate and form a temporary barrier that allows the steam pressure below to increase. The rising pressure forces some of the bubbles upward past the constriction and short bursts of steam and water spurt from the geyser. This lowers the steam pressure at the constriction, causing the hot water to vaporize, blowing steam and hot water skyward (Fig. 15–22).

The most famous geyser in North America is Old Faithful in Yellowstone Park, which erupts on the average of once every 65 minutes. Old Faithful is not as regular as people like to believe; the intervals between eruptions vary from about 30 to 95 minutes.

▶ 15.8 GEOTHERMAL ENERGY

Hot ground water can be used to drive turbines and generate electricity, or it can be used directly to heat homes and other buildings. Energy extracted from the Earth's heat is called **geothermal energy**. In January 1995, 70 geothermal plants in California, Hawaii, Utah, and Nevada had a generating capacity of 2500 megawatts, enough to supply over one million people with electricity and equivalent to the power output of 2 1/2 large nuclear reactors. However, this amount of energy is minuscule compared with the potential of geothermal energy.

The major problem with current methods of extracting energy from the Earth is that they work only where deep ground water is heated naturally. Unfortunately, only a limited number of "wet" sites exist where abundant ground water and hot rock are found together at shallow depths. However, many "dry" sites occur where rising magma has heated rocks close to the surface but little ground water is available. Technology is being developed to harness energy from dry sites. For example, imported water can be circulated through wells drilled in dry, hot rock and then extracted and reused. Easily accessible dry geothermal sites in the United States have enough energy to supply all U.S. consumption for nearly 8000 years.

Scientists and engineers are developing methods for extracting energy from dry Earth heat at a pilot project at Fenton Hill, New Mexico. They drilled two separate wells side by side and pumped water down the injection well to a depth of about 4 kilometers (Fig. 15–23). The pump forces the water into hot, fractured granite at the bottom of the well and then into the extraction well, where it returns to the surface. The scientists have succeeded in pumping water into the well at 20°C and extracting it 12 hours later at 190°C.

With 1996 technology, construction of a dry geothermal plant is about 3.6 times as expensive as the cost of a new gas-fired generating plant. A large part of the high cost results from the expense of drilling large-diameter holes 2 or more kilometers into hard rock. As long as coal and petroleum prices are low, there is little incentive to develop geothermal projects to compete with coal- and oil-fired electric generating plants.

Figure 15–22 Shallow magma heats ground water, causing numerous geyser eruptions in Yellowstone National Park. (Corel Photos)

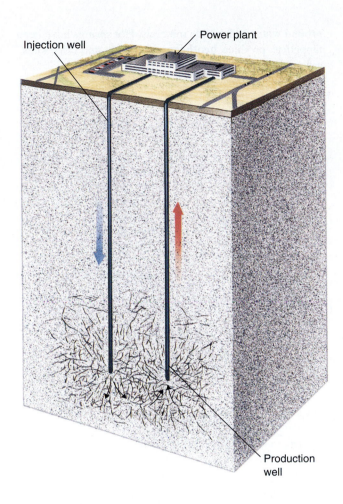

Injection well

Power plant

Production well

Figure 15–23 A schematic view of the Fenton Hill, New Mexico, dry geothermal energy plant.

SUMMARY

Much of the rain that falls on land seeps into soil and bedrock to become **ground water**. Ground water saturates the upper few kilometers of soil and bedrock to a level called the **water table**. **Porosity** is the proportion of rock or soil that consists of open space. **Permeability** is the ability of rock or soil to transmit water. An **aquifer** is a body of rock that can yield economically significant quantities of water. An aquifer is both porous and permeable.

Most ground water moves slowly, about 4 centimeters per day. In humid environments, the water table follows the topography of the land and ground water flows into **effluent** (or **gaining**) streams. In a desert, the water table may be below the stream bed, and the **influent** (or **losing**) stream seeps into the desert aquifer. **Springs** occur where the water table intersects the surface of the land. Dipping layers of permeable and impermeable rock can produce an **artesian aquifer**.

If water is withdrawn from a well faster than it can be replaced by the aquifer, a **cone of depression** forms. If rapid withdrawal continues, the water table falls. Other effects of excessive removal of ground water include **subsidence** of the land and **salt-water intrusion** near a seacoast.

Ground-water pollution can originate from both **point sources** and **non-point sources**. A pollutant normally spreads slowly into an aquifer as a contaminant **plume**. Because many pollutants persist in an aquifer and render the water unfit for use, expensive and difficult **remediation** efforts are commonly undertaken to cleanse a polluted aquifer.

Caverns form where ground water dissolves limestone. A **sinkhole** forms when the roof of a limestone cavern collapses. **Karst topography**, with numerous caves, sinkholes, and subterranean streams, is characteristic of limestone regions. **Hot springs** develop when hot ground water rises to the surface. Ground water can be heated by (1) the geothermal gradient, (2) shallow magma or a cooling pluton, or (3) chemical reactions between ground water and sulfide minerals. Hot springs have been tapped to produce **geothermal energy**, and "dry sites" are now being explored.

KEY WORDS

ground water 258	aquifer 261	salt-water intrusion 267	stalactite 273
porosity 260	effluent (gaining) stream 262	point and non-point source pollution 268	stalagmite 273
permeability 260	influent (losing) stream 262	plume (of contamination) 268	column 273
zone of saturation 261	perched water table 263	remediation (of contamination) 268	sinkhole 273
water table 261	artesian aquifer 263	bioremediation 271	karst topography 273
zone of aeration 261	artesian well 263	cavern 273	hot spring 275
capillary action 261	cone of depression 263	speleothem 273	geyser 276
soil moisture belt 261	subsidence 266		geothermal energy 276
well 261			
recharge 261			

REVIEW QUESTIONS

1. (a) Describe what happens to rain that falls in an arid region and then continues to fall heavily for several days. (b) Describe what happens to this water when no more rain falls for a month.

2. Describe the difference between porosity and permeability. Can soil or rock be porous and not permeable? Permeable but not porous?

3. (a) Draw a cross section showing the soil moisture belt, zone of saturation, water table, and zone of aeration. (b) Explain each of these terms.

4. What is the capillary fringe, and how does it form?

5. What is an aquifer, and how does water reach it?

6. What does water level in a well tell you about the location of the water table?

7. Explain why bedrock or regolith must be both porous and permeable to be an aquifer.

8. Compare the movement of ground water in an aquifer with that of water in a stream.

9. Why does the water table below hills usually remain elevated above the level of adjacent streams and lakes?

10. Explain how a temperate-climate stream continues to flow during a prolonged drought.

11. How does an artesian aquifer differ from a normal one? How does water from an artesian well rise without being pumped?

12. Describe three reasons why ground water is a particularly valuable resource.

13. Describe three problems that can arise from excessive use of ground water.

14. Why is ground-water depletion likely to have longer-lasting effects than depletion of a surface reservoir?

15. Explain how land subsides when ground water is depleted. If the removal of ground water is stopped, will the land rise again to its original level? Explain your answer.

16. Discuss the differences between point and non-point pollution sources. Give examples of each.

17. Describe the steps in cleaning polluted ground water.

18. Give two reasons why ground water purifies itself slowly.

19. Explain how remediation efforts can cleanse a contaminated aquifer.

20. Explain how caverns, speleothems, and sinkholes form.

21. What is karst topography? How can it be recognized? How does it form?

22. Describe three types of heat source for a hot spring.

DISCUSSION QUESTIONS

1. Imagine that you live on a hill 25 meters above a nearby stream. You drill a well 40 meters deep and do not reach water. Explain.

2. The ancient civilization of Mesopotamia fell after its agricultural system collapsed due to problems resulting from a failure of irrigation techniques. In his book *Cadillac Desert*, author Marc Reisner describes the transformation of the western United States from its natural semidesert condition to its modern agricultural wealth. However, he argues that this system, like others that preceded it, cannot be sustained indefinitely. Argue for or against Reisner's hypothesis.

3. How does a desert aquifer become recharged?

4. Would you expect to find a cavern in granite? Would you expect to discover a cavern in shale? Defend your answers.

5. Why can't stalactites or stalagmites form when a cavern is filled with water?

6. Discuss differences in problems of ground-water pollution in a region of karst topography in contrast to a region with a sandstone aquifer.

7. Imagine that a high percentage of newly born infants in a small town had birth defects and that you were called in to study the problem. Local health officials suspected that polluted ground water might be the cause. Outline a study project to determine whether or not polluted ground water was responsible. Discuss the role of the precautionary principle and cost–benefit analysis in any policy decisions you might make.

8. A neighbor's underground fuel tank develops a leak. You live downhill from the neighbor, and fuel oil floats to the surface in your yard. The tank was buried 3 meters deep; your well is 50 meters deep. Your neighbor's lawyer makes the following three statements: (a) The fuel oil in your yard may have come from a more distant source and may not be his client's responsibility; (b) even if his client's (your neighbor's) tank did leak, the tank capacity was only 1000 liters, so the total quantity of oil spilled would be relatively small; (c) because oil floats, any spilled fuel would be unlikely to contaminate your drinking water. Evaluate each of these statements.

Deserts

Deserts evoke an image of thirsty travelers crawling across lifeless sand dunes. This image accurately depicts some deserts, but not others. Many deserts are rocky and even mountainous, with colorful cliffs or peaks towering over plateaus and narrow canyons. Rains may punctuate the long hot summers, and in winter a thin layer of snow may cover the ground. Although plant life in deserts is not abundant, it is diverse. Cactus, sage, grasses, and other plants may dot the landscape. After a rainstorm, millions of flowers bloom.

A **desert** is any region that receives less than 25 centimeters (10 inches) of rain per year and consequently supports little or no vegetation.[1] Most deserts are surrounded by **semiarid** zones that receive 25 to 50 centimeters of rainfall, more moisture than a true desert but less than surrounding regions.

Deserts cover 25 percent of the Earth's land surface outside of the polar regions and make up a significant proportion of every continent. If you were to visit the great deserts of the Earth, you might be surprised by their geologic and topographic variety. You would see coastal deserts along the beaches of Chile, shifting dunes in the Sahara, deep red sandstone canyons in southern Utah, and stark granite mountains in Arizona. The world's deserts are similar to one another only in that they all receive scant rainfall.

Throughout human history, cultures have adapted to the low water and sparse vegetation of desert ecosystems. Traditionally, many desert societies were nomadic, taking advantage of resources where and when they were available. Other desert cultures developed irrigation systems to water crops close to rivers and wells. Modern irrigation systems have improved human adaptation to dry environments and enabled 13 percent of the world's population to live in deserts. Two thirds of the world's crude oil lies beneath the deserts of the Middle East, transforming some of the poorest nations of the world into the richest. In the future, vast arrays of solar cells may convert desert sunlight to electricity.

[1]The definition of *desert* is most directly linked to soil moisture and depends on temperature and amount of sunlight in addition to rainfall. Therefore, the 25-centimeter criterion is approximate.

Deserts can be warm, cold, sandy, or rocky. This photograph shows sand dunes in a southwest African desert.   (E. D. McKee/USGS)

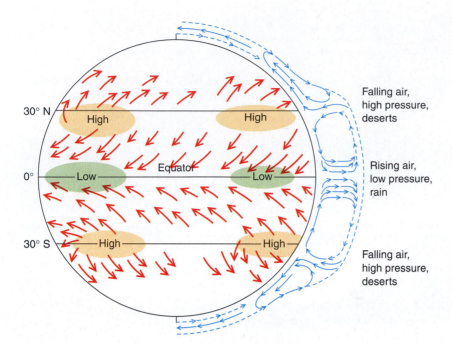

Figure 16–1 Falling air creates high pressure and deserts at 30° north and south latitudes. The arrows drawn inside the globe indicate surface winds. The arrows to the right show both vertical and horizontal movement of air on the surface and at higher elevations.

▶ 16.1 WHY DO DESERTS EXIST?

THE EFFECT OF LATITUDE

The Sun shines most directly near the equator, warming air near the Earth's surface. The air absorbs moisture from the equatorial oceans and rises because it is less dense than surrounding air. This warm, wet air cools as it ascends, and the water vapor condenses and falls as rain. For this reason, vast tropical rainforests grow near the equator. The rising equatorial air, which is now drier because of the loss of moisture, flows northward and southward at high altitudes. This air cools, becomes denser, and sinks back toward the Earth's surface at about 30° north and south latitudes (Fig. 16–1). As the air falls,

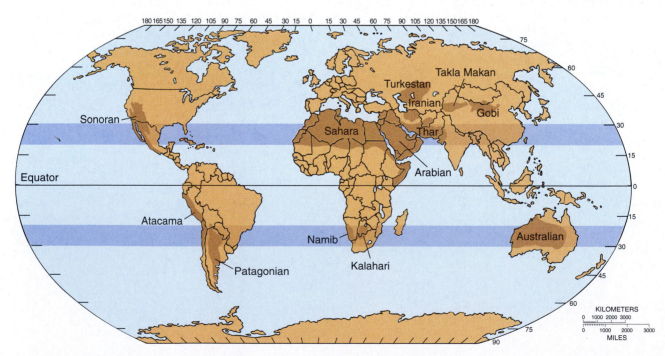

Figure 16–2 The major deserts of the world. Note the global concentration of deserts at 30° north and south latitudes. Most of the deserts are surrounded by semiarid lands.

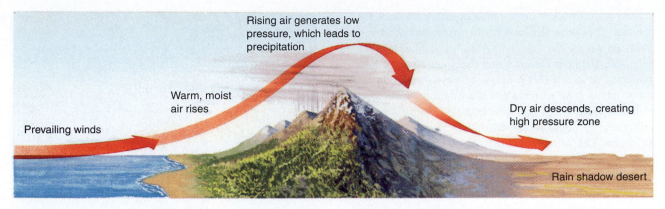

Figure 16–3 Formation of a rain shadow desert. Warm, moist air from the ocean rises. As it rises, it cools and water vapor condenses to form rain. The dry, descending air on the lee side absorbs moisture, forming a desert.

it is compressed and becomes warmer, which enables it to hold more water vapor. As a result, water evaporates from the land surface into the air. Because the sinking air absorbs water, the ground surface is dry and rainfall is infrequent. Thus, many of the world's largest deserts lie at about 30° north and south latitudes (Fig. 16–2).

EFFECT OF TOPOGRAPHY: RAIN SHADOW DESERTS

When moisture-laden air flows over a mountain range, it rises. As the air rises, it cools and its ability to hold water decreases. As a result, the water vapor condenses into rain or snow, which falls as precipitation on the wind-

ward side and on the crest of the range (Fig. 16–3). This cool air flows down the leeward (or downwind) side and sinks. As in the case of sinking air at 30° latitude, the air is compressed and warmed as it falls, and it has already lost much of its moisture. This warm, dry air creates an arid zone called a **rain shadow desert** on the leeward side of the range. Figure 16–4 shows the rainfall distribution in California. Note that the leeward valleys are much drier than the mountains to the west.

CONTINENTAL COASTLINES AND INTERIORS

Because most evaporation occurs over the oceans, one might expect that coastal areas would be moist and cli-

Figure 16–4 Rainfall patterns in the state of California. Note that rain shadow deserts lie east of the mountain ranges. Rainfall is reported in centimeters per year.

mates would become drier with increasing distance from the sea. This generalization is often true, but notable exceptions exist.

The Atacama Desert along the west coast of South America is so dry that portions of Peru and Chile have received no rainfall for a decade or more. Cool ocean currents flow along the west coast of South America. When the cool marine air encounters warm land, the air is heated. The warm, expanding air holds relatively little water vapor and absorbs moisture from the ground, creating a coastal desert.

The Gobi Desert is a broad, arid region in central Asia. The center of the Gobi lies at about 40°N latitude, and its eastern edge is a little more than 400 kilometers from the Yellow Sea. As a comparison, Pittsburgh, Pennsylvania, lies at about the same latitude and is 400 kilometers from the Atlantic Ocean. If latitude and distance from the ocean were the only factors, the two regions would have similar climates. However, the Gobi is a barren desert and western Pennsylvania receives enough rainfall to support forests and rich farmland. The Gobi is bounded by the Himalayas to the south and the Urals to the west, which shadow it from the prevailing winds. In contrast, winds carry abundant moisture from the Gulf of Mexico, the Great Lakes, and the Atlantic Ocean to western Pennsylvania.

Thus, in some regions deserts extend to the seashore and in other regions the interior of a continent is humid. The climate at any particular place on the Earth results from a combination of many factors. Latitude and proximity to the ocean are important, but the direction of prevailing winds, the direction and temperature of ocean currents, and the positions of mountain ranges also control climate.

▶ 16.2   DESERT LANDFORMS

Landforms in humid climates are commonly smooth and rounded because abundant rain promotes chemical weathering. In a desert, however, chemical weathering is slower because less rain falls. Instead, mechanical weathering predominates and intermittent desert streams undercut rock outcrops. As a result, steep cliffs and angular landforms dominate many deserts (Fig. 16–5).

DESERT STREAMS

Large rivers flow through some deserts. For example, the Colorado River crosses the arid southwestern United States, and the Nile River flows through North African deserts (Fig. 16–6). Desert rivers receive most of their water from wetter, mountainous areas bordering the arid lands.

Figure 16–5 Mechanical weathering, here at Castleton Tower in the Utah desert, produces angular landforms and steep cliffs in many deserts.

Figure 16–6 The Colorado River flows through the Utah desert. Most of the water flows from mountains to the east.

In a desert, the water table is often so low that water seeps out of the stream bed into the ground. As a result, many desert streams flow for only a short time after a rainstorm or during the spring, when winter snows are melting. A stream bed that is dry for most of the year is called a **wash** (Fig. 16–7).

DESERT LAKES

While most lakes in wetter environments intersect the water table and are fed, in part, by ground water, many desert lake beds lie above the water table. During the wet season, water enters a desert lake bed by stream flow and to a lesser extent by direct precipitation. Some desert lakes are drained by outflowing streams, while many lose water only by evaporation and seepage. During the dry season, water loss may be so great that the lake dries up completely. An intermittent desert lake is called a **playa lake**, and the dry lake bed is called a **playa** (Fig. 16–8).

Recall from Chapters 6 and 14 that water dissolves ions from rock and soil. When this slightly salty water evaporates, the ions precipitate to deposit salts and other evaporite minerals on the playa. Over many years, economically valuable evaporite deposits, such as those of Death Valley, may accumulate (Fig. 16–9).

FLASH FLOODS AND DEBRIS FLOWS

Bedrock or tightly compacted soil covers the surface of many deserts, and little vegetation is present to absorb moisture. As a result, rainwater runs over the surface to collect in gullies and washes. During a rainstorm, a dry stream bed may fill with water so rapidly that a **flash flood** occurs. Occasionally, novices to desert camping pitch their tents in a wash, where they find soft sand to sleep on and shelter from the wind. However, if a thunderstorm occurs upstream during the night, a flash flood may fill the wash with a wall of water mixed with rocks and boulders, creating disaster for the campers. By mid-

(a)

(b)

Figure 16–7 Courthouse wash, (a) in the spring when rain and melting snow fill the channel with water and (b) in mid-summer, when the creek bed is a dry wash.

Figure 16–8 Mud cracks pattern the floor of a playa in Utah.

Figure 16–10 Alluvial fans form where steep mountain streams deposit sediment where they enter a valley. This photograph shows a fan in Death Valley.

morning of the next day, the wash may contain only a tiny trickle, and within 24 hours it may be completely dry again.

When rainfall is unusually heavy and prolonged, the desert soil itself may become saturated enough to flow. Viscous wet mud flows downslope in a debris flow that carries boulders and anything else in its path. Some of the most expensive homes in Phoenix, Arizona, and other desert cities are built on the slopes of mountains, where they have good views but are prone to debris flows during wet years.

PEDIMENTS AND BAJADAS

When a steep, flooding mountain stream empties into a flat valley, the water slows abruptly and deposits most of

its sediment at the mountain front, forming an alluvial fan. Although fans form in all climates, they are particularly conspicuous in deserts (Fig. 16–10). A large fan may be several kilometers across and rise a few hundred meters above the surrounding valley floor.

If the mouths of several canyons are spaced only a few kilometers apart, the alluvial fans extending from each canyon may merge. A **bajada** is a broad depositional surface formed by merging alluvial fans and extending into the center of the desert valley. Typically, the alluvial fans merge incompletely, forming an undulating surface that may follow the mountain front for tens of kilometers. The sediment that forms the bajada may fill the valley to a depth of several thousand meters.

A **pediment** is a nearly flat, gently sloping surface eroded into bedrock. Pediments commonly form along the front of desert mountains. The bedrock surface of a pediment is covered with a thin veneer of gravel that is in the process of being transported from the mountains, across the pediment to the bajada (Figs. 16–11 and 16–12).

Figure 16–9 Borax and other valuable minerals are abundant in the evaporite deposits of Death Valley. Mule teams hauled the ore from the valley in the 1800s. (U.S. Borax)

Figure 16–11 The bajada in the foreground merges with the gently sloping pediment to form a continuous surface east of Reno, Nevada.

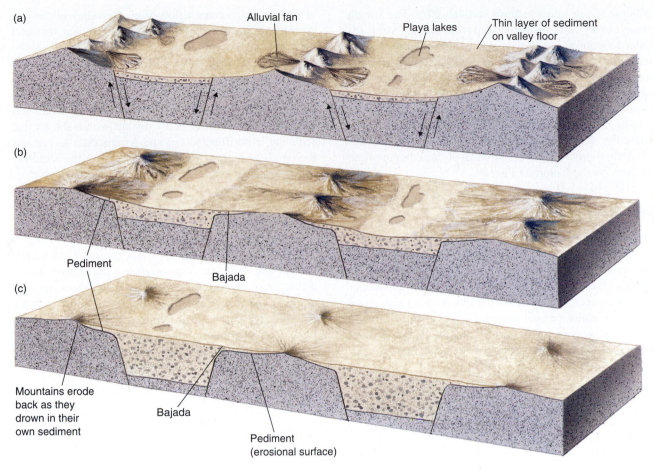

Figure 16–12 One scenario for the formation of bajadas and pediments. (a) The mountains and valleys were formed by block faulting. Desert streams deposit sediment to form alluvial fans. (b) Alluvial fans merge, creating a broad undulating surface called a bajada. (c) The mountain range erodes backward, creating a pediment along the mountain front.

Several different hypotheses have been suggested to explain pediment development. According to one, the streams flowing from desert mountains carry so much sediment that they develop braided channels that shift back and forth across the mountain front. Slowly, the streams erode a flat surface into bedrock. With time, the streams cut the pediment backward into the mountains. Another hypothesis suggests that pediments formed when the climate was wetter than it is today. The greater stream discharge eroded the mountain front rather than depositing alluvial fans.

It is commonly difficult to distinguish a pediment from a bajada because they form a continuous surface from the mountain front to the center of the valley. In addition, both are slightly concave upward and are covered with sand and gravel. To tell the difference, you would have to dig or drill a hole. If you were on a pediment, you would strike bedrock after only a few meters, but on a bajada, bedrock may be buried beneath hundreds or even thousands of meters of sediment.

▶ 16.3 TWO DESERT LANDSCAPES IN THE UNITED STATES

THE COLORADO PLATEAU

The Colorado Plateau covers a broad region across portions of Utah, Colorado, Arizona, and New Mexico. Through Earth history this region has been alternately covered by shallow seas, lakes, and deserts. Sediment accumulated, sedimentary rocks formed, and the land was later uplifted without much faulting or deformation, forming the Colorado Plateau. The Colorado River cut through the rock as it rose, to form a 1.6-kilometer-deep canyon, called Grand Canyon. Grand Canyon is intersected by smaller canyons formed by tributary streams.

A stream forms a canyon by downcutting and transporting sediment downslope. If the stream reaches a resistant rock layer, it erodes laterally, widening the canyon. Vertical joints occur in many of the rocks of the Colorado Plateau. As lateral erosion undercuts the cliffs, the walls collapse along joints to form vertical cliffs. The stream continuously removes the eroded rock, leaving steep angular mountains called mesas and buttes. A **plateau** is a large elevated area of fairly flat land. The term plateau is used for regions as large as the Colorado Plateau as well as for smaller elevated flat surfaces. A **mesa** is smaller than a plateau and is a flat-topped mountain shaped like a table. A **butte** is also a flat-topped mountain characterized by steep cliff faces and is smaller and more tower-like than a mesa (Fig. 16–13). Each of these features can be seen on the Colorado Plateau.

DEATH VALLEY

Death Valley is a deep depression in southeastern California, with a maximum depth of 82 meters below sea level (Fig. 16–14). It is a classic rain shadow desert and receives only a scant 5 centimeters of rainfall per year. However, the mountains to the west receive more abundant moisture, and during the winter rainy season, streams flow from the mountains into the Valley, eroding the rock to form broad pediments. Because Death Valley

is a basin, rivers cannot flow from the valley to the sea. As a result, sediment collects to form alluvial fans and bajadas along the mountain front and sand dunes on the valley floor. Stream water collects in broad playa lakes that dry up under the hot summer sun.

Like Death Valley, many desert regions in the American West have no through-flowing drainage. Because intermittent streams dry up within the basins, sediment is not flushed out and accumulates to become thousands of meters thick. In some cases, the sediment fills the valleys nearly to the tops of the mountains.

▶ 16.4 WIND IN DESERTS

When wind blows through a forest or across a prairie, the trees or grasses protect the soil from wind erosion. In addition, rain accompanies most windstorms in wet climates; the water dampens the soil and binds particles together. Therefore, little wind erosion occurs. In contrast, a desert commonly has little or no vegetation and rainfall, so wind erodes bare, unprotected soil. One can

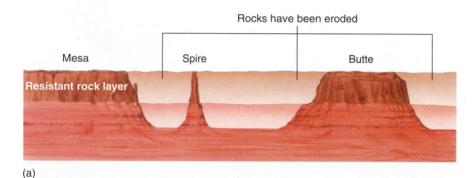

(a)

(b)

Figure 16–13 (a) Spires and buttes form when streams reach a temporary base level and erode laterally. (b) A landscape in Monument Valley, Arizona.

Figure 16–14   Sediment eroded from surrounding mountains is slowly filling Death Valley.

hardly think about deserts without imagining a hide-behind-your-camel-type sandstorm.

WIND EROSION

Wind erosion, called **deflation**, is a selective process. Because air is much less dense than water, wind moves only small particles, mainly silt and sand. (Clay particles usually stick together, and consequently wind does not erode clay effectively.) Imagine bare soil containing silt, sand, pebbles, and cobbles (Fig. 16–15a). When wind blows, it removes only the silt and sand, leaving the pebbles and cobbles as a continuous cover of stones called **desert pavement** (Fig. 16–15b). Desert pavement prevents the wind from eroding additional sand and silt, even though they may be abundant beneath the layer of stones. Thus, most deserts are rocky and covered with gravel, and sandy deserts are relatively rare.

TRANSPORT AND ABRASION

Because sand grains are relatively heavy, wind blows sand near the surface (usually less than 1 meter above the surface) and carries it only a short distance. In a windstorm, the sand grains bounce over the ground by saltation. (Recall from Chapter 14 that sand also moves by saltation in a stream bed.) In contrast, wind carries fine silt in suspension. Skiers in the Alps commonly encounter a silty surface on the snow, blown from the Sahara Desert and carried across the Mediterranean Sea.

Windblown sand is abrasive and erodes bedrock. Because wind carries sand close to the surface, wind erosion occurs near ground level. If you see a tall desert pinnacle topped by a delicately perched cap, you know that the top was not carved by wind erosion because it is too

Wind removes
surface sand

Formation of desert
pavement complete–
No further wind
erosion

(a)

(b)

Figure 16–15   (a) Wind erodes silt and sand but leaves larger rocks behind to form desert pavement. (b) Desert pavement is a continuous cover of stones left behind when wind blows silt and sand away.

Figure 16–16 Wind abrasion (here in Lago Poopo, Bolivia) selectively eroded the base of this rock because windblown sand moves mostly near the surface.

high above the ground. However, if the base of a pinnacle is sculpted, wind may be the responsible agent (Fig. 16–16). (Salt cracking at ground level also contributes to the weathering of desert rocks.)

Cobbles and boulders lying on the desert surface often have faces worn flat by windblown sand. Such rocks, called **ventifacts**, often have two or three flat faces because of changing wind directions (Fig. 16–17).

DUNES

A **dune** is a mound or ridge of wind-deposited sand (Fig. 16–18). As explained earlier, wind removes sand from the surface in many deserts, leaving behind a rocky desert

Figure 16–18 Dunes near Lago Poopo, Bolivia.

pavement. The wind then deposits the sand in a topographic depression or other place where the wind slows down. Approximately 80 percent of the world's desert area is rocky and only 20 percent is covered by dunes. Although some desert dune fields cover only a few square kilometers, the largest is the Rub Al Khali (Empty Quarter) in Arabia, which covers 560,000 square kilometers, larger than the state of California.

Dunes also form where glaciers have recently melted and along sandy coastlines. A glacier deposits large quan-

Figure 16–17 A ventifact shows flat faces scoured by windblown sand. (Hubbard Scientific Co.)

Figure 16–19 Many blowouts are only a meter or two deep, but some can be much larger. The grassy surface on top of the hummock is the level of the prairie before wind erosion occurred. (Courtesy of N. H. Darton, USGS)

tities of bare, unvegetated sediment. A sandy beach is commonly unvegetated because sea salt prevents plant growth. Thus, both of these environments contain the essentials for dune formation: an abundant supply of sand and a windy environment with sparse vegetation.

Dunes form when wind erodes sand from one location and deposits it nearby. A saucer or trough-shaped hollow formed by wind erosion in sand is called a **blowout**. In the 1930s, intense, dry winds eroded large areas of the Great Plains and created the Dust Bowl. Deflation formed tens of thousands of blowouts, many of which remain today. Some are small, measuring only 1 meter deep and 2 or 3 meters across, but others are much larger (Fig. 16–19). One of the deepest blowouts in the world is the Qattara Depression in western Egypt. It is more than 100 meters deep and 10 kilometers in diameter. Ultimately, the lower limit for a blowout is the water table. If the bottom of the depression reaches moist soil near the water table, where water binds the sand grains, wind erosion is no longer effective.

If wind-transported sand moves over a rock, a natural depression, or a small clump of vegetation, the wind slows down in the downwind, or **lee**, side of the obstacle. Sand settles out in this protected zone. The growing mound of sand creates a larger windbreak, and more sand accumulates, forming a dune. Dunes commonly grow to heights of 30 to 100 meters, and some giants exceed 500 meters. In places, they are tens or even hundreds of kilometers long.

Most dunes are asymmetrical. Wind erodes sand from the windward side of a dune, and then the sand slides down the sheltered leeward side. In this way, dunes migrate in the downwind direction (Fig. 16–20). The leeward face of a dune is called the **slip face**. Typically, the slip face is about twice as steep as the windward face.

Migrating dunes overrun buildings and highways. For example, near the town of Winnemucca, Nevada, dunes advance across U.S. Highway 95 several times a year. Highway crews must remove as much as 4000 cubic meters of sand to reopen the road. Engineers often attempt to stabilize dunes in inhabited areas. One method is to plant vegetation to reduce deflation and stop dune migration. The main problem with this approach is that desert dunes commonly form in regions that are too dry to support vegetation. Another solution is to build artificial windbreaks to create dunes in places where they do the least harm. For example, a fence traps blowing sand and forms a dune, thereby protecting areas downwind. Fencing is a temporary solution, however, because eventually the dune covers the fence and resumes its migration. In Saudi Arabia, dunes are sometimes stabilized by covering them with tarry wastes from petroleum refining.

Fossil Dunes

When dunes are buried by younger sediment and lithified, the resulting sandstone retains the original sedimentary structures of the dunes. Figure 16–21 shows a rock face in Zion National Park in Utah. The sloping sedimentary layering is not evidence of tectonic tilting but is the original steeply dipping layers of the dune slip face. The beds dip in the direction in which the wind was blowing when it deposited the sand. Notice that the planes dip in different directions, indicating changes in wind direction. The layering is an example of **cross-bedding**, described in Chapter 7.

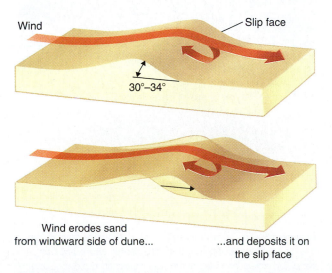

Figure 16–20 The formation and migration of a sand dune.

Figure 16–21 Cross-bedded sandstone in Zion National Park preserves the sedimentary bedding of ancient sand dunes.

Types of Sand Dunes

Wind speed and sand supply control the shapes and orientation of dunes (Fig. 16–22). **Barchan dunes** form in rocky deserts with little sand. The center of the dune grows higher than the edges (Fig. 16–23a). When the dune migrates, the edges move faster because there is less sand to transport. The resulting barchan dune is crescent shaped with its tips pointing downwind (Fig. 16–23b). Barchan dunes are not connected to one another, but instead migrate independently. In a rocky desert, barchan dunes cover only a small portion of the land; the remainder is bedrock or desert pavement.

If sand is plentiful and evenly dispersed, it accumulates in long ridges called **transverse dunes** that align perpendicular to the prevailing wind (Figs. 16–22b and 16–24). They are shaped like sand ripples, although they are much larger.

If desert vegetation is plentiful, the wind may form a blowout in a bare area among the desert plants. As sand is carried out of the blowout, it accumulates in a **parabolic dune**, the tips of which are anchored by plants on

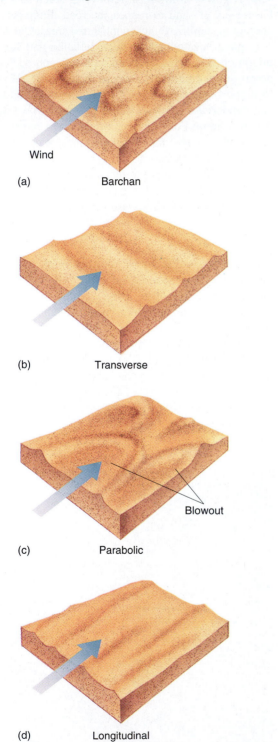

(a) Barchan

(b) Transverse

(c) Parabolic

Blowout

(d) Longitudinal

Figure 16–22 Sand dunes take on different shapes depending on sand supply and variations in wind direction. (a) Barchan dunes. (b) Transverse dunes. (c) Parabolic dunes. (d) Longitudinal dunes.

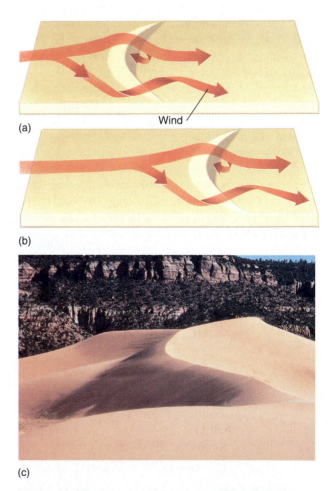

(a)

Wind

(b)

(c)

Figure 16–23 (a and b) When sand supply is limited, the tips of a dune travel faster than the center and point downwind, forming a barchan dune. (c) A barchan dune in Coral Pinks, Utah.

Figure 16–24 Transverse dunes form perpendicularly to the prevailing wind direction in regions with abundant sand, such as this part of the Oregon coast.   (Galen Rowell/Mountain Light)

Figure 16–26 Longitudinal dunes on the Oregon coast form where the wind is erratic and the sand supply is limited. (Albert Copley/Visuals Unlimited)

each side of the blowout (Figs. 16–22c and 16–25). A parabolic dune is similar in shape to a barchan dune, except that the tips of the parabolic dune point into the wind. Parabolic dunes are common in moist semidesert regions and along seacoasts.

If the wind direction is erratic but prevails from the same general quadrant of the compass and the supply of sand is limited, then long, straight **longitudinal dunes** form parallel to the prevailing wind direction (Figs.

16–22d and 16–26). In portions of the Sahara Desert, longitudinal dunes reach 100 to 200 meters in height and are as much as 100 kilometers long.

LOESS

Wind can carry silt for hundreds or even thousands of kilometers and then deposit it as **loess** (pronounced luss). Loess is porous, uniform, and typically lacks layering. Often the angular silt particles interlock. As a result, even though the loess is not cemented, it typically forms vertical cliffs and bluffs (Fig. 16–27).

The largest loess deposits in the world, found in central China, cover 800,000 square kilometers and are more than 300 meters thick. The silt was blown from the Gobi and the Takla Makan deserts of central Asia. The parti-

Figure 16–25 A parabolic dune in Death Valley, California, has formed where wind blows sand from a blowout, and grass or shrubs anchor the dune tips.   (Martin G. Miller/Visuals Unlimited)

Figure 16–27 Villagers in Askole, Pakistan, have dug caves in these vertical loess cliffs.

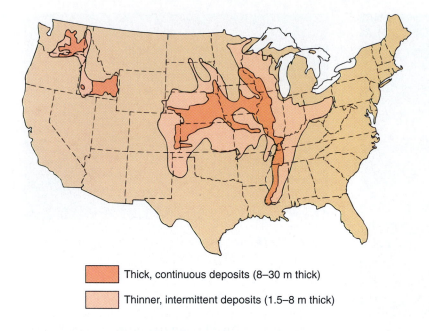

Thick, continuous deposits (8–30 m thick)

Thinner, intermittent deposits (1.5–8 m thick)

Figure 16–28 Loess deposits in the United States.

cles interlock so effectively that people have dug caves into the loess cliffs to make their homes. However, in 1920 a great earthquake caused the cave system to collapse, burying and killing an estimated 100,000 people.

Large loess deposits accumulated in North America during the Pleistocene Ice Age, when continental ice sheets ground bedrock into silt. Streams carried this fine sediment from the melting glaciers and deposited it in vast plains. These zones were cold, windy, and devoid of vegetation, and wind easily picked up and transported the silt, depositing thick layers of loess as far south as Vicksburg, Mississippi.

Loess deposits in the United States range from about 1.5 meters to 30 meters thick (Fig. 16–28). Soils formed on loess are generally fertile and make good farmland. Much of the rich soil of the central plains of the United States and eastern Washington State formed on loess.

SUMMARY

Deserts have an annual precipitation of less than 25 centimeters (10 inches). The world's largest deserts occur near 30° north and south latitudes, where warm, dry, descending air absorbs moisture from the land. Deserts also occur in rain shadows of mountains, continental interiors, and coastal regions adjacent to cold ocean currents.

Chemical weathering is slow in deserts because water is scarce, and mechanical processes may form angular landscapes. Desert streams are often dry for much of the year but may develop **flash floods** when rainfall occurs. **Playa lakes** are desert lakes that dry up periodically, leaving abandoned lake beds called **playas**. Alluvial fans are common in desert environments. A **bajada** is a broad depositional surface formed by merging alluvial fans. A **pediment** is a planar erosional surface that may lie at the base of a mountain front in arid and semiarid regions.

The Colorado Plateau desert is distinguished by through-flowing streams. The water carries sediment away, forming canyons. The plateaus have been eroded to form **mesas** and **buttes**. Death Valley has no external drainage, and as a result, the valley is filling with sediment eroded from the surrounding mountains.

Deflation is erosion by wind. Silt and sand are removed selectively, leaving larger stones on the surface and creating **desert pavement**. Sand grains are carried short distances and a meter or less above the ground by **saltation**, but silt can be transported great distances at higher elevations. Wind erosion forms **blowouts**. Windblown particles are abrasive, but because the heaviest grains travel close to the surface, abrasion occurs mainly near ground level.

A **dune** is a mound or ridge of wind-deposited sand. Most dunes are asymmetrical, with gently sloping windward sides and steeper **slip faces** on the lee sides. Dunes migrate. The various types of dunes include **barchan dunes**, **transverse dunes**, **longitudinal dunes**, and **parabolic dunes**. Wind-deposited silt is called **loess**.

KEY WORDS

desert *280*
rain shadow desert *283*
wash *285*
playa lake *285*
playa *285*
flash flood *285*

bajada *286*
pediment *286*
plateau *288*
mesa *288*
butte *288*
deflation *289*

desert pavement *289*
ventifact *290*
dune *290*
blowout *291*
slip face *291*
cross-bedding *291*

barchan dune *292*
transverse dune *292*
parabolic dune *292*
longitudinal dune *293*
loess *293*

REVIEW QUESTIONS

1. Why are many deserts concentrated along zones at 30° latitude in both the Northern and Southern Hemispheres?

2. List three conditions that produce deserts.

3. Explain why angular topography is common in desert regions.

4. Why do flash floods and debris flows occur in deserts?

5. Why are alluvial fans more prominent in deserts than in humid environments?

6. Compare and contrast floods in deserts with those in more humid environments.

7. Compare and contrast pediments and bajadas.

8. Why is wind erosion more prominent in desert environments than it is in humid regions?

9. Describe the formation of desert pavement.

10. Describe the evolution and shape of a dune.

11. Describe the differences among barchan dunes, transverse dunes, parabolic dunes, and longitudinal dunes. Under what conditions does each type of dune form?

12. Compare and contrast desert plateaus, mesas, and buttes. Describe the formation of each.

13. Compare the effects of stream erosion and deposition in the Colorado Plateau and Death Valley.

DISCUSSION QUESTIONS

1. Coastal regions boast some of the wettest and some of the driest environments on Earth. Briefly outline the climatological conditions that produce coastal rainforests versus coastal deserts.

2. Explain why soil moisture content might be more useful than total rainfall in defining a desert. How could one region have a higher soil moisture content and lower rainfall than another region?

3. Discuss two types of tectonic change that could produce deserts in previously humid environments.

4. Imagine that you lived on a planet in a distant solar system. You had no prior information on the topography or climate of the Earth and were designing an unmanned spacecraft to land on Earth. The spacecraft had arms that could reach out a few meters from the landing site to collect material for chemical analysis. It also had instruments to measure the immediate meteorological conditions and cameras that could focus on anything within a range of 100 meters. The batteries on your radio transmitter had a life expectancy of two weeks. The spacecraft landed and you began to receive data. What information would convince you that the spacecraft had landed in a desert?

5. Deserts are defined as areas with low rainfall, yet water is an active agent of erosion in desert landscapes. Explain this apparent contradiction.

6. Compare and contrast erosion, transport, and deposition by wind with erosion and deposition by streams.

7. Imagine that someone told you that an alluvial fan had formed by wind deposition. What evidence would you look for to test this statement?

8. What type of dunes form under the following conditions? (a) Relatively high vegetation cover, sand supply, and wind strength. (b) Low vegetation and sand supply.

9. What type of environment would produce fossilized seashells embedded in lithified sand dunes?

Glaciers and Ice Ages

We often think of glaciers as features of high mountains and the frozen polar regions. Yet anyone who lives in the northern third of the United States is familiar with glacial landscapes. Many low, rounded hills of upper New York State, Wisconsin, and Minnesota are piles of gravel deposited by great ice sheets as they melted. In addition, people in this region swim and fish in lakes that were formed by recent glaciers.

Glaciers have advanced and retreated at least five times during the past 2 million years. Before the most recent major glacial advance, beginning about 100,000 years ago, the world was free of ice except for the polar ice caps of Antarctica and Greenland. Then, in a relatively short time—perhaps only a few thousand years—the Earth's climate cooled by a few degrees. As winter snow failed to melt in summer, the polar ice caps grew and spread into lower latitudes. At the same time, glaciers formed near the summits of high mountains, even near the equator. They flowed down mountain valleys into nearby lowlands. When the glaciers reached their maximum size 18,000 years ago, they covered one third of the Earth's continents. About 15,000 years ago, Earth's climate warmed again and the glaciers melted rapidly.

Although 18,000 years is a long time when compared with a single human lifetime, it is a blink of an eye in geologic time. In fact, humans lived through the most recent glaciation. In southwestern France and northern Spain, humans developed sophisticated spearheads and carved body ornaments between 40,000 and 30,000 years ago. People first began experimenting with agriculture about 10,000 years ago.

Alpine glaciers sculpt mountain landscapes, Bugaboo Mountains, British Columbia.

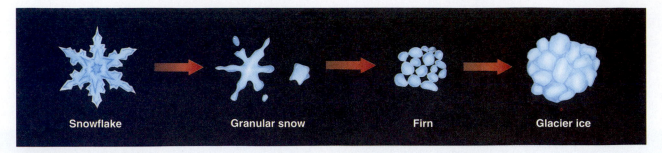

Figure 17–1 The change of newly fallen snow through several stages to form glacier ice.

▶ 17.1 FORMATION OF GLACIERS

In most temperate regions, winter snow melts in spring and summer. However, in certain cold, wet environments, only a portion of the winter snow melts and the remainder accumulates year after year. During summer, snow crystals become rounded as the snowpack is compressed and alternately warmed during daytime and cooled at night. Temperature changes and compaction make the snow denser. If snow survives through one summer, it converts to rounded ice grains called **firn** (Fig. 17–1). Mountaineers like firn because the sharp points of their ice axes and crampons sink into it easily and hold firmly. If firn is buried deeper in the snowpack, it converts to closely packed ice crystals.

A **glacier** is a massive, long-lasting, moving mass of compacted snow and ice. Glaciers form only on land, wherever the amount of snow that falls in winter exceeds the amount that melts in summer. Glaciers in mountain regions flow downhill. Glaciers on level land flow outward under their own weight, just as cold honey poured onto a tabletop spreads outward.

Glaciers form in two environments. Alpine glaciers form at all latitudes on high, snowy mountains. Continental ice sheets form at all elevations in the cold polar regions.

ALPINE GLACIERS

Mountains are generally colder and wetter than adjacent lowlands. Near the mountain summits, winter snowfall is deep and summers are short and cool. These conditions create **alpine glaciers** (Fig. 17–2). Alpine glaciers exist on every continent—in the Arctic and Antarctica, in temperate regions, and in the tropics. Glaciers cover the summits of Mount Kenya in Africa and Mount Cayambe in South America, even though both peaks are near the equator.

Some alpine glaciers flow great distances from the peaks into lowland valleys. For example, the Kahiltna Glacier, which flows down the southwest side of Denali (Mount McKinley) in Alaska, is about 65 kilometers long, 12 kilometers across at its widest point, and about 700 meters thick. Although most alpine glaciers are smaller than the Kahiltna, some are larger.

The growth of an alpine glacier depends on both temperature and precipitation. The average annual temperature in the state of Washington is warmer than that

Figure 17–2 This alpine glacier flows around granite peaks in British Columbia, Canada.

Figure 17–3 The Beardmore glacier is a portion of the Antarctic ice sheet. (Kevin Killelea)

in Montana, yet alpine glaciers in Washington are larger and flow to lower elevations than those in Montana. Winter storms buffet Washington from the moisture-laden Pacific. Consequently, Washington's mountains receive such heavy winter snowfall that even though summer melting is rapid, large quantities of snow accumulate every year. In much drier Montana, snowfall is light enough that most of it melts in the summer, and thus Montana's mountains have no or only very small glaciers.

CONTINENTAL GLACIERS

Winters are so long and cold and summers so short and cool in polar regions that glaciers cover most of the land regardless of its elevation. An **ice sheet**, or **continental glacier**, covers an area of 50,000 square kilometers or more (Fig. 17–3).[1] The ice spreads outward in all directions under its own weight.

Today, the Earth has only two ice sheets, one in Greenland and the other in Antarctica. These two ice sheets contain 99 percent of the world's ice and about three fourths of the Earth's fresh water. The Greenland sheet is more than 2.7 kilometers thick in places and covers 1.8 million square kilometers. Yet it is small compared with the Antarctic ice sheet, which blankets about 13 million square kilometers, almost 1.5 times the size of the United States. The Antarctic ice sheet covers entire mountain ranges, and the mountains that rise above its surface are islands of rock in a sea of ice. If the Antarctic ice sheet melted, the meltwater would create a

---

[1] A continental glacier with an area of less than 50,000 square kilometers is called an ice cap.

river the size of the Mississippi that would flow for 50,000 years.

Whereas the South Pole lies in the interior of the Antarctic continent, the North Pole is situated in the Arctic Ocean. Only a few meters of ice freeze on the relatively warm sea surface, and the ice fractures and drifts with the currents. As a result, no ice sheet exists at the North Pole.

▶ 17.2 GLACIAL MOVEMENT

Imagine that you set two poles in dry ground on opposite sides of a glacier, and a third pole in the ice to form a straight line with the other two. After a few months, the center pole would have moved downslope, and the three poles would form a triangle. This simple experiment shows us that the glacier moved downhill.

Rates of glacial movement vary with slope steepness, precipitation, and air temperature. In the coastal ranges of Alaska, where annual precipitation is high and average temperature is relatively high (for glaciers), some glaciers typically move 15 centimeters to a meter a day. In contrast, in the interior of Alaska where conditions are generally cold and dry, glaciers move only a few centimeters a day. At these rates, ice flows the length of an alpine glacier in a few hundred to a few thousand years. In some instances, a glacier may *surge* at a speed of 10 to 100 meters per day.

Glaciers move by two mechanisms: basal slip and plastic flow. In **basal slip**, the entire glacier slides over bedrock in the same way that a bar of soap slides down a tilted board. Just as wet soap slides more easily than dry soap, an accumulation of water between bedrock and the base of a glacier accelerates basal slip.

Several factors cause water to accumulate near the base of a glacier. The Earth's heat melts ice near bedrock. Friction from glacial movement also generates heat. Water occupies less volume than an equal amount of ice. As a result, pressure from the weight of overlying ice favors melting. Finally, during the summer, water melted from the surface of a glacier may seep downward to its base.

A glacier also moves by **plastic flow**, in which it deforms as a viscous fluid. Plastic flow is demonstrated by two experiments. In one, scientists set a line of poles in the ice (Fig. 17–4). After a few years, the ice moves downslope so that the poles form a U-shaped array. This experiment shows us that the center of the glacier moves faster than the edges. Frictional resistance with the valley walls slows movement along the edges and glacial ice flows plastically, allowing the center to move faster than the sides.

Figure 17–4 If a line of pipes is set into the ice, the pipes near the center of the glaciers move downslope faster than those near the margins.

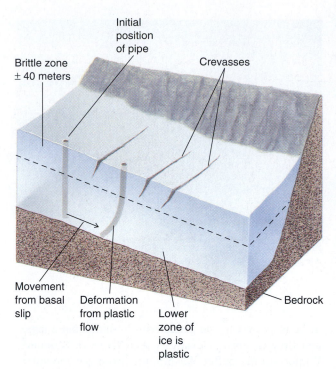

Figure 17–5 In this experiment, a pipe is driven through a glacier until it reaches bedrock. The entire pipe moves downslope. The pipe becomes curved because the center of the glacier deforms plastically and moves faster than the base. The top 40 meters of ice are brittle, and the pipe remains straight in this section.

As another experiment, imagine driving a straight but flexible pipe downward into the glacier to study the flow of ice at depth (Fig. 17–5). At a later date, you notice that not only has the entire pipe moved downslope, but the pipe has become curved. At the surface of a glacier, the ice acts as a brittle solid, like an ice cube or the ice found on the surface of a lake. In contrast, at depths greater than about 40 meters, the pressure is sufficient to allow ice to deform in a plastic manner. The curvature in the pipe shows that the ice has moved plastically and that middle levels of the glacier moved faster than the lower part. The base of the glacier is slowed by friction against bedrock, so it moves more slowly than the plastic portion above it.

The relative rates of basal slip and plastic flow depend on the steepness of the bedrock underlying the glacier and on the thickness of the ice. A small alpine glacier on steep terrain moves mostly by basal slip. In contrast, the bedrock beneath portions of the Antarctica and Greenland ice sheets is relatively level, so the ice cannot slide downslope. Thus, these continental glaciers are huge plastic masses of ice (with a thin rigid cap) that ooze outward mainly under the forces created by their own weight.

When a glacier flows over uneven bedrock, the deeper plastic ice bends and flows over bumps, stretch-ing the brittle upper layer of ice so that it cracks, forming **crevasses** (Fig. 17–6). Crevasses form only in the brittle upper 40 meters of a glacier, not in the lower plastic zone. Crevasses open and close slowly as a glacier moves. An **ice fall** is a section of a glacier consisting of crevasses and towering ice pinnacles. The pinnacles form where ice blocks break away from the crevasse walls and rotate as the glacier moves. With crampons, ropes, and ice axes, a skilled mountaineer might climb into a crevasse. The walls are a pastel blue, and sunlight filters through the narrow opening above. The ice shifts and cracks, making creaking sounds as the glacier advances. Many mountaineers have been crushed by falling ice while traveling through ice falls.

THE MASS BALANCE OF A GLACIER

Consider an alpine glacier flowing from the mountains into a valley (Fig. 17–7). At the upper end of the glacier, snowfall is heavy, temperatures are below freezing for much of the year, and avalanches carry large quantities of snow from the surrounding peaks onto the ice. Thus, more snow accumulates in winter than melts in sum-

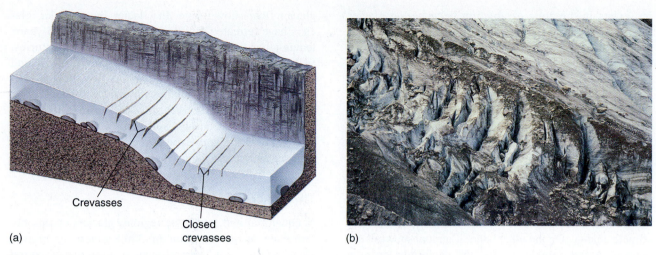

Crevasses

Closed
crevasses

(a)

(b)

Figure 17–6 (a) Crevasses form in the upper, brittle zone of a glacier where the ice flows over uneven bedrock. (b) Crevasses in the Bugaboo Mountains of British Columbia.

mer, and snow piles up from year to year. This higher-elevation part of the glacier is called the **zone of accumulation**. There the glacier's surface is covered by snow year round.

Lower in the valley, the temperature is higher throughout the year, and less snow falls. This lower part of a glacier, where more snow melts in summer than accumulates in winter, is called the **zone of ablation**. When the snow melts, a surface of old, hard glacial ice is left behind. The **snowline** is the boundary between permanent snow and seasonal snow. The snowline shifts up and down the glacier from year to year, depending on weather.

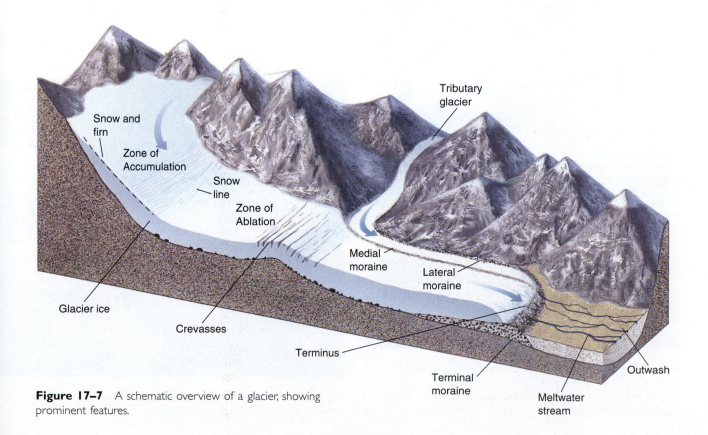

Snow and firn

Zone of Accumulation

Snow line

Zone of Ablation

Glacier ice

Crevasses

Tributary glacier

Medial moraine

Lateral moraine

Terminus

Terminal moraine

Meltwater stream

Outwash

Figure 17–7 A schematic overview of a glacier, showing prominent features.

Ice exists in the zone of ablation because the glacier flows downward from the accumulation area. Farther down-valley, the rate of glacial flow cannot keep pace with melting, so the glacier ends at its **terminus** (Fig. 17–8).

Glaciers grow and shrink. If annual snowfall increases or average temperature drops, more snow accumulates; then the snowline of an alpine glacier descends to a lower elevation, and the glacier grows thicker. At first the terminus may remain stable, but eventually it advances farther down the valley. The lag time between a change in climate and a glacial advance may range from a few years to several decades depending on the size of the glacier, its rate of motion, and the magnitude of the climate change. On the other hand, if annual snowfall decreases or the climate warms, the accumulation area shrinks and the glacier retreats.

When a glacier retreats, its ice continues to flow downhill, but the terminus melts back faster than the glacier flows downslope. In Glacier Bay, Alaska, glaciers have retreated 60 kilometers in the past 125 years, leaving barren rock and rubble. Over the centuries, seabird droppings will mix with windblown silt and weathered rock to form thin soil. At first, lichens will grow on the bare rock, and then mosses will take hold in sheltered niches that contain soil. The mosses will be followed by grasses, bushes, and, finally, trees, as vegetation reclaims the landscape. Eventually, the glacier may advance again, destroying the vegetation.

TIDEWATER GLACIERS

In equatorial and temperate regions, glaciers commonly terminate at an elevation of 3000 meters or higher. However, in a cold, wet climate, a glacier may extend into the sea to form a **tidewater glacier**. The terminus of a tidewater glacier is often a steep ice cliff dropping abruptly into the sea (Fig. 17–9). Giant chunks of ice break off, or **calve**, forming **icebergs**.

The largest icebergs in the world are those that calve from the Antarctic ice shelf. In January 1995, the edge of the 300-meter-thick Larson Ice Shelf cracked and an iceberg almost as big as Rhode Island broke free and floated into the Antarctic Ocean. The tallest icebergs in the world calve from tidewater glaciers in Greenland. Some extend 150 meters above sea level; since the visible portion of an iceberg represents only about 10 to 15 percent of its mass, these bergs may be as much as 1500 meters from base to tip.

Figure 17–8 The terminus of an alpine glacier on Baffin Island, Canada, in mid-summer. Dirty, old ice forms the lower part of the glacier below the firn line, and clean snow lies higher up on the ice above the firn line. (Steve Sheriff)

Figure 17–9 A kayaker paddles among small icebergs that calved from the Le Conte glacier, Alaska.

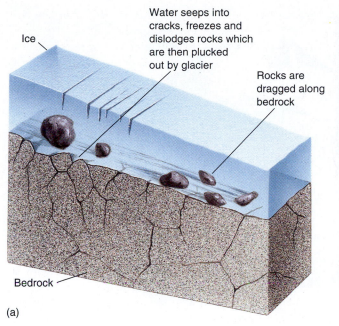

Water seeps into
cracks, freezes and
dislodges rocks which
are then plucked
out by glacier

Ice

Rocks are
dragged along
bedrock

Bedrock

(a)

(b)

Figure 17–10 (a) A glacier plucks rocks from bedrock and then drags them along, abrading both the loose rocks and the bedrock. (b) Plucking formed these crescent-shaped depressions in granite at Le Conte Bay, Alaska.

▶ 17.3 GLACIAL EROSION

Rock at the base and sides of a glacier may have been fractured by tectonic forces and may be loosened by weathering processes, such as frost wedging or pressure-release fracturing. The moving ice then dislodges the loosened rock in a process called **plucking** (Fig. 17–10). Ice is viscous enough to pick up and carry particles of all sizes, from silt-sized grains to house-sized boulders. Thus glaciers erode and transport huge quantities of rock and sediment.

Ice itself is not abrasive to bedrock because it is too soft. However, rocks embedded in the ice scrape across bedrock like a sheet of rough sandpaper pushed by a giant's hand. This process cuts deep, parallel grooves and scratches in bedrock called **glacial striations** (Fig. 17–11). When glaciers melt and striated bedrock is exposed, the markings show the direction of ice movement. Glacial striations are used to map the flow directions of glaciers. Rocks that were embedded in the base of a glacier also commonly show striations.

Sand and silt embedded in a glacier polish bedrock to a smooth, shiny finish. The abrasion grinds rocks into fine silt-sized grains called **rock flour**. Characteristically, a glacial stream is so muddy with rock flour that it is gritty and brown or gray in color. Sometimes the suspended silt scatters sunlight to make alpine streams and lakes appear turquoise, blue, or green.

EROSIONAL LANDFORMS CREATED BY ALPINE GLACIERS

Let's take an imaginary journey through a mountain range that was glaciated in the past but is now mostly ice

Figure 17–11 Stones embedded in the base of a glacier gouged these striations in bedrock in British Columbia.

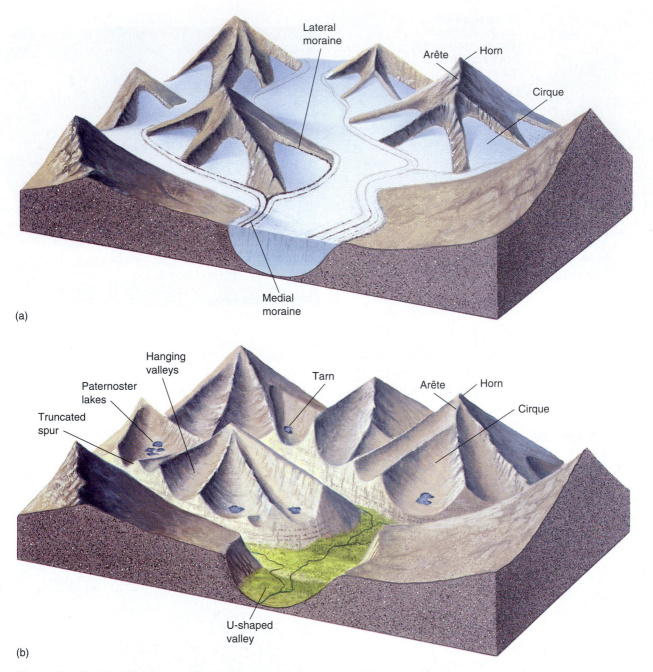

(a)

(b)

Figure 17–12 Glacial landscapes. (a) A landscape as it appears when it is covered by glaciers. (b) The same landscape as it appears after the glaciers have melted.

free. We start with a helicopter ride to the summit of a high, rocky peak. Our first view from the helicopter is of sharp, jagged mountains rising steeply above smooth, rounded valleys (Fig. 17–12).

A mountain stream commonly erodes downward into its bed, cutting a steep-sided, V-shaped valley. A glacier, however, is not confined to a narrow stream bed but instead fills its entire valley. As a result, it scours the

sides of the valley as well as the bottom, carving a broad, rounded, **U-shaped valley** (Fig. 17–13).

We land on one of the peaks and step out of the helicopter. Beneath us, a steep cliff drops off into a horseshoe-shaped depression in the mountainside called a **cirque**. A small glacier at the head of the cirque reminds us of the larger mass of ice that existed in a colder, wetter time (Fig. 17–14a).

Figure 17–13 Pleistocene glaciers carved this U-shaped valley in the Canadian Rockies, Alberta.

To understand how a glacier creates a cirque, imagine a gently rounded mountain. As snow accumulates and a glacier forms, the ice flows down the mountainside (Fig. 17–14b). The ice plucks a small depression that grows slowly as the glacier flows (Fig. 17–14c). With time, the cirque walls become steeper and higher. The glacier carries the eroded rock from the cirque to lower parts of the valley (Fig. 17–14d). When the glacier finally melts, it leaves a steep-walled, rounded cirque.

Streams and lakes are common in glaciated mountain valleys. As a cirque forms, the glacier commonly erodes a depression into the bedrock beneath it. When the glacier melts, this depression fills with water, forming a small lake, or **tarn**, nestled at the base of the cirque. If we hike down the valley below the high cirques, we may encounter a series of lakes called **paternoster lakes**, which are commonly connected by rapids and waterfalls

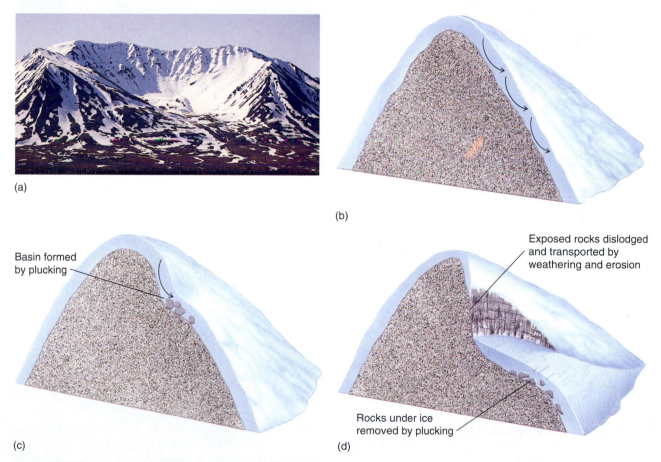

Figure 17–14 (a) A glacier eroded this concave depression, called a cirque, into a mountainside in the Alaska Range. (McCutcheon/Visuals Unlimited) (b) Snow accumulates, and a glacier begins to flow downslope from the summit of a peak. (c) Glacial plucking erodes a small depression in the mountainside. (d) Continued glacial erosion and weathering enlarge the depression. When the glacier melts, it leaves a cirque carved in the side of the peak, as in the photograph.

Figure 17–15 Glaciers cut this string of paternoster lakes in the Sierra Nevada.

Figure 17–16 The Matterhorn formed as three glaciers eroded cirques into the peak from three different sides. (Swiss Tourist Board)

(Fig. 17–15). Paternoster lakes are a sequence of small basins plucked out by a glacier. The name *paternoster* refers to a string of rosary beads or in this case, a string of small lakes strung out across a valley. When the glacier recedes, the basins fill with water.

If glaciers erode three or more cirques into different sides of a peak, they may create a steep, pyramid-shaped rock summit called a **horn**. The Matterhorn in the Swiss Alps is a famous horn (Fig. 17–16). Two glaciers flowing along opposite sides of a mountain ridge may erode both sides of the ridge, forming a sharp, narrow **arête** between adjacent valleys.

Looking downward from our peak, we may see a waterfall pouring from a small, high valley into a larger, deeper one. A small glacial valley lying high above the floor of the main valley is called a **hanging valley** (Fig. 17–17). The famous waterfalls of Yosemite Valley in California cascade from hanging valleys. To understand how a hanging valley forms, imagine these mountain valleys filled with glaciers, as they were several millennia ago (Fig. 17–12). The main glacier gouged the lower valley deeply. In contrast, the smaller tributary glacier did not scour its valley as deeply, creating an abrupt drop

where the small valley joins the main valley. If the main valley glacier cuts off the lower portion of an arête, a triangular-shaped rock face called a **truncated spur** forms.

Figure 17–17 Two hanging valleys in Yosemite National Park. (Science Graphics/Ward's Natural Science Establishment, Inc.)

Figure 17–18 A steep-sided fjord bounded by thousand-meter-high cliffs in Baffin Island, Canada.

Figure 17–19 A glacier flowed from right to left over a bedrock hill to carve this streamlined roche moutonnée in Rocky Mountain National Park, Colorado. (Don Dickson/ Visuals Unlimited)

Deep, narrow inlets called **fjords** extend far inland on many high-latitude seacoasts. Most fjords are glacially carved valleys that were later flooded by rising seas as the glaciers melted (Fig. 17–18).

EROSIONAL LANDFORMS CREATED BY A CONTINENTAL GLACIER

A continental glacier erodes the landscape just as an alpine glacier does. However, a continental glacier is considerably larger and thicker and is not confined to a valley. As a result, it covers vast regions, including entire mountain ranges.

If a glacier flows over a bedrock knob, it carves an elongate, streamlined hill called a **roche moutonnée**. (This term is derived from the French words *roche*, for "rock," and *mouton*, for "sheep." Clusters of roches moutonnées resemble herds of grazing sheep.) The upstream side of the roche moutonnée is typically gently inclined, rounded, and striated by abrasion. As the ice rides over the bedrock, it plucks rocks from the downstream side, producing a steep, jagged face (Fig. 17–19). Both alpine and continental glaciers form roches moutonnées.

▶ 17.4 GLACIAL DEPOSITS

In the 1800s, geologists recognized that the large deposits of sand and gravel found in some places had been transported from distant sources. A popular hypothesis at the time explained that this material had drifted in on icebergs during catastrophic floods. The deposits were called "drift" after this inferred mode of transport.

Today we know that continental glaciers covered vast parts of the land only 10,000 to 20,000 years ago, and these glaciers carried and deposited drift. Although the term *drift* is a misnomer, it remains in common use. Now geologists define **drift** as all rock or sediment transported and deposited by a glacier. Glacial drift averages 6 meters thick over the rocky hills and pastures of New England and 30 meters thick over the plains of Illinois.

Drift is divided into two categories. **Till** was deposited directly by glacial ice. **Stratified drift** was first carried by a glacier and then transported and deposited by a stream.

LANDFORMS COMPOSED OF TILL

Ice is so much more viscous than water that it carries particles of all sizes together. When a glacier melts, it deposits its entire sediment load; fine clay and huge boulders end up mixed together in an unsorted, unstratified mass (Fig. 17–20). Within a glacier, each rock or grain of sediment is protected by the ice that surrounds it. Therefore, the pieces do not rub against one another, and glacial transport does not round sediment as a stream does. If you find rounded gravel in till, it became rounded by a stream before the glacier picked it up.

If you travel in country that was once glaciated, you occasionally find large boulders lying on the surface. In

Figure 17–20 Unsorted glacial till. Note that large cobbles are mixed with smaller sediment. The cobbles were rounded by stream action before they were transported and deposited by the glacier.

Figure 17–21 An end moraine is a ridge of till piled up at a glacier's terminus.

many cases the boulders are of a rock type different from the bedrock in the immediate vicinity. Boulders of this type are called **erratics** and were transported to their present locations by a glacier. The origins of erratics can be determined by exploring the terrain in the direction the glacier came from until the parent rock is found. Some erratics were carried 500 or even 1000 kilometers from their point of origin and provide clues to the movement of glaciers. Plymouth Rock, where the pilgrims allegedly landed, is a glacial erratic.

Moraines

A **moraine** is a mound or a ridge of till. Think of a glacier as a giant conveyor belt. An airport conveyor belt carries suitcases to the end of the belt and dumps them in a pile. Similarly, a glacier carries sediment and deposits it at its terminus. If a glacier is neither advancing nor retreating, its terminus may remain in the same place for years. During that time, sediment accumulates at the terminus to form a ridge called an **end moraine**. An end moraine that forms when a glacier is at its greatest advance, before beginning to retreat, is called a **terminal moraine** (Fig. 17–21).

If warmer conditions prevail, the glacier recedes. If the glacier stabilizes again during its retreat and the terminus remains in the same place for a year or more, a new end moraine, called a **recessional moraine**, forms.

When ice melts, till is deposited in a relatively thin layer over a broad area, forming a **ground moraine**. Ground moraines fill old stream channels and other low spots. Often this leveling process disrupts drainage patterns. Many of the swamps in the northern Great Lakes region lie on ground moraines formed when the most recent continental glaciers receded.

End moraines and ground moraines are characteristic of both alpine and continental glaciers. An end moraine deposited by a large alpine glacier may extend for several kilometers and be so high that even a person in good physical condition would have to climb for an hour to reach the top. Moraines may be dangerous to hike over if their sides are steep and the till is loose. Large boulders are mixed randomly with rocks, cobbles, sand, and clay. A careless hiker can dislodge boulders and send them tumbling to the base.

Terminal moraines leave record of the maximum extent of Pleistocene continental glaciers. In North America they lie in a broad, undulating front extending across the northern United States. Enough time has passed since the glaciers retreated that soil and vegetation have stabilized the till and most of the hills are now wooded (Fig. 17–22).

When an alpine glacier moves downslope, it erodes the valley walls as well as the valley floor. Therefore, the edges of the glacier carry large loads of sediment.

Figure 17–22 This terminal moraine in New York State marks the southernmost extent of glaciers in that region. (Science Graphics/Ward's Natural Science, Inc.)

Additional debris falls from the valley walls and accumulates on and near the sides of mountain glaciers. Sediment near the glacial margins forms a **lateral moraine** (Fig. 17–23).

If two glaciers converge, the lateral moraines along the edges of the two glaciers merge into the middle of the larger glacier. This till forms a visible dark stripe on the surface of the ice called a **medial moraine** (Fig. 17–24).

Figure 17–24 Three separate medial moraines formed by merging lateral moraines from coalescing glaciers, Baffin Island, Canada.

Figure 17–23 A lateral moraine lies against the valley wall in the Bugaboo Mountains, British Columbia.

Drumlins

Elongate hills, called **drumlins**, cover parts of the northern United States and are well exposed across the rolling farmland in Wisconsin (Fig. 17–25). Each one looks like a whale swimming through the ground with its back in the air. Drumlins are usually about 1 to 2 kilometers long and about 15 to 50 meters high. Most are made of till, while others consist partly of till and partly of bedrock. In either case, the elongate shape of a drumlin develops when a glacier flows over a mound of sediment. The flow of the ice creates the streamlined shape, which is elongated in the same direction as the glacial flow.

Landforms Composed of Stratified Drift

Because of the great amount of sediment eroded by a glacier, streams flowing from a glacier are commonly laden with silt, sand, and gravel. The stream deposits this sediment beyond the glacier terminus as **outwash**

Figure 17–25 Crop patterns emphasize glacially streamlined drumlins in Wisconsin. (Kevin Horan/Tony Stone Images)

(Fig. 17–26). Glacial streams carry such a heavy load of sediment that they often become braided, flowing in multiple channels. Outwash deposited in a narrow valley is called a **valley train**. If the sediment spreads out from the confines of the valley into a larger valley or plain, it forms an **outwash plain** (Fig. 17–27). Outwash plains are also characteristic of continental glaciers.

During the summer, when snow and ice melt rapidly, streams form on the surface of a glacier. Many are too wide to jump across. Some of these streams flow off the front or sides of the glacier. Others plunge into crevasses and run beneath the glacier over bedrock or drift. These streams commonly deposit small mounds of sediment, called **kames**, at the margin of a receding glacier or where the sediment collects in a crevasse or other depression in the ice. An **esker** is a long, sinuous ridge that forms as the channel deposit of a stream that flowed within or beneath a melting glacier.

Because kames, eskers, and other forms of stratified drift are stream deposits and were not deposited directly by ice, they show sorting and sedimentary bedding, which distinguishes them from unsorted and unstratified till. In addition, the individual cobbles or grains are usually rounded.

Figure 17–26 Streams flowing from the terminus of a glacier filled this valley on Baffin Island with outwash.

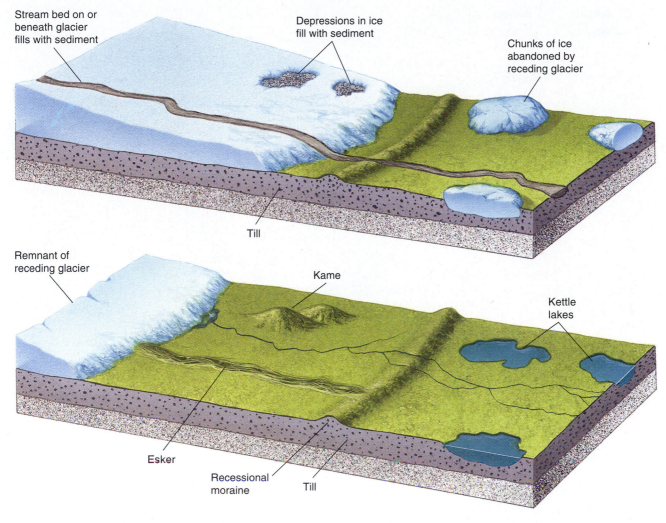

Figure 17–27 Landforms created as a glacier retreats.

Large blocks of ice may be left behind in a moraine or an outwash plain as a glacier recedes. When such an ice block melts, it leaves a depression called a **kettle**. Kettles fill with water, forming kettle lakes. A kettle lake is as large as the ice chunks that melted to form the hole. The lakes vary from a few tens of meters to a kilometer or so in diameter, with a typical depth of 10 meters or less.

▶ 17.5   THE PLEISTOCENE ICE AGE

The extent of glacial landforms is proof that glaciers once covered much larger areas than they do today. A time when alpine glaciers descend into lowland valleys and continental glaciers spread over land in high latitudes is called an **ice age**. Geologic evidence shows that the Earth has been warm and relatively ice free for at least 90 percent of the past 1 billion years. However, at least five major ice ages occurred during that time (Fig. 17–28). Each one lasted from 2 to 10 million years.

Glacial landforms created during older ice ages have been mostly obliterated by erosion and tectonic processes. However, in a few places till from older glaciers has been lithified into a glacial conglomerate called **tillite**. Geologists know that tillites were deposited by glaciers because the cobbles in tillite are often angular and striated.

The most recent ice age took place mainly during the Pleistocene Epoch and is called the **Pleistocene Ice Age**. It began about 2 million years ago (although evidence of an earlier beginning has been found in the Southern Hemisphere). However, the Earth has not been glaciated continuously during the Pleistocene Ice Age; instead, climate has fluctuated and continental glaciers grew and then melted away several times (Fig. 17–28). Although the climate has been relatively warm for the most recent 15,000 years, most climate models indicate

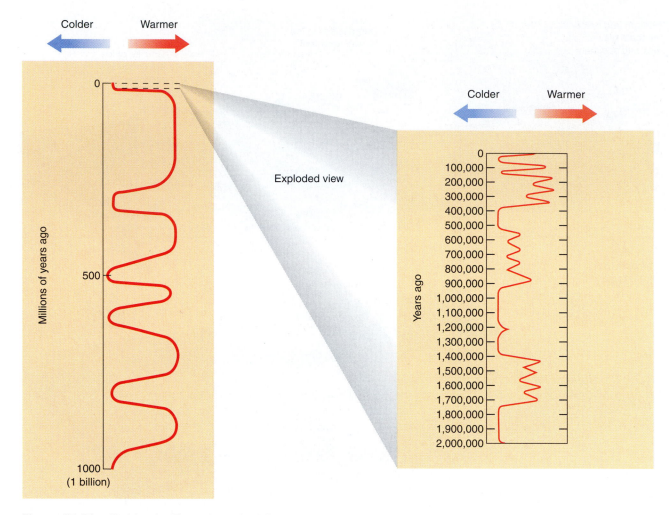

Figure 17–28 Glacial cycles. The scale on the left shows an approximate curve for average global temperature variation during the past billion years. Times of lowest temperature are thought to coincide with major ice ages. The scale on the right is an exploded view of the Pleistocene Ice Age. We are probably still living within the Pleistocene Ice Age, and continental ice sheets will advance again.

that we are still in the Pleistocene Ice Age and the glaciers will advance again.

PLEISTOCENE GLACIAL CYCLES

Most scientists now think that the relatively rapid climate fluctuations that caused the Pleistocene glacial cycles resulted from periodic variations in the Earth's orbit and spin axis (Fig 17–29). Astronomers have detected three types of variations:

1. The Earth's orbit around the Sun is elliptical rather than circular. The shape of the ellipse is called **eccentricity**. The eccentricity varies in a regular cycle lasting about 100,000 years.

2. The Earth's axis is currently tilted at about 23.5° with respect to a line perpendicular to the plane of

its orbit around the Sun. The **tilt** oscillates by 2.5° on about a 41,000-year cycle.

3. The Earth's axis, which now points directly toward the North Star, circles like that of a wobbling top. This circling, called **precession**, completes a full cycle every 23,000 years.

Although these changes do not appreciably affect the total solar radiation received by the Earth, they do affect the distribution of solar energy with respect to latitude and season. Therefore, these changes influence the duration of the seasons. Seasonal changes in sunlight reaching higher latitudes can cause an onset of glaciation by reducing summer temperature. If summers are cool and short, winter snow and ice persist, leading to growth of glaciers.

Early in the 20th century, a Yugoslavian astronomer, Milutin Milankovitch, calculated the combined effects of

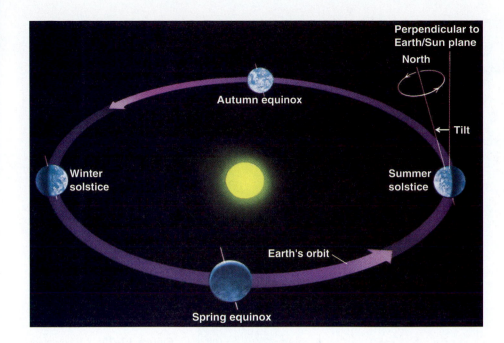

Figure 17–29 Earth orbital variations may explain the temperature oscillations and glacial advances and retreats during the Pleistocene Epoch. Orbital variations occur over time spans of tens of thousands of years.

the three orbital variations on climate. His calculations showed that they should interact to generate alternating cool and warm climates in the mid- and high latitudes. Moreover, the timing of the calculated high-latitude cooling coincided with that of Pleistocene glacial advances.

EFFECTS OF PLEISTOCENE CONTINENTAL GLACIERS

At its maximum extent about 18,000 years ago, the most recent North American ice sheet covered 10 million square kilometers—most of Alaska, Canada, and parts of the northern United States (Fig. 17–30). At the same time, alpine glaciers flowed from the mountains into the lowland valleys.

The erosional features and deposits left by these glaciers dominate much of the landscape of the northern states. Today terminal moraines form a broad band of rolling hills from Montana across the Midwest and eastward to the Atlantic Ocean. Long Island and Cape Cod are composed largely of terminal moraines. Kettle lakes or lakes dammed by moraines are abundant in northern Minnesota, Wisconsin, and Michigan. Drumlins dot the landscape in the northern states. Ground moraines, outwash, and loess (windblown glacial silt) cover much of the northern Great Plains. These deposits have weathered to form the fertile soil of North America's "breadbasket."

Pleistocene glaciers advanced when mid- and high-latitude climates were colder and wetter than today. When the glaciers melted, the rain and meltwater flowed through streams and collected in numerous lakes. Later,

as the ice sheets retreated and the climate became drier, many of these streams and lakes dried up.

Today, extinct stream channels and lake beds are common in North America. The extinct lakes are called **pluvial lakes**, a term derived from the Latin word *pluvia*, meaning "rain." The basin that is now Death Valley was once filled with water to a depth of 100 meters or more. Most of western Utah was also covered by a pluvial lake called Lake Bonneville. As drier conditions returned, Lake Bonneville shrank to become Great Salt Lake, west of Salt Lake City.

When glaciers grow, they accumulate water that would otherwise be in the oceans, and sea level falls. When glaciers melt, sea level rises again. When the Pleistocene glaciers reached their maximum extent 18,000 years ago, global sea level fell to about 130 meters below its present elevation. As submerged continental shelves became exposed, the global land area increased by 8 percent (although about one third of the land was ice covered).

When the ice sheets melted, much of the water returned to the oceans, raising sea level again. At the same time, portions of continents rebounded isostatically as the weight of the ice was removed. The effect along any specific coast depends upon the relative amounts of sea level rise and isostatic rebound. Some coastlines were submerged by the rising seas. Others rebounded more than sea level rose. Today, beaches in the Canadian Arctic lie tens to a few hundred meters above the sea. Portions of the shoreline of Hudson's Bay have risen 300 meters.

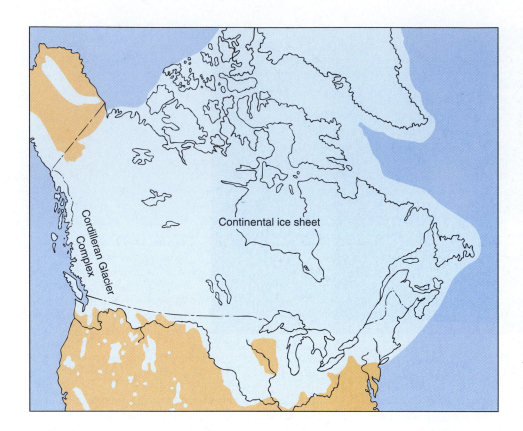

Figure 17–30 Maximum extent of the continental glaciers in North America during the latest glacial advance, approximately 18,000 years ago.

Labels on map: Cordilleran Glacier Complex; Continental ice sheet

SUMMARY

If snow survives through one summer, it becomes a relatively hard, dense material called **firn**. A **glacier** is a massive, long-lasting accumulation of compacted snow and ice that forms on land and creeps downslope or outward under the influence of its own weight. **Alpine glaciers** form in mountainous regions; **continental glaciers** cover vast regions. Glaciers move by both **basal slip** and **plastic flow**. The upper 40 meters of a glacier is too brittle to flow, and large cracks called **crevasses** develop in this layer.

In the **zone of accumulation** of a glacier, the annual rate of snow accumulation is greater than the rate of melting, whereas in the **zone of ablation**, melting exceeds accumulation. The **snowline** is the boundary between permanent snow and seasonal snow. The end of the glacier is called the **terminus**.

Glaciers erode bedrock by **plucking** and by abrasion. Glaciated mountains often contained U-shaped valleys and other landforms eroded by flowing ice. A knob of bedrock streamlined by glacial erosion is called a **roche moutonnée**.

Drift is any rock or sediment transported and deposited by a glacier. The unsorted drift deposited directly by a glacier is **till**. Most glacial terrain is characterized by large mounds of till known as **moraines**. **Terminal moraines**, **ground moraines**, **recessional moraines**, **lateral moraines**, **medial moraines**, and **drumlins** are all depositional features formed by glaciers. **Stratified drift** consists of sediment first carried by a glacier and then transported, sorted, and deposited by streams flowing on, under, or within a glacier. **Valley trains**, **outwash plains**, **kames**, and **eskers** are formed from stratified drift. **Kettles** are depressions created by melting of large blocks of ice abandoned by a retreating glacier.

During the past 1 billion years, at least five major ice ages have occurred. The most recent occurred during the **Pleistocene Epoch**, when continental glaciers created many topographic features that are prominent today. One theory contends that Pleistocene advances and retreats were caused by climate changes induced by changes in the Earth's orbit and the orientation of its rotational axis. **Pluvial lakes** formed in the wetter climate of those times. Ice sheets isostatically depress continents, which later rebound when the ice melts. Sea level falls when continental ice sheets form and rises again when the ice melts.

KEY WORDS

firn   *298*

glacier   *298*

alpine glacier   *298*

ice sheet   *299*

continental glacier   *299*

basal slip   *299*

plastic flow   *299*

crevasse   *300*

ice fall   *300*

zone of accumulation   *301*

zone of ablation   *301*

snowline   *301*

terminus   *302*

tidewater glacier   *302*

iceberg   *302*

plucking   *303*

glacial striation   *303*

rock flour   *303*

U-shaped valley   *304*

cirque   *304*

tarn   *305*

paternoster lake   *305*

horn   *306*

arête   *306*

hanging valley   *306*

truncated spur   *306*

fjord   *307*

roche moutonnée   *307*

drift   *307*

till   *307*

stratified drift   *307*

erratic   *308*

moraine   *308*

end moraine   *308*

terminal moraine   *308*

recessional moraine   *308*

ground moraine   *308*

lateral moraine   *309*

medial moraine   *309*

drumlin   *309*

outwash   *309*

valley train   *310*

outwash plain   *310*

kame   *310*

esker   *310*

kettle   *311*

ice age   *311*

tillite   *311*

Pleistocene Ice Age   *311*

pluvial lake   *313*

REVIEW QUESTIONS

1. Outline the major steps in the metamorphism of newly fallen snow to glacial ice.

2. Differentiate between alpine glaciers and continental glaciers. Where are alpine glaciers found today? Where are continental glaciers found today?

3. Distinguish between basal slip and plastic flow.

4. Why are crevasses only about 40 meters deep, even though many glaciers are much thicker?

5. Describe the surface of a glacier in the summer and in the winter in (a) the zone of accumulation and (b) the zone of ablation.

6. How do icebergs form?

7. Describe how glacial erosion can create (a) a cirque, (b) striated bedrock, and (c) smoothly polished bedrock.

8. Describe the formation of arêtes, horns, hanging valleys, and truncated spurs.

9. Distinguish among ground, recessional, terminal, lateral, and medial moraines.

10. Why are kames and eskers features of receding glaciers? How do they form?

11. What topographic features were left behind by the continental ice sheets? Where can they be found in North America today?

12. How do geologists recognize the existence and movement of continental glaciers that advanced hundreds of millions of years ago?

DISCUSSION QUESTIONS

1. Compare and contrast the movement of glaciers with stream flow.

2. Outline the changes that would occur in a glacier if (a) the average annual temperature rose and the precipitation decreased;   (b) the temperature remained constant but the precipitation increased; and   (c) the temperature decreased and the precipitation remained constant.

3. Explain why plastic flow is a minor mechanism of movement for thin glaciers but is likely to be more important for a thick glacier.

4. In some regions of northern Canada, both summer and winter temperatures are cool enough for glaciers to form, but there are no glaciers. Speculate on why continental glaciers are not forming in these regions.

5. If you found a large boulder lying in a field, how would you determine whether or not it was an erratic?

6. A bulldozer can only build a pile of dirt when it is moving forward. Yet a glacier can build a terminal moraine when it is neither advancing nor retreating. Explain.

7. Imagine you encountered some gravelly sediment. How would you determine whether it was a stream deposit or a ground moraine?

8. Explain how medial moraines prove that glaciers move.

9. If you were hiking along a wooded hill in Michigan, how would you determine whether or not it was a moraine?

Coastlines

The seashore is an attractive place to live or visit. Because the ocean moderates temperature, coastal regions are cooler in summer and warmer in winter than continental interiors. People enjoy the salt air and find the rhythmic pounding of surf soothing and relaxing. Vacationers and residents sail, swim, surf, and fish along the shore. In addition, the sea provides both food and transportation. For all of these reasons, coastlines have become heavily urbanized and industrialized. In the United States, 75 percent of the population, 40 percent of the manufacturing plants, and 65 percent of the electrical power generators are located within 80 kilometers of the oceans or the Great Lakes.

Coastlines are also one of the most geologically active environments on Earth. Rivers deposit great amounts of sediment on coastal deltas. Waves and currents erode the shore and transport sediment. Converging tectonic plates buckle many coastal regions, creating mountain ranges, earthquakes, and volcanic eruptions. Over geologic time, sea level rises and falls, flooding some beaches and raising others high above sea level.

Coastlines are among the most changeable landforms. Montauk Point, Long Island, is composed of glacial till deposited during the last Ice Age. Waves and currents are eroding the Point and carrying the sediment westward.

▶ 18.1 WAVES, TIDES, AND CURRENTS

OCEAN WAVES

Most waves develop when wind blows across the water. Waves vary from gentle ripples on a pond to destructive giants that can topple beach houses during a hurricane. In deep water, the size of a wave depends on (1) the wind speed, (2) the length of time that the wind has blown, and (3) the distance that the wind has traveled (sailors call this last factor **fetch**). A 25-kilometer-per-hour wind blowing for 2 to 3 hours across a 15-kilometer-wide bay will generate waves about 0.5 meter high. But if a Pacific storm blows at 90 kilometers per hour for several days over a fetch of 3500 kilometers, it can generate 30-meter-high waves, as tall as a ship's mast.

The highest part of a wave is called the **crest**; the lowest is the **trough** (Fig. 18–1). The **wavelength** is the distance between successive crests. The **wave height** is the vertical distance from the crest to the trough.

If you tie one end of a rope to a tree and shake the other end, a wave travels from your hand to the tree, but any point on the rope just moves up and down (Fig. 18–2). In a similar manner, a single water molecule in a water wave does not travel in the same direction as the wave. The water molecule moves in circles, as shown in Figure 18–3. Water at the surface completes relatively small circles with little forward motion. That is why if you are sitting in a boat on the ocean, you bob up and down and sway back and forth as the waves pass beneath you, but you do not travel along with the waves. In addition, the circles of water movement become smaller with depth. At a depth equal to about one half the wavelength, the movement becomes negligible. Thus, if you dive deep enough, you escape wave motion.

In deep water, therefore, the bottom of a wave does not contact the sea floor. But when a wave enters shallow water, the deepest circles interact with the bottom

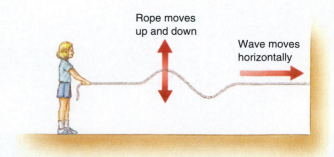

Figure 18–2 When a wave moves along a rope, any point moves up and down, but the wave travels horizontally.

and are compressed into ellipses. This deformation slows the lower part of the wave, so that the upper part moves more rapidly than the lower. As a result, the front of the wave steepens until it collapses forward, or **breaks** (Fig. 18–4). Chaotic, turbulent waves breaking along a shore are called **surf**.

Wave Refraction

Most waves approach the shore obliquely rather than directly. When this happens, one end of the wave encounters shallow water and slows down, while the rest of the wave is still in deeper water and continues to advance at a constant speed. As a result, the wave bends. This effect is called **refraction**. Consider the analogy of a sled gliding down a snowy hill onto a cleared road. If the sled hits the road at an angle, one runner will reach it before the other. The runner that hits the pavement first slows down, while the other, which is still on the snow, continues to travel rapidly (Fig. 18–5). As a result, the sled turns abruptly.

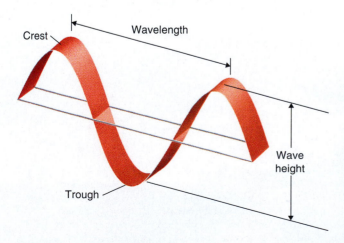

Figure 18–1 Terminology used to describe waves.

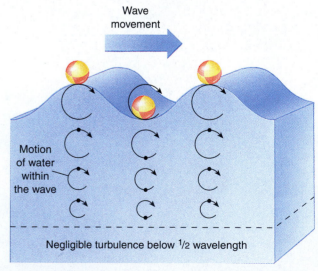

Figure 18–3 Movement of a wave and the movement of water within the wave.

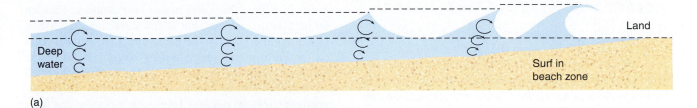

Deep water

Land

Surf in beach zone

(a)

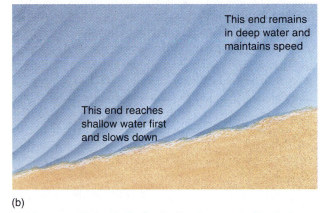

(b)

Figure 18–4   (a) When a wave approaches the shore, the circular motion flattens out and becomes elliptical. The wavelength shortens, and the wave steepens until it finally breaks, creating surf. The dashed line shows the changes in wavelength and wave height as the wave approaches shore. (b) Surf breaks along the beach in Hawaii.   (Corel Photos)

Sled analogy

Snowy hill

Road

(a)

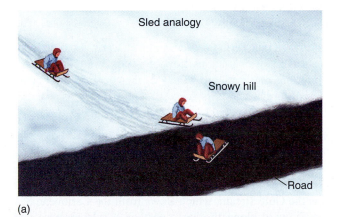

This end remains in deep water and maintains speed

This end reaches shallow water first and slows down

(b)

(c)

Figure 18–5   (a) A sled turns upon striking a paved roadway at an angle because one runner hits the roadway and slows down before the other does. (b) When a water wave strikes the shore at an angle, one end slows down, causing the wave to refract, or bend. (c) Wave refraction on a lakeshore.

TIDES

Even the most casual observer will notice that on any beach the level of the ocean rises and falls on a cyclical basis. If the water level is low at noon, it will reach its maximum height about 6 hours later, at 6 o'clock, and be low again near midnight. These vertical displacements are called **tides**. Most coastlines experience two high tides and two low tides approximately every 24 hours.

Tides are caused by the gravitational pull of the Moon and Sun. Although the Moon is much smaller than the Sun, it is so much closer to the Earth that its influence predominates. At any time, one region of the Earth (marked A in Fig. 18–6) lies directly under the Moon. Because gravitational force is greater for objects that are closer together, the part of the ocean nearest to the Moon is attracted with the strongest force. The water rises, resulting in a high tide in that region. (Although land also experiences a gravitational attraction to the Moon, it is too rigid to rise perceptibly.)

But now our simple explanation runs into trouble. As the Earth spins on its axis, a given point on the Earth passes directly under the Moon approximately once every 24 hours, but the period between successive high tides is only 12 hours. Why are there ordinarily two high tides in a day? The tide is high not only when a point on Earth is directly under the Moon, but also when it is 180° away. To understand this, we must consider the Earth–Moon orbital system. Most people visualize the Moon orbiting around the Earth, but it is more accurate to say that the Earth and the Moon orbit around a common center of gravity. The two celestial partners are locked together like dancers spinning around in each other's arms. Just as the back of a dancer's dress flies outward as she twirls, the oceans on the opposite side of the Earth from the Moon bulge outward. This bulge is the high tide 180° away from the Moon (point B in Fig. 18–6). Thus, the tides rise and fall twice daily.

High and low tides do not occur at the same time each day, but are delayed by approximately 50 minutes every 24 hours. The Earth makes one complete rotation on its axis in 24 hours, but at the same time, the Moon is orbiting the Earth in the same direction. After a point on the Earth makes one complete rotation in 24 hours, that point must spin for an additional 50 minutes to catch up with the orbiting Moon. This is why the Moon rises approximately 50 minutes later each night and the tides are approximately 50 minutes later each day.

Although the Sun's gravitational pull on the Earth's oceans is smaller than the Moon's, it does affect ocean tides. When the Sun and Moon are directly in line, their gravitational fields are added together, creating a strong tidal bulge. During these times, the variation between high and low tides is large, producing **spring tides** (Fig. 18–7a). When the Sun and Moon are 90° out of alignment, each partially offsets the effect of the other and the differences between the levels of high and low tide are smaller. These relatively small tides are called **neap tides** (Fig. 18–7b).

Tidal variations differ from place to place. For example, in the Bay of Fundy, the tidal variation is as much as 15 meters during a spring tide, while in Santa Barbara, California, it is less than 2 meters. Mariners consult tide tables that give the time and height of the tides in any area on any day.

OCEAN CURRENTS

A wave is a periodic oscillation of water. In contrast, a current is a continuous flow of water in a particular direction. Currents are found everywhere in the ocean, from its surface to its greatest depths.

Surface Currents

Prevailing winds push the sea surface to generate broad, slow, surface currents that are deflected into circular paths by the Earth's rotation. One familiar ocean current is the Gulf Stream, which flows from the Caribbean Sea northward along the east coast of North America and

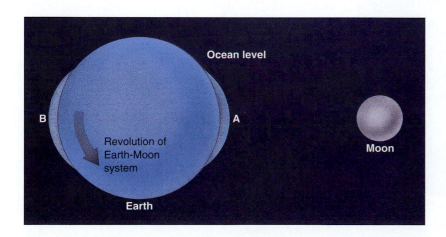

Figure 18–6 Schematic view of tide formation. (Magnitudes and sizes are exaggerated for emphasis.)

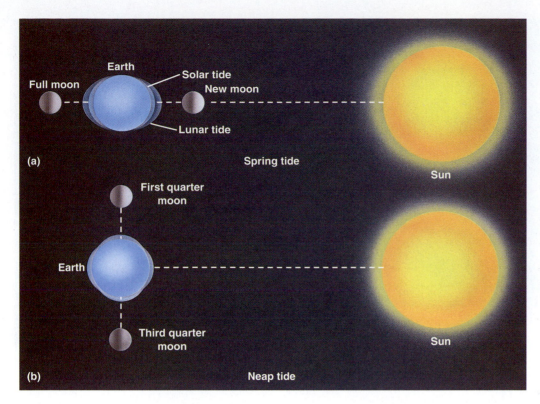

Figure 18–7 Formation of spring and neap tides.

then across the Atlantic Ocean to Europe. Ocean currents affect climates by carrying warm water from the equator toward higher latitudes or cold water from the Arctic and Antarctic toward lower latitudes.

Currents Generated by Tides

When tides rise and fall along an open coastline, the water moves in and out as a broad sheet. If the tidal flow is constricted by a narrow bay, a fjord, islands, or other obstructions, the moving water is funneled into **tidal currents**. Tidal currents can be intense where large differences exist between high and low tide and where narrow constrictions occur in the shoreline. In parts of the west coast of British Columbia, a diesel-powered fishing boat cannot make headway against tidal currents flowing between closely spaced islands. Fishermen must wait until the tide and tidal currents reverse direction before passing through the constrictions.

Currents Generated by Near-Shore Waves

After a wave breaks and washes onto the beach, the water flows back toward the sea. This outward flow creates a current called a **rip current**, or **undertow**, that can be strong enough to carry swimmers out to sea.

If waves regularly strike shore at an angle, they create a **longshore current** that flows parallel to the beach (Fig. 18–8). A longshore current flows in the surf zone and a little farther out to sea and may travel for tens or even hundreds of kilometers. When waves strike shore at an angle, they wash sand onto the beach in the direction that they are moving. However, the water then flows straight back down the beach slope, taking some of the sand with it. The zig-zag motion carries sand along the coast in a process called **beach drift**.

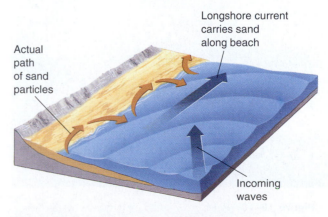

Figure 18–8 Formation of a longshore current.

(a)

Figure 18–9 (a) Sandy beaches are common near Santa Barbara, California. (b) Big Sur, to the north, is dominated by rocky beaches.

(b)

▶ 18.2 THE WATER'S EDGE

BEACHES

When most people think about going to the beach, they think of gently sloping expanses of sand. However, a **beach** is any strip of shoreline that is washed by waves and tides. Most beaches are covered with sediment. Although many beaches are sandy, others are swampy, rocky, or bounded by cliffs (Fig. 18–9).

A beach is divided into two zones, the **foreshore** and the **backshore**. The foreshore, called the **intertidal zone** by biologists, lies between the high and low tide lines and is alternately exposed to the air at low tide and covered by water at high tide. The backshore is usually dry but is washed by waves during storms. Many terrestrial plants cannot survive even occasional inundation by salt water, so specialized, salt-resistant plants live in the backshore. The backshore can be wide or narrow depending on its topography and the frequency and intensity of storms. In a region where the land rises steeply, the backshore may be a narrow strip. In contrast, if the coast consists of low-lying plains and if coastal storms occur regularly, the backshore may extend several kilometers inland.

REEFS

A **reef** is a wave-resistant ridge or mound built by corals, oysters, algae, or other marine organisms. Because corals need sunlight and warm, clear water to thrive, coral reefs develop in shallow tropical seas where little suspended clay or silt muddies the water (Fig. 18–10). As the corals die, their offspring grow on their remains. Oyster reefs form in temperate estuaries and can grow in more turbid water.

The South Pacific and portions of the Indian Ocean are dotted with numerous islands called atolls. An **atoll** is a circular coral reef that forms a ring of islands around a lagoon. Atolls vary from 1 to 130 kilometers in diameter and are surrounded by deep water of the open sea. If corals live only in shallow water, how did atolls form

Figure 18–10 Reefs grow in the clear, shallow water near Vanuatu and many other South Pacific islands.

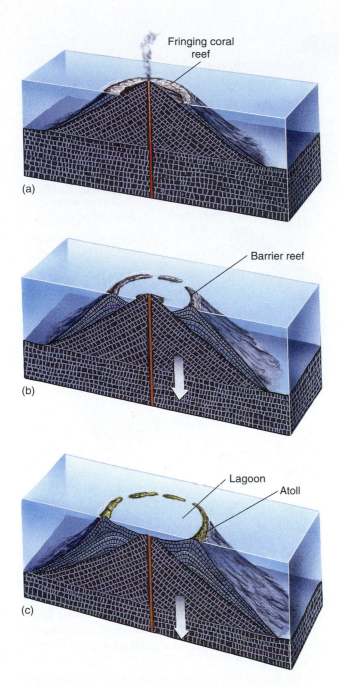

Fringing coral reef

(a)

Barrier reef

(b)

Lagoon

Atoll

(c)

Figure 18–11   (a) When a volcanic island is rising or static, the reef remains attached to the beach and is called a fringing reef. (b) As the island sinks, the reef continues to grow upward to form a barrier reef. (c) Finally the island becomes submerged and the reef forms a circular atoll.

in the deep sea? Charles Darwin studied this question during his famous voyage on the *Beagle* from 1831 to 1836. He reasoned that a coral reef must have formed in shallow water on the flanks of a volcanic island. Eventually the island sank, but the reef continued to grow upward, so that the living portion always remained in shallow water (Fig. 18–11). This proposal

was not accepted at first because scientists could not explain how a volcanic island could sink. However, when scientists drilled into a Pacific atoll shortly after World War II and found volcanic rock hundreds of meters beneath the reef, Darwin's original hypothesis was reconsidered. Today we know that the weight of a volcano causes the lithosphere to sink. In addition, the hot lithosphere beneath a volcanic island cools after the volcano becomes extinct. As a result, it becomes denser and contracts, adding to the sinking effect.

Reefs around the world have suffered severe epidemics of disease and predation within the last decade. Studies of fossils show that epidemics and mass extinctions have affected reefs periodically for hundreds of millions of years. However, some oceanographers have suggested that human activity has provoked the recent epidemics. One suggested cause is that sewage provides nutrients for algae and other organisms that smother reef organisms. Another is that chemical pollutants are altering the species balance in aquatic ecosystems. A third is that seawater temperature has risen in response to global warming and that reef organisms are adversely affected by warmer seawater.

▶ 18.3   EMERGENT AND SUBMERGENT COASTLINES

Geologists have found drowned river valleys and fossils of land animals on continental shelves beneath the sea. They have also found fossils of fish and other marine organisms in continental interiors. As a result, we infer that sea level has changed, sometimes dramatically, throughout geologic time. An **emergent coastline** forms when a portion of a continent that was previously under water becomes exposed as dry land. Falling sea level or rising land can cause emergence. In contrast, a **submergent coastline** develops when the sea floods low-lying land and the shoreline moves inland (Fig. 18–12). Submergence occurs when sea level rises or coastal land sinks.

FACTORS THAT CAUSE COASTAL EMERGENCE AND SUBMERGENCE

Tectonic processes, such as mountain building or basin formation, can cause a coastline to rise or sink. Isostatic adjustment can also depress or elevate a portion of a coastline. About 18,000 years ago, a huge continental glacier covered most of Scandinavia, causing it to sink isostatically. As the crust settled, the displaced asthenosphere flowed southward, causing the Netherlands to rise. When the ice melted, the process reversed. Today, Scandinavia is rebounding and the Netherlands is sinking. These tectonic and isostatic processes cause local or

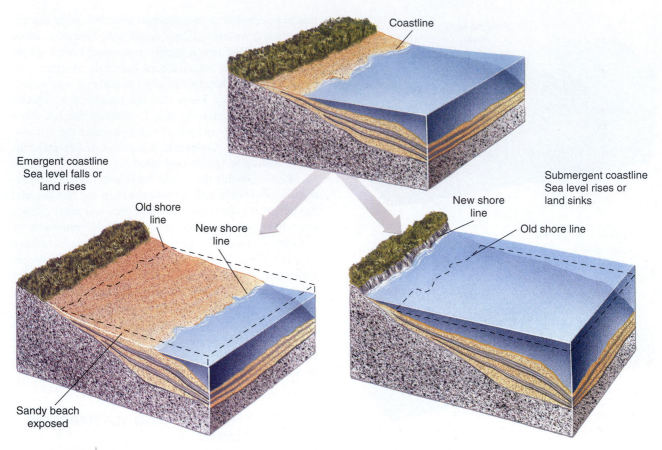

Coastline

Emergent coastline
Sea level falls or
land rises

Old shore
line

New shore
line

Sandy beach
exposed

Submergent coastline
Sea level rises or
land sinks

New shore
line

Old shore line

Figure 18–12 Emergent and submergent coastlines. If sea level falls or if the land rises, offshore sand is exposed to form a sandy beach. If coastal land sinks or sea level rises, areas that were once land are flooded. Irregular shorelines develop and beaches are commonly sediment poor.

regional sea level changes but do not affect global sea level.

Sea level can also change globally. A global sea-level change, called **eustatic change**, occurs by three mechanisms: changes in water temperature, changes in the volume of the mid-oceanic ridge, and growth and melting of glaciers.

Water expands or contracts when its temperature changes. Although this change is not noticeable in a glass of water, the volume of the oceans is so great that a small temperature change can alter sea level measurably. Water is most dense at 4°C. Because most temperate and tropical oceans are warmer than 4°C, most ocean water expands when warmed and contracts when cooled. Thus, global warming causes a sea-level rise and cooling leads to falling sea level.

As explained in Chapter 11, changes in the volume of the mid-oceanic ridge can also affect sea level. The mid-oceanic ridge displaces seawater. When lithospheric plates spread slowly from the mid-oceanic ridge, the new lithosphere cools and shrinks before it travels far. Thus, slow sea-floor spreading creates a narrow ridge that dis-

places relatively small amounts of seawater and leads to low sea level. In contrast, rapidly spreading plates produce a high-volume ridge, causing a global sea-level rise just as the water level in the bathtub rises when you settle into a bath. At times in Earth history, spreading has been relatively rapid, and as a result, global sea level has been high.

During an ice age, vast amounts of water move from the sea to continental glaciers, and sea level falls globally, resulting in emergence. At the same time, the weight of a glacier can isostatically depress a coastline, causing local or regional submergence. Similarly, when glaciers melt, sea level rises globally, causing submergence, but the melting ice allows the unburdened continent to rise isostatically, resulting in local or regional emergence. The net result along any particular coast is determined by the balance between global sea-level change and the local or regional isostatic adjustments.

Temperature changes and glaciation are linked. When global temperature rises, seawater expands and glaciers melt; when temperature falls, seawater contracts and glaciers grow.

▶ 18.4 SANDY AND ROCKY COASTLINES

Coastal weathering and erosion occur by many processes (Fig. 18–13). Waves hurl sand and gravel against sea cliffs, wearing them away. Salt water dissolves soluble minerals. Salt water also soaks into cracks in the bedrock; when the water evaporates, the growing salt crystals pry the rock apart. In addition, when a wave strikes fractured rock, it compresses air in the cracks. This compressed air can enlarge the cracks and dislodge rocks. Storm waves create forces as great as 25 to 30 tons per square meter and can dislodge and lift large boulders. Engineers built a breakwater of house-sized rocks weighing 80 to 100 tons each in Wick Bay, Scotland. The rocks were bound together with steel rods set in concrete and topped by a steel-reinforced concrete cap weighing over 800 tons. One large storm broke the cap and scattered the rocks. On the Oregon coast, waves tossed a 60-kilogram rock over a 25-meter-high lighthouse. The rock then crashed through the roof of the keeper's cottage, startling the inhabitants.

If weathering and erosion occur along all coastlines, why are some beaches sandy and others rocky? The answer is that most coastal sediment is not formed by weathering and erosion at the beach itself, but is trans-ported from other places. Major rivers carry large quantities of sand, silt, and clay to the sea and deposit it on deltas that may cover thousands of square kilometers. In some coastal regions, waves and currents erode sediment from glacial till that was deposited along coastlines during the Pleistocene Ice Age. In some tropical regions, eroding reefs supply sediment. Sandy coastlines occur where sediment from any of these sources is abundant; rocky coastlines occur where sediment is scarce.

SANDY COASTLINES

Most of the sand along coasts accumulates in shallow water offshore from the beach. If a coastline rises or sea level falls, this vast supply of sand is exposed. Thus, sandy beaches are abundant on emergent coastlines.

Longshore currents and waves erode, transport, and deposit sand along a coast. Much of the sand found at Cape Hatteras, North Carolina, originated from the mouth of the Hudson River and from glacial deposits on Long Island and southern New England. Midway along this coast, at Sandy Hook, New Jersey, an average of 2000 tons a day move along the beach. As a result of this process, beaches have been called "rivers of sand."

A long ridge of sand or gravel extending out from a beach is called a **spit** (Fig. 18–14). A spit may block the

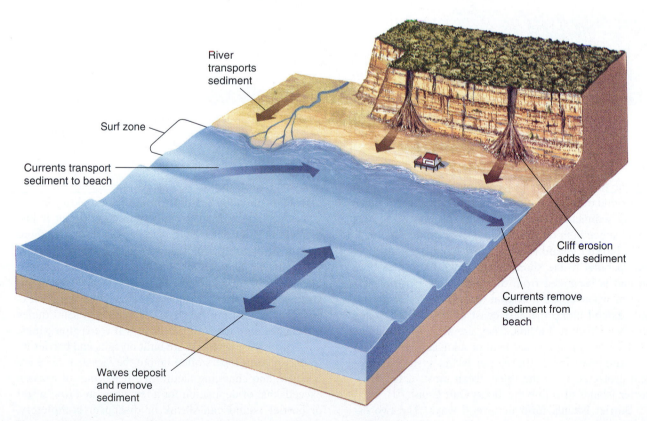

Figure 18–13 The growth and shrinkage of a beach depend on the sum total of erosion and deposition.

River transports sediment

Surf zone

Currents transport sediment to beach

Waves deposit and remove sediment

Cliff erosion adds sediment

Currents remove sediment from beach

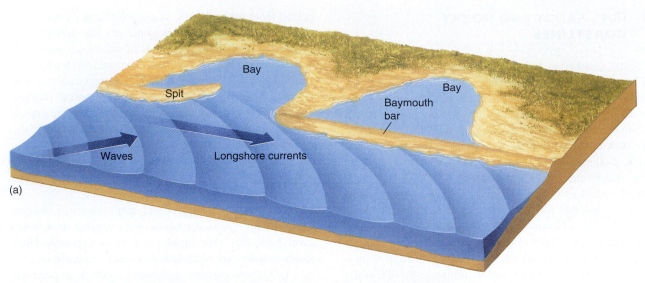

Figure 18–14 (a) A spit forms where sediment is carried away from the shore and deposited. If a spit closes the mouth of a bay, it becomes a baymouth bar. (b) Aerial view of a spit that formed along a low-lying coast in northern Siberia.

entrance to a bay, forming a **baymouth bar**. A spit may also extend outward into the sea, creating a trap for other moving sediment. A well-developed spit may be several meters above high tide level and tens of kilometers long.

A **barrier island** is a long, low-lying island that extends parallel to the shoreline. It looks like a beach or spit and is separated from the mainland by a sheltered body of water called a **lagoon** (Fig. 18–15). Barrier islands extend along the east coast of the United States from New York to Florida. They are so nearly continuous that a sailor in a small boat or a kayak can navigate the entire coast inside the barrier island system and remain protected from the open ocean most of the time. Barrier islands also line the Texas Gulf Coast.

Barrier islands form in several ways. The two essential ingredients are a large supply of sand and waves or currents to transport it. If a coast is shallow for several kilometers outward from shore, breaking storm waves may carry sand toward shore and deposit it just offshore as a barrier island. Alternatively, if a longshore current veers out to sea, it slows down and deposits sand where it reaches deeper water. Waves may then pile up the sand to form a barrier island. Other mechanisms involve sea-level change. Underwater sand bars may be exposed as a coastline emerges. Alternatively, sand dunes or beaches may form barrier islands if a coastline sinks.

Many seaside resorts are built on spits and barrier islands, and developers often ignore the fact that these are transient and changing landforms. If the rate of erosion exceeds that of deposition for a few years in a row, a spit or barrier island can shrink or disappear completely, leading to destruction of beach homes and resorts. In ad-

Figure 18–15   An aerial view of a barrier island along the south coast of Long Island. The sheltered lagoon is on the left side of the island.

dition, barrier islands are especially vulnerable to hurricanes, which can wash over low-lying islands and move enormous amounts of sediment in a very brief time.

ROCKY COASTLINES

In contrast to the sandy beaches typical of emergent coastlines, submergent coastlines are commonly sediment poor and are characterized by steep, rocky shores. In many areas on land, bedrock is exposed or covered by a thin layer of soil. If this type of sediment-poor terrain is submerged, and if rivers do not supply large amounts of sand, the coastline is rocky.

A submergent coast is commonly irregular, with many bays and headlands. The coast of Maine, with its numerous inlets and rocky bluffs, is a submergent coastline (Fig. 18–16). Small sandy beaches form in protected coves, but most of the headlands are rocky and steep.

As waves approach an irregular, rocky coast, they reach the headlands first, breaking against the rocks and eroding the cliffs. The waves then refract around the headland and break against its sides. Thus, most of the wave energy is spent on the headlands. As a result, the waves inside the adjacent bay are less energetic and deposit the sediment eroded from the headland. As the headlands erode and the interiors of bays fill with sediment, an irregular coastline eventually straightens (Fig. 18–17).

A **wave-cut cliff** forms when waves erode a rocky headland into a steep profile. As the cliff erodes, it leaves a flat or gently sloping **wave-cut platform** (Fig. 18–18). If waves cut a cave into a narrow headland, the cave may eventually erode all the way through the headland, forming a scenic **sea arch**. When an arch collapses or when the inshore part of a headland erodes faster than the tip, a pillar of rock called a **sea stack** forms (Fig. 18–19). As waves continue to batter the rock, eventually the sea stack crumbles.

Figure 18–16   The Maine coast is a rocky, irregular, submergent coastline.

(a)

(b)

(c)

Figure 18–17 (a, b, and c) A three-step sequence in which an irregular coastline is straightened. Erosion is greatest at the points of the headlands, and sediment is deposited inside the bays, leading to a gradual straightening of the shoreline.

If the sea floods a long, narrow, steep-sided coastal valley, a sinuous bay called a **fjord** is formed. Fjords are common at high latitudes, where rising sea level flooded glacially scoured valleys submerged as the Pleistocene glaciers melted (refer back to Fig. 17–18). Fjords may be hundreds of meters deep, and often the cliffs drop straight into the sea.

An **estuary** forms where rising sea level or a sink-ing coastline submerges a broad river valley or other basin. Estuaries are ordinarily shallow and have gentle, sloping beaches. Streams transport nutrients to the bay, and the shallow water provides habitats for marine organisms. Estuaries also make excellent harbors and therefore are prime sites for industrial activity. As a result, many estuaries have become seriously polluted in recent years.

Figure 18–18 Waves hurl sand and gravel against solid rock to erode cliffs and wave-cut platforms along the Oregon coast.

Figure 18–19 Sea stacks are common along the rocky Oregon coast.

18.5 DEVELOPMENT OF COASTLINES

CASE STUDY

Long Island

Long Island extends eastward from New York City and is separated from Connecticut by Long Island Sound (Fig. 18–20). Narrow, low barrier islands line the southern coast of Long Island. Longshore currents flow westward, eroding sand from glacial deposits at the eastern end of the island and carrying it past beaches and barrier islands toward New Jersey.

Over geologic time, the beaches and barrier islands of Long Island are unstable. The glacial deposits at the eastern end of the island will become exhausted and the flow of sand will cease. Then the entire coastline will erode and the barrier islands and beaches will disappear. However, this change will not occur in the near future because a vast amount of sand is still available at the eastern end of the island. Thus, the beaches are stable over a period of hundreds of years. Over this time, longshore currents move sand continuously. At any point along the beach, the currents erode and deposit sand at approximately the same rate.

If we narrow our time perspective further and look at a Long Island beach over a season or during a single storm, it may shrink or expand. Over such short times, the rates of erosion and deposition are not equal. In the winter violent waves and currents erode beaches, whereas sand accumulates on the beaches during the calmer summer months. In an effort to prevent these seasonal fluctuations and to protect their personal beaches, Long Island property owners have built stone barriers called **groins** from shore out into the water. A groin intercepts the steady flow of sand moving from the east and keeps that particular part of the beach from eroding (Fig. 18–21). But the groin impedes the overall flow of sand. West of the groin the beach erodes as usual, but the sand is not replenished because the upstream groin traps it. As a result, beaches downcurrent from the groin erode away (Fig. 18–22). The landowner living downcurrent from a groin may then decide to build another groin to protect his or her beach (Fig. 18–21c). The situation has a domino effect, with the net result that millions of dollars are spent in ultimately futile attempts to stabilize a system that was naturally stable in its own dynamic manner.

Storms pose another dilemma. Hurricanes strike Long Island in the late summer and fall, generating storm waves that completely overrun the barrier islands, flattening dunes and eroding beaches. When the storms are over, gentler waves and longshore currents carry sediment back to the beaches and rebuild them. As the sand accumulates again, salt marshes rejuvenate and the dune grasses grow back within a few months.

These short-term fluctuations are incompatible with human ambitions, however. People build houses, resorts, and hotels on or near the shifting sands. The owner of a home or resort hotel cannot allow the buildings to be flooded or washed away. Therefore, property owners construct large sea walls along the beach. When a storm wave rolls across an undeveloped low-lying beach, it dissipates its energy gradually as it flows over the dunes and transports sand. The beach is like a judo master who defeats an opponent by yielding to the attack, not countering it head on. A sea wall interrupts this gradual absorption of wave energy. The waves crash violently against the barrier and erode sediment at its base until

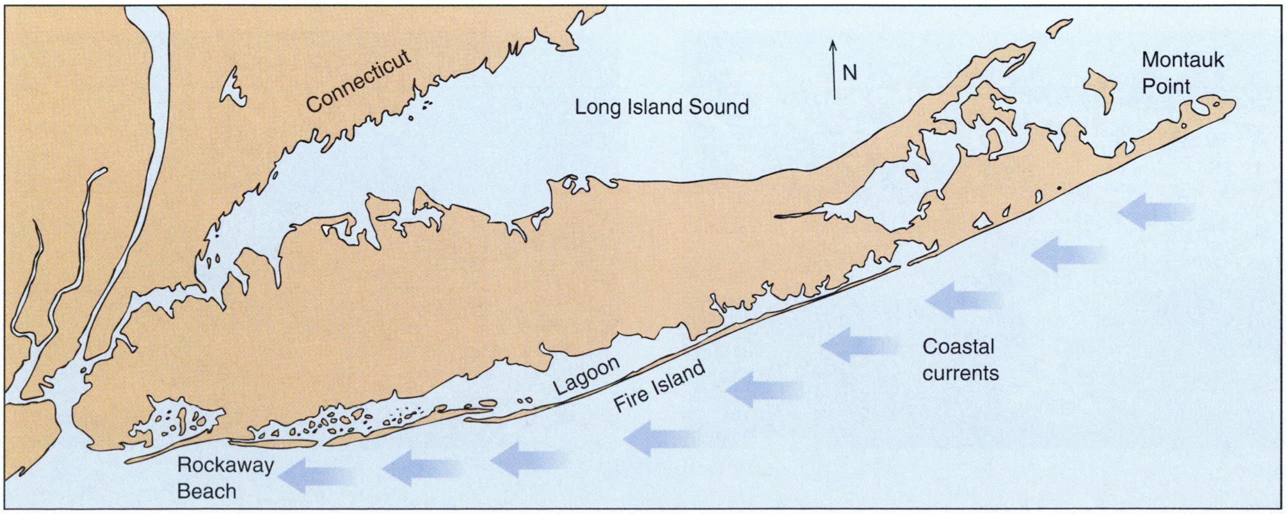

Figure 18–20 Longshore currents carry sand westward along the south shore of Long Island.

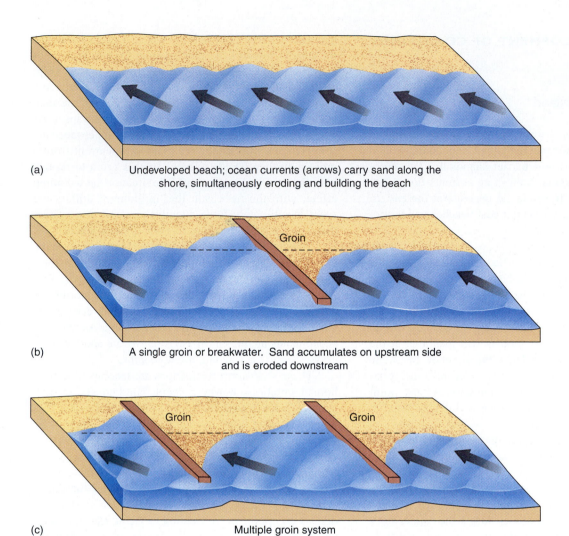

(a) Undeveloped beach; ocean currents (arrows) carry sand along the shore, simultaneously eroding and building the beach

(b) A single groin or breakwater. Sand accumulates on upstream side and is eroded downstream

(c) Multiple groin system

Figure 18–21 (a) Longshore currents simultaneously erode and deposit sand along an undeveloped beach. (b) A single groin or breakwater traps sand on the upstream side, resulting in erosion on the downstream side. (c) A multiple groin system propagates the uneven distribution of sand along the entire beach.

(a)   (b)

Figure 18–22 (a) This aerial photograph of a Long Island beach shows sand accumulating on the upstream side of a groin, and erosion on the downstream side. (b) A closeup of the house in (a) shows waves lapping against the foundation.

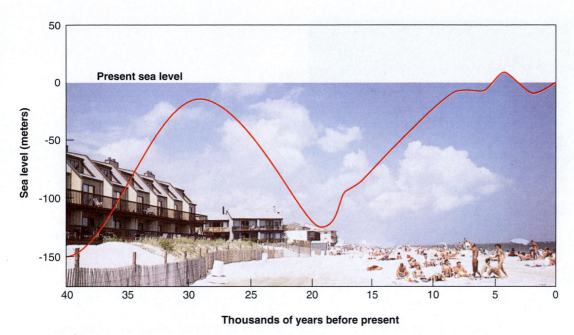

Figure 18–23 Sea level has fluctuated more than 150 meters during the past 40,000 years. (Data from J. D. Hansom, *Coasts.* Cambridge, U.K.: Cambridge University Press, 1988)

the wall collapses. It may seem surprising that a reinforced concrete sea wall is *more* likely to be permanently destroyed than a beach of grasses and sand dunes, yet this is often the case.

▶ **18.6 GLOBAL WARMING AND SEA-LEVEL RISE**

Sea level has risen and fallen repeatedly in the geologic past, and coastlines have emerged and submerged throughout Earth history. During the past 40,000 years, sea level has fluctuated by 150 meters, primarily in response to growth and melting of glaciers (Fig. 18–23). The rapid sea-level rise that started about 18,000 years ago began to level off about 7000 years ago. By coincidence, humans began to build cities about 7000 years ago. Thus, civilization has developed during a short time when sea level has been relatively constant.

Global sea level started rising again about 75 years ago, at a rate of about 1.5 to 2.5 millimeters per year (Fig. 18–24). The change in a single year is small, but it is half as fast as the dramatic postglacial sea-level rise. Many climatologists predict that the greenhouse effect will raise global temperature during the next century. If global warming occurs, sea level will rise because of melting polar ice sheets and expansion of seawater. Although estimates vary, many scientists predict a 1-meter rise in sea level by the year 2100.

Consequences of a 1-meter sea-level rise vary with location and economics. The wealthy, developed nations

would build massive barriers to protect cities and harbors. In regions where global sea-level rise is compounded by local tectonic sinking, dikes are already in

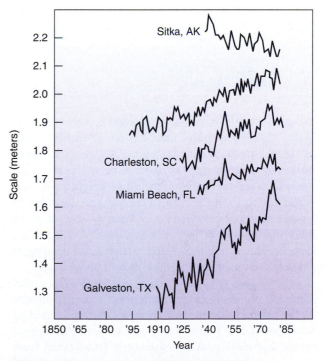

Figure 18–24 Coastal emergence and submergence at several locations in the United States. Land subsidence in Galveston has led to rapid submergence, and tectonic uplift along the Alaskan coast has led to local emergence in Sitka. (Stephen H. Schneider, *Global Warming*, p. 164)

Figure 18–25 A 1-meter sea-level rise would flood 17 percent of the land area of Bangladesh and displace 38 million people. (Wide World Photos)

place or planned. Portions of Holland lie below sea level, and the land is protected by a massive system of dikes. In London, where the high-tide level has risen by 1 meter in the past century, multimillion-dollar storm gates have been built on the Thames River. A similar system is now planned to protect Venice from further flooding.

If sea level rises as predicted, people in the United States will spend about $10 billion per year to protect developed coastlines. The cost will exceed that of any construction project in history. Wetlands, farms, and houses that are not valuable enough to be protected will be lost.

If sea level rises by 1 meter, 20,000 square kilometers of dry land and 17,000 square kilometers of coastal wetlands in the United States will be flooded. In addition, storm damage and coastal erosion will increase.

Many poor countries cannot afford coastal protection. A 1-meter rise in sea level would flood portions of the Nile delta, displacing 10 million people and decreasing Egypt's agricultural productivity by 15 percent. Seventeen percent of the land area of Bangladesh would be flooded, displacing 38 million inhabitants (Fig. 18–25).

SUMMARY

In deep water, the size of a wave depends on (1) the wind speed, (2) the amount of time that the wind has blown, and (3) the **fetch**. The highest part of a wave is the **crest**; the lowest, the **trough**. The distance between successive crests is called the **wavelength**. **Wave height** is the vertical distance from the crest to the trough. The water in a wave moves in circular paths. When a wave nears the shore, the bottom of the wave is slowed down and the wave **breaks**, creating **surf**. **Refraction** is the bending of a wave when it strikes the shore at an oblique angle.

Surface currents are driven by wind and affect global climate. **Longshore currents** transport sediment along a shore. A **beach** is a strip of shoreline that is washed by waves and tides. Weathering produces sediment along a beach, but most coastal sediment is transported from river deltas and glacial deposits. Reefs also add sediment in certain areas.

A **reef** is a wave-resistant ridge or mound built by corals, algae, oysters, or other marine organisms.

An **atoll** forms when an island, surrounded by a reef, sinks.

If land rises or sea level falls, the coastline migrates toward the open ocean and old beaches are abandoned above the sea, forming an **emergent coastline**. Emergent coastlines are sediment rich and are characterized by sandy beaches, **spits**, **baymouth bars**, and **barrier islands**. In contrast, a **submergent coastline** forms when land sinks or sea level rises. Submergent coastlines are often sediment poor. **Wave-cut cliffs**, **wave-cut platforms**, **arches**, and **stacks** are common in this environment. Irregular coastlines are straightened by erosion and deposition. **Fjords** are submerged glacial valleys. **Estuaries** are submerged river beds and flood plains.

Human intervention may upset the natural movement of coastal sediment and alter patterns of erosion and deposition on beaches. Sea level has been rising over the past century, and it may continue to rise into the next.

KEY WORDS

fetch *318*
crest *318*
trough *318*
wavelength *318*
wave height *318*
surf *318*
refraction *318*
spring tide *320*
neap tide *320*

tidal current *321*
rip current *321*
undertow *321*
longshore current *321*
beach drift *321*
beach *322*
foreshore *322*
backshore *322*
intertidal zone *322*

reef *322*
atoll *322*
emergent coastline *323*
submergent coastline *323*
eustatic change *324*
spit *325*
baymouth bar *326*
barrier island *326*

lagoon *326*
wave-cut cliff *327*
wave-cut platform *327*
sea arch *327*
sea stack *327*
fjord *328*
estuary *328*
groin *329*

REVIEW QUESTIONS

1. List the three factors that determine the size of a wave.

2. Draw a picture of a wave and label the crest, the trough, the wavelength, and the wave height.

3. Describe the motion of both the surface and the deeper layers of water that is disturbed by waves.

4. Explain how surf forms.

5. What is refraction? How does it affect coastal erosion?

6. Explain the differences among a mid-oceanic current, a tidal current, a rip current, a longshore current, and beach drift.

7. List four different sources of coastal sediment.

8. Discuss the most important weathering processes along a coastline.

9. What is an emergent coastline and how does it form? Are emergent coastlines sediment rich or sediment poor? Why?

10. What is a submergent coastline and how does it form? Are submergent coastlines sediment rich or sediment poor? Why?

11. Compare and contrast a beach, a barrier island, and a spit.

12. Explain how an irregular coastline is straightened by coastal processes.

13. Describe some dominant features of a sediment-poor coastline.

14. What is a groin? How does it affect the beach in its immediate vicinity? How does it affect the entire shoreline?

15. Explain how greenhouse warming could lead to a rise in sea level.

DISCUSSION QUESTIONS

1. Earthquake waves were discussed in Chapter 10. Compare and contrast earthquake waves with water waves.

2. How can a ship survive 30-meter-high storm waves, while a house along the beach will be smashed by waves of the same size?

3. Explain why very large waves cannot strike a beach directly in shallow coastal waters.

4. During World War II, few maps existed of the underwater profile of shore lines. When planning amphibious attacks on beaches of the islands in the Pacific, Allied commanders needed to know how deep the water was adjacent to the shore. Explain how this information could be deduced from aerial photographs of breaking waves and surf.

5. Imagine that an oil spill occurred from a tanker accident. Discuss the effects of mid-ocean currents, longshore currents, storm waves, and tides on the dispersal of the oil.

6. In Section 18.4 we explained how erosion and deposition tend to smooth out an irregular coastline by eroding headlands and depositing sediment in bays. If coastlines are affected in this manner, why haven't they all been smoothed out in the 4.6-billion-year history of the Earth?

7. Prepare a three-way debate. Have one side argue that the government should support the construction of groins. Have the second side argue that the government should prohibit the construction of groins. The third position defends the argument that groins should be permitted, but not supported.

8. Prepare another debate to argue whether or not government funding should be used to repair storm damage to property on barrier islands.

9. In evaluating flood danger, hydrologists use the concept of the 100-year flood. Would a similar concept be useful in planning coastal development?

Geologic Resources

With time, the human use of geologic resources has become increasingly sophisticated. Prehistoric people used flint and obsidian to make weapons and hide scrapers. About 7000 B.C., people learned to shape and fire clay to make pottery. Archeologists have found copper ornaments in Turkey from 6500 B.C.; 1500 years later, Mesopotamian farmers used copper farm implements. Today, the silicon chip that operates your computer, the titanium valves in a space probe, and the gasoline that powers your car are all derived from Earth resources.

A miner shovels rubble to clear the track in an underground coal mine. (Mike Abrahams/Tony Stone Worldwide)

▶ 19.1 GEOLOGIC RESOURCES

Humans use two different types of geologic resources: mineral resources and energy resources. **Mineral resources** include all useful rocks and minerals. Mineral resources fall into two groups: nonmetallic resources and metals. A **nonmetallic resource** is any useful rock or mineral that does not have metallic properties, such as salt or sand and gravel. A **metal** is any chemical element with a metallic luster, ductility, and the ability to conduct electricity and heat. About 40 metals are commercially important. Some, such as iron, lead, copper, aluminum, silver, and gold, are familiar (Fig. 19–1). Others, such as vanadium, titanium, and tellurium, are less well known but are vital to industry. All mineral resources are **nonrenewable**: We use them up at a much faster rate than natural processes create them.

We use **energy resources** for heat, light, work, and data transmission. Petroleum, coal, and natural gas are called **fossil fuels** because they formed from the remains of plants and animals. **Nuclear fuels** are radioactive isotopes used to generate electricity in nuclear reactors. Uranium is the most commonly used nuclear fuel. These energy resources, like mineral resources, are nonrenewable. **Alternative energy resources**, such as solar, wind, and geothermal energy, are renewable.

Figure 19–1 In the early 1900s, miners extracted gold, copper, and other metals from underground mines such as this one 600 meters below the surface in Butte, Montana. (Montana Historical Society)

▶ 19.2 NONMETALLIC MINERAL RESOURCES

When we think about striking it rich from mining, we usually think of gold. However, more money has been made mining sand and gravel than gold. For example, in the United States in 1994, sand and gravel produced $4.26 billion in revenue, but gold produced $4.1 billion. Sand and gravel are mined from stream and glacial deposits, sand dunes, and beaches.

Portland cement is made by heating a mixture of crushed limestone and clay. Concrete is a mixture of cement, sand, and gravel. Reinforced with steel, it is used to build roads, bridges, and buildings.

Many buildings are faced with stone—usually granite or limestone, although marble, slate, sandstone, and other rocks are also used. Stone is mined from **quarries** cut into bedrock.

▶ 19.3 METALS AND ORE

If you picked up any rock and sent it to a laboratory for analysis, the report would probably show that the rock contains measurable amounts of iron, gold, silver, aluminum, and other valuable metals. However, the concentrations of these metals are so low in most rocks that the extraction cost would be much greater than the income gained by selling the metals. In certain locations, however, geologic processes have enriched metals many times above their normal concentrations (Table 19–1).

A **mineral deposit** is a local enrichment of one or more minerals. **Ore** is rock sufficiently enriched in one or more minerals to be mined profitably. Geologists usually use the term *ore* to refer to metallic mineral deposits, and the term is commonly accompanied by the name of the metal—for example, iron ore or silver ore. Table 19–1 shows that the concentration of a metal in ore may exceed its average abundance in ordinary rock by a factor of more than 100,000. **Mineral reserves** are the known supply of ore in the ground. The term can refer to the amount of ore remaining in a particular mine, or it can be used on a global or national scale.

▶ 19.4 HOW ORE FORMS

One of the primary objectives of many geologists is to find new ore deposits. Successful exploration requires an understanding of the processes that concentrate metals to form ore. For example, platinum concentrates in certain types of igneous rocks. Therefore, if you were exploring for platinum, you would focus on those rocks rather than on sandstone or limestone.

Table 19–1 • COMPARISON OF CONCENTRATIONS OF SPECIFIC ELEMENTS IN EARTH'S CRUST WITH CONCENTRATIONS NEEDED TO OPERATE A COMMERCIAL MINE

ELEMENT	NATURAL CONCENTRATION IN CRUST (% BY WEIGHT)	CONCENTRATION REQUIRED TO OPERATE A COMMERCIAL MINE (% BY WEIGHT)	ENRICHMENT FACTOR
Aluminum	8	24–32	3–4
Iron	5.8	40	6–7
Copper	0.0058	0.46–0.58	80–100
Nickel	0.0072	1.08	150
Zinc	0.0082	2.46	300
Uranium	0.00016	0.19	1,200
Lead	0.00010	0.2	2,000
Gold	0.0000002	0.0008	4,000
Mercury	0.000002	0.2	100,000

MAGMATIC PROCESSES

Magmatic processes form mineral deposits as liquid magma solidifies to form an igneous rock. These processes create metal ores as well as some gems and valuable sulfur deposits.

Layered Plutons

Some large bodies of igneous rock, particularly those of mafic (basaltic) composition, solidify in layers. Each layer contains different minerals and is of a different chemical composition from adjacent layers. Some of the layers may contain rich ore deposits.

The layering can develop by at least three processes:

1. Recall from Chapter 4 that cooling magma does not solidify all at once. Instead, higher-temperature minerals crystallize first, and lower-temperature minerals form later as the temperature drops. Most minerals are denser than magma. Consequently, early-formed crystals may sink to the bottom of a magma chamber in a process called **crystal settling** (Fig. 19–2). In some instances, ore minerals crystallize with other early-formed minerals and consequently accumulate in layers near the bottom of the pluton.

2. Some large bodies of mafic magma crystallize from the bottom upward. Thus, early-formed ore minerals become concentrated near the base of the pluton by this process.

3. In some cases, a large body of magma may begin to develop layering by either of the two processes just described. Then, additional magma of a different composition or temperature may flow into the magma chamber. As a result of these changes, dif-

ferent minerals may crystallize at different times to create layering in the pluton.

The largest ore deposits found in mafic layered plutons are the rich chromium and platinum reserves of South Africa's Bushveld intrusion. The pluton is about 375 by 300 kilometers in area—roughly the size of the state of Maine—and about 7 kilometers thick. The Bushveld deposits contain more than 20 billion tons of chromium and more than 10 billion grams of platinum, the greatest reserves in any known deposit on Earth. The

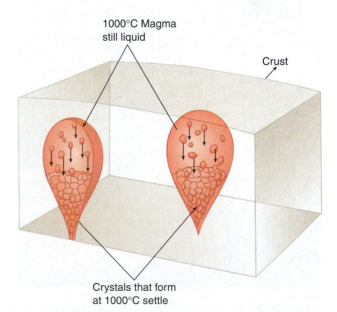

Figure 19–2 Early-formed crystals settle and concentrate near the bottom of a magma chamber.

world's largest known nickel deposit occurs in a layered mafic pluton at Sudbury, Ontario, and rich platinum ores are mined from layered plutons in southern Montana and Norilsk, Russia.

Kimberlites

In Chapter 5, we described rare igneous rocks called kimberlites that originate in the mantle and are the world's main source of diamonds. Diamonds are used both for jewelry and as industrial abrasives because of their great hardness. Now, however, most industrial diamonds are produced synthetically.

Volcanic Vent Deposits

Sulfur, used primarily for sulfuric acid in industrial applications, precipitates as a pure yellow deposit from gases escaping from some volcanic vents (Fig. 19–3).

Figure 19–3 Yellow sulfur coats the vent of Ollagüe volcano, southern Bolivia.

Such deposits are sometimes mined even as the sulfur-rich fumes continue to escape from the volcano.

HYDROTHERMAL PROCESSES

Hydrothermal processes are probably responsible for the formation of more ore deposits, and a larger total quantity of ore, than all other processes combined. To form a hydrothermal ore deposit, hot water (hence the roots *hydro* for water and *thermal* for hot) dissolves metals from rock or magma. The metal-bearing solutions then seep through cracks or through permeable rock until they precipitate to form an ore deposit.

Three main sources provide water for hydrothermal activity.

1. Many magmas, particularly those of granitic composition, leave behind a water-rich residual fluid after most of the magma has solidified. Under certain conditions, that fluid crystallizes to form pegmatite, as described in Chapter 4. Under other conditions, the water and dissolved ions escape from the magma chamber to form hydrothermal solutions. For this reason, hydrothermal ore deposits are commonly associated with granite and similar igneous rocks.

2. Ground water can seep into the crust where it is heated and forms a hydrothermal solution. This is particularly true in areas of active volcanism where hot rock or magma heats ground water at shallow depths. For this reason, hydrothermal ore deposits are also common in volcanic regions.

3. In the oceans, seawater is heated as it seeps into cracks along the mid-oceanic ridge and near submarine volcanoes.

As you learned in Chapter 6, water by itself is capable of dissolving some minerals. The dissolved salts and high temperature of hydrothermal solutions greatly increase their ability to dissolve minerals. Thus, hot, salty hydrothermal water is a very powerful solvent, capable of dissolving and transporting metals.

Table 19–1 shows that tiny amounts of all metals are found in the average rocks of the Earth's crust. For example, gold makes up 0.0000002 percent of the crust, while copper makes up 0.0058 percent and lead 0.0001 percent. As hydrothermal solutions migrate through rock, they dissolve these metals. Although the metals are present in very low concentrations in country rock, hydrothermal solutions percolate through vast volumes of rock, dissolving and accumulating the metals. The solutions then deposit the metals when they encounter changes in temperature, pressure, or chemical environment (Fig. 19–4). In this way, hydrothermal solutions

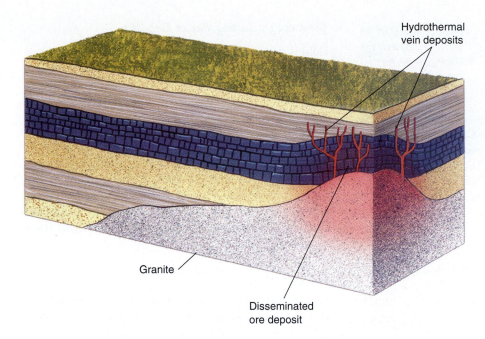

Hydrothermal
vein deposits

Granite

Disseminated
ore deposit

Figure 19–4   Hydrothermal ore deposits form when hot water deposits metals in bedrock.

scavenge metals from large volumes of average crustal rocks and then deposit them locally to form ore. In addition, some magmas also contain metals, which concentrate with the hydrothermal solutions that form as the magma solidifies.

Types of Hydrothermal Ore Deposits

A **hydrothermal vein deposit** forms when dissolved metals precipitate in a fracture in rock. Ore veins range from less than a millimeter to several meters in width. A single vein can yield several million dollars worth of gold or silver. The same hydrothermal solutions may also soak into country rock surrounding the vein to create a large but much less concentrated **disseminated ore deposit**. Because they commonly form from the same solutions, rich ore veins and disseminated deposits are often found together. The history of many mining districts is one in which early miners dug shafts and tunnels to follow the rich veins. After the veins were exhausted, later miners used huge power shovels to extract low-grade ore from disseminated deposits surrounding the veins.

Disseminated copper deposits, with accompanying veins, are abundant along the entire western margin of North and South America (Fig. 19–5). They are most commonly associated with large plutons and are called **porphyry copper deposits**. (*Porphyry* is a term for an igneous rock in which large crystals, usually feldspar, are set in a finer-grained matrix. Most porphyry copper

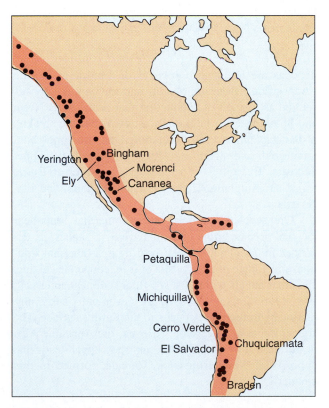

Figure 19–5   Porphyry copper deposits in the Western Hemisphere lie along modern or ancient tectonic plate boundaries.

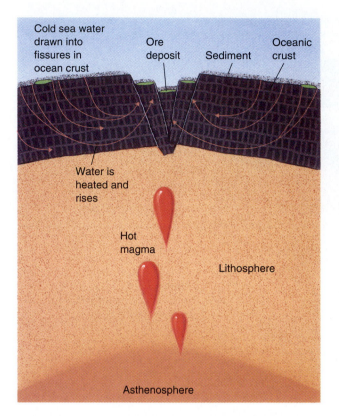

Figure 19–6 Submarine hydrothermal ore deposits precipitate from circulating seawater near the mid-oceanic ridge.

deposits are found in porphyrys of granitic to dioritic composition.) Both the plutons and the copper deposits formed as a result of subduction that occurred as North and South America migrated westward after the breakup of Pangea. Other metals, including lead, zinc, molybdenum, gold, and silver, are found with porphyry copper deposits. Examples of such deposits occur at Butte, Montana; Bingham, Utah; Morenci, Arizona; and Ely, Nevada.

Ore deposits also form at hydrothermal vents near the mid-oceanic ridge and submarine volcanoes, as described in Chapter 11. Metal-bearing hydrothermal solutions precipitate huge deposits of iron, copper, lead, zinc, and other metals within the sea-floor sediment and basalt (Fig. 19–6). Tectonic activity may eventually carry submarine hydrothermal deposits to the Earth's surface. The copper deposits of Jerome, Arizona, the Appenine Alps of northern Italy and Cyprus, and the copper–lead–zinc deposits of New Brunswick, Canada, formed in this manner.

SEDIMENTARY PROCESSES

Two types of sedimentary processes form ore deposits: sedimentary sorting and precipitation.

Sedimentary Sorting: Placer Deposits

Gold is denser than any other mineral. Therefore, if you swirl a mixture of water, gold dust, and sand in a gold pan, the gold falls to the bottom first (Fig. 19–7). Differential settling also occurs in nature. Many streams carry silt, sand, and gravel with an occasional small grain of gold. The gold settles first when the current slows down. Over years, currents agitate the sediment and the heavy grains of gold work their way into cracks and crevices in the stream bed. Thus, grains of gold concentrate near bedrock or in coarse gravel, forming a **placer deposit** (Fig. 19–8). It was primarily placer deposits that brought prospectors to California in the Gold Rush of 1849.

Precipitation

Ground water dissolves minerals as it seeps through soil and bedrock. In most environments, ground water eventually flows into streams and then to the sea. Some of these dissolved ions, such as sodium and chloride, make seawater salty. In deserts, however, lakes develop with no outlet to the ocean. Water flows into the lakes but can escape only by evaporation. As the water evaporates, the

Figure 19–7 Jeffery Embrey panning for gold near his cabin in Park City, Montana, in 1898. (Maud Davis Baker/Montana Historical Society)

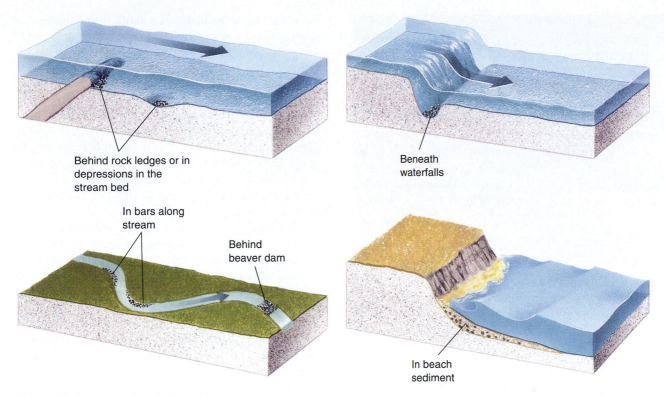

Behind rock ledges or in
depressions in the
stream bed

In bars along
stream

Behind
beaver dam

Beneath
waterfalls

In beach
sediment

Figure 19–8   Placer deposits form where water currents
slow down and deposit heavy minerals.

dissolved ions concentrate until they precipitate to form
evaporite deposits.

You can perform a simple demonstration of evaporation and precipitation. Fill a bowl with warm water and
add a few teaspoons of table salt. The salt dissolves and
you see only a clear liquid. Set the bowl aside for a few
days until the water evaporates. The salt precipitates and
encrusts the sides and bottom of the bowl.

Evaporite deposits formed in desert lakes include
table salt, borax, sodium sulfate, and sodium carbonate.
These salts are used in the production of paper, soap, and
medicines and for the tanning of leather.

Several times during the past 500 million years, shallow seas covered large regions of North America and all
other continents. At times, those seas were so poorly
connected to the open oceans that water did not circulate
freely between them and the oceans. Consequently, evaporation concentrated the dissolved ions until salts precipitated as marine evaporites. Periodically, storms
flushed new seawater from the open ocean into the shallow seas, providing a new supply of salt. Thick marine
evaporite beds formed in this way underlie nearly 30 percent of North America. Table salt, gypsum (used to manufacture plaster and sheetrock), and potassium salts (used
in fertilizer) are mined extensively from these deposits.

About 1 billion tons of iron are mined every year, 90
percent from sedimentary rocks called **banded iron formations**, which consist of layers of iron-rich minerals
sandwiched between beds of silicates. The alternating
layers are a few centimeters thick and give the rocks their
banded appearance (Fig. 19–9). The most abundant and
economically important banded iron formations developed between 2.6 and 1.9 billion years ago when iron
precipitated from seawater as a result of rising atmospheric oxygen concentration.

WEATHERING PROCESSES

In environments with high rainfall, the abundant water
dissolves and removes most of the soluble ions from soil
and rock near the Earth's surface. This process leaves the
relatively insoluble ions in the soil to form **residual
deposits**. Both aluminum and iron have very low
solubilities in water. **Bauxite**, the principal source of
aluminum, forms as a residual deposit, and in some instances iron also concentrates enough to become ore.
Most bauxite deposits form in warm, rainy, tropical or
subtropical environments where chemical weathering
occurs rapidly, such as those of modern Jamaica, Cuba,
Guinea, Australia, and parts of the southeastern United

Figure 19–9 In this banded iron formation from Michigan, the red bands are iron minerals and the dark layers are chert. (Barbara Gerlach/Visuals Unlimited)

States (Fig. 19–10). Some bauxite deposits are found today in regions with dry, cool climates. Most of them, however, formed when the regions had a warm, wet climate, and they reflect climatic change since their origin.

Weathering processes also can enrich metal concentrations in mineral deposits formed by other processes. For example, a disseminated hydrothermal deposit may contain copper, lead, zinc, and silver, but in concentra-

tions too low to be mined at a profit. Over millions of years, ground water and rain can weather the deposit and carry off most of the dissolved rocks and minerals in solution. In some cases, however, the valuable metals react to form new minerals in the zone of weathering and are not removed. In this way, the metals may become concentrated by factors of tens or hundreds to create rich **supergene ore** lying above a low-grade mineral deposit. These mineral deposits are easily mined because they lie at the Earth's surface. Supergene ore caps once covered many of the great porphyry copper deposits of the United States, but because they were so rich and easily mined, most are now gone.

Weathering also forms immense amounts of clay, a nonmetallic resource described in Chapter 6. One clay mineral, called kaolin, is the primary constituent of porcelain and other ceramics. Another, smectite, is mixed with water and other minerals to make drilling mud, a clay-rich slurry used to cool and lubricate drill bits in the drilling of deep wells. Several types of clay are used to line sanitary landfills and irrigation ditches and ponds because wet clay is impermeable.

METAMORPHIC PROCESSES

Recall from Chapter 8 that a metamorphic rock forms when heat and pressure alter the mineralogy and texture of any preexisting rock. Metamorphism can also expel water from rocks to create hydrothermal fluids, which, in turn, deposit metal ores. Thus, some hydrothermal ore deposits are of metamorphic origin.

Metamorphic processes also form several types of nonmetallic mineral resources. Graphite, the main component of the "lead" in pencils, forms when metamorphism alters the carbon in some organic-rich rocks. Asbestos is a commercial name for two different minerals formed by metamorphism. Metamorphism also forms marble, a valuable building stone and sculptor's material, from limestone.

▶ 19.5 MINERAL RESERVES

Mining depletes mineral reserves by decreasing the amount of ore remaining in the ground; but reserves may increase in two ways. First, geologists may discover new mineral deposits, thereby adding to the known amount of ore. Second, subeconomic mineral deposits—those in which the metal is not sufficiently concentrated to be mined at a profit—can become profitable if the price of that metal increases, or if improvements in mining or refining technology reduce extraction costs.

Consider an example of the changing nature of reserves. In 1966 geologists estimated that global reserves

Figure 19–10 Bauxite forms by intense weathering of aluminum-bearing rocks. (H. E. Simpson/USGS)

of iron were about 5 billion tons.[1] At that time, world consumption of iron was about 280 million tons per year. Assuming that consumption continued at the 1966 rate, the global iron reserves identified in 1966 would have been exhausted in 18 years (5 billion tons/280 million tons per year = 18 years), and we would have run out of iron ore in 1984. But iron ore is still plentiful and cheap today because new and inexpensive methods of processing lower-grade iron ore were developed. Thus, deposits that were subeconomic in 1966, and therefore not counted as reserves, are now ore.

THE GEOPOLITICS OF METAL RESOURCES

The Earth's mineral resources are unevenly distributed, and no single nation is self-sufficient in all minerals. For example, almost two thirds of the world's molybdenum reserves and more than one third of the lead reserves are located in the United States. More than half of the aluminum reserves are found in Australia and Guinea. The United States uses 40 percent of all aluminum produced in the world, yet it has no large bauxite deposits. Zambia and Zaire supply half of the world's cobalt, although neither nation uses the metal for its own industry.

Five nations—the United States, Russia, South Africa, Canada, and Australia—supply most of the mineral resources used by modern societies. Many other nations have few mineral resources. For example, Japan has almost no metal or fuel reserves; despite its thriving economy and high productivity, it relies entirely on imports for both.

Developed nations consume most of the Earth's mineral resources. Four nations—the United States, Japan, Germany, and Russia—consume about 75 percent of the

most intensively used metals, although they account for only 14 percent of world population.

Currently, the United States depends on 25 other countries for more than half of its mineral resources. Some must be imported because we have no reserves of our own. We have reserves of others, but we consume them more rapidly than we can mine them, or we can buy them more cheaply than we can mine them.

▶ 19.6 COAL

The three major fossil fuels are coal, petroleum, and natural gas. All form from the partially decayed remains of living organisms. Humans began using coal first because it is easily mined and can be burned without refining.

Coal-fired electric generating plants burn about 60 percent of the coal consumed in the United States. The remainder is used to make steel or to produce steam in factories. Although it is easily mined and abundant in many parts of the world, coal emits air pollutants that can be removed only with expensive control devices.

Large quantities of coal formed worldwide during the Carboniferous Period, between 360 and 285 million years ago, and later in Cretaceous and Paleocene times, when warm, humid swamps covered broad areas of low-lying land. Coal is probably forming today in some places, such as in the Ganges River delta in India, but the process is much slower than the rate at which we are consuming coal reserves. As shown in Figure 19–11, widespread availability of this fuel is expected at least until the year 2200.

COAL FORMATION

When plants die in forests and grasslands, organisms consume some of the litter, and chemical reactions with oxygen and water decompose the remainder. As a result,

[1]B. Mason, *Principles of Geochemistry*, 3rd ed. New York: John Wiley, 1996, Appendix III.

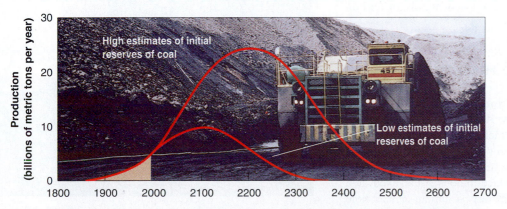

Figure 19–11   Past and predicted global coal supplies based on two different estimates of reserves. Shaded area shows coal already consumed.   (Adapted from M. King Hubbard)

(a) Litter falls to floor of stagnant swamp

(b) Debris accumulates, barrier forms, decay is incomplete

(c) Sediment accumulates, organic matter is converted to peat

(d) Peat is lithified to coal

Figure 19–12 Peat and coal form as sediment buries organic litter in a swamp.

little organic matter accumulates except in the topsoil. In some warm swamps, however, plants grow and die so rapidly that newly fallen vegetation quickly buries older plant remains. The new layers prevent atmospheric oxygen from penetrating into the deeper layers, and decomposition stops before it is complete, leaving brown, partially decayed plant matter called **peat.** Commonly, peat is then buried by mud deposited in the swamp.

Plant matter is composed mainly of carbon, hydrogen, and oxygen and contains large amounts of water. During burial, rising pressure expels the water and chemical reactions release most of the hydrogen and oxygen, and the proportion of carbon increases. The result is **coal,** a combustible rock composed mainly of carbon (Fig. 19–12).

▶ **19.7 MINES AND MINING**

Miners extract both coal and ore from **underground mines** and **surface mines.** A large underground mine may consist of tens of kilometers of interconnected passages that commonly follow ore veins or coal seams (Fig. 19–13). The lowest levels may be several kilometers deep. In contrast, a surface mine is a hole excavated into the Earth's surface. The largest human-created hole is the open-pit copper mine at Bingham Canyon, Utah. It is 4 kilometers in diameter and 0.8 kilometer deep.

The mine produces 230,000 tons of copper a year and smaller amounts of gold, silver, and molybdenum (Fig. 19–14). Most modern coal mining is done by large power shovels that extract coal from huge surface mines (Fig. 19–15).

Figure 19–13 Machinery extracts coal from an underground coal mine.

Figure 19–14 An aerial view of the Bingham Canyon, Utah, open-pit copper mine. (Agricultural Stabilization and Conservation Service/USDA)

In the United States, the Surface Mining Control and Reclamation Act requires that mining companies restore mined land so that it can be used for the same purposes for which it was used before mining began. In ad-

dition, a tax is levied to reclaim land that was mined and destroyed before the law was enacted. However, the government has been unable to enforce the act fully. More than 6000 unrestored coal and metal surface mines cover an area of about 90,000 square kilometers, slightly smaller than the state of Virginia. This figure does not include abandoned sand and gravel mines and rock quarries, which probably account for an even larger area.

Although underground mines do not directly disturb the land surface, many abandoned mines collapse, and occasionally houses have fallen into the holes (Fig. 19–16). Over 800,000 hectares (2 million acres) of land in central Appalachia have settled into underground mine shafts.

Both metal ore and coal are commonly covered and surrounded by soil and rock that is not of marketable quality. As a result, miners must dig up and discard large amounts of waste rock as they expose and extract the resource. Before pollution-control laws were enacted, the waste rock from both surface and underground mines was usually piled up near the mine. The wastes were easily eroded, and the muddy runoff poured into nearby streams, destroying aquatic habitats. Heavy metals, such as lead, cadmium, zinc, and arsenic, are common in many metal ores, and rain can leach them from the mine wastes and carry them to streams and ground water. In addition, sulfur is abundant in many metal ores as well as in coal. The sulfur reacts with water in the presence of air to produce sulfuric acid (H_2SO_4). If pollution control is inadequate, the sulfuric acid then runs off into streams and ground water below the mine or mill.

Today, responsible mining companies use several methods to stabilize mine and mill wastes. For example,

Figure 19–15 A huge power shovel dwarfs a person standing inside the Navajo Strip Mine in New Mexico. (H. E. Malde/USGS)

Figure 19–16 This house is being torn in half and tilted as it sinks into an abandoned underground coal mine. (Chuck Meyers/U.S. Department of the Interior)

they use crushed limestone to neutralize acid waters; they build well-designed settling ponds to trap silt; and they backfill abandoned mines and settling ponds. These measures can be costly, but they greatly reduce the quantity of pollutants that escape into streams and ground water.

ORE SMELTERS AND COAL BURNING

When sulfide ore minerals are refined or when sulfur-bearing coal is burned without pollution control devices, the sulfur escapes into the atmosphere, where it forms hydrogen sulfide and sulfuric acid. These gases contribute to acid precipitation. Most of the sulfur can be removed by pollution control devices, which are required by law in the United States. Other toxic elements and compounds, including heavy metals such as lead, cadmium, zinc, and arsenic, can escape from smelters into the atmosphere and water. They too can be removed by pollution control devices.

▶ 19.8 PETROLEUM AND NATURAL GAS

The first commercial petroleum well was drilled in the United States in 1859, ushering in a new energy age.

Petroleum is the most versatile of the fossil fuels. Crude oil, as it is pumped from the ground, is a gooey, viscous, dark liquid made up of thousands of different chemical compounds. It is then refined to produce propane, gasoline, heating oil, and other fuels. During refining, the crude oil is treated chemically and heated under pressure to break apart its large molecules. The mixture is then separated in multistory distillation columns (Fig. 19–17). Many petroleum products are used to manufacture plastics, nylon, and other useful materials.

Natural gas, or methane (CH_4), forms naturally when crude oil is heated above 100°C during burial. Many oil wells contain natural gas floating above the heavier liquid petroleum. In other instances, the lighter, more mobile gas escaped and was trapped elsewhere in a separate reservoir.

Natural gas is extracted as a nearly pure compound and is used without refining for home heating, cooking, and to fuel large electrical generating plants. Because natural gas contains few impurities, it releases no sulfur and other pollutants when it burns. This fuel has a higher net energy yield, produces fewer pollutants, and is less expensive to produce than petroleum. At current consumption rates, global natural gas supplies will last for 80 to 200 years.

THE ORIGIN OF PETROLEUM

Streams carry organic matter from decaying land plants and animals to the sea and to some large lakes, and deposit it with mud in shallow coastal waters. Marine plants and animals die and settle to the sea floor, adding more organic matter to the mud. Younger sediment then buries this organic-rich mud. Rising temperature and pressure resulting from burial convert the mud to shale. At the same time, the elevated temperature and pressure convert

Figure 19–17 An oil refinery converts crude oil to useful products such as gasoline.

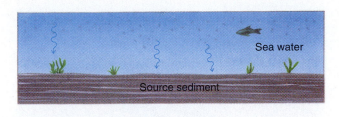

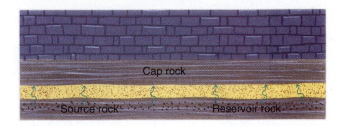

Figure 19–18   Most oil forms in shaly source rock. It must migrate to a permeable reservoir in order to be recovered from an oil well.

the organic matter to liquid **petroleum** that is finely dispersed in the rock (Fig. 19–18). The activity of bacteria may enhance the process. Typically petroleum forms in the temperature range from 50 to 100°C. At temperatures above about 100°C, oil begins to convert to natural gas. Consequently, many oil fields contain a mixture of oil and gas.

The shale or other sedimentary rock in which oil originally forms is called the **source rock**. Oil dispersed in shale cannot be pumped from an oil well because shale is relatively impermeable; that is, liquids do not flow through it rapidly. But under favorable conditions, petroleum migrates slowly to a nearby layer of permeable rock—usually sandstone or limestone—where it can flow readily. Because petroleum is less dense than water or rock, it then rises through the permeable rock until it is trapped within the rock or escapes onto the Earth's surface.

Many **oil traps** form where impermeable **cap rock** prevents the petroleum from rising further. Oil or gas then accumulates in the trap as a petroleum **reservoir**. The cap rock is commonly impermeable shale. Folds and faults create several types of oil traps (Fig. 19–19). In some regions, large, lightbulb-shaped bodies of salt have flowed upward through solid rocks to form salt domes. The ris-

ing salt folded the surrounding rock to form an oil trap (Fig. 19–19d). The salt originated as a sedimentary bed of marine evaporite, and it rose because salt is less dense than the surrounding rocks. An oil reservoir is not an underground pool or lake of oil. It consists of oil-saturated permeable rock that is like an oil-soaked sponge.

Geologic activity can destroy an oil reservoir as well as create one. A fault may fracture the cap rock, or tectonic forces may uplift the reservoir and expose it to erosion. In either case, the petroleum escapes once the trap is destroyed. Sixty percent of all oil wells are found in relatively young rocks that formed during the Cenozoic Era. Undoubtedly, much petroleum that had formed in older Mesozoic and Paleozoic rocks escaped long ago and decomposed at the Earth's surface.

PETROLEUM EXTRACTION, TRANSPORT, AND REFINING

To extract petroleum, an oil company drills a well into a reservoir and pumps the oil to the surface. Fifty years ago, many reservoirs lay near the surface and oil was easily extracted from shallow wells. But these reserves have been exploited, and modern oil wells are typically deeper. For example, in 1949, the average oil well drilled in the United States was 1116 meters deep. In 1994, the average well was 1629 meters deep. The average cost of drilling a new oil well in 1960 was slightly more than $200,000; by 1993, the cost had risen to about $350,000. After the hole has been bored, the expensive drill rig is removed and replaced by a pumper that slowly extracts the petroleum. In 1989, the USGS reported that oil reserves in the United States were being rapidly depleted. Petroleum companies found half as much oil as they did in the 1950s for every meter of exploratory drilling. However, in 1995 the USGS revised these earlier assessments.[2] According to the new report, due to improved extraction techniques, oil reserves increased by 41 percent between 1989 and 1995.

On the average, more than half of the oil in a reservoir is too viscous to be pumped to the surface by conventional techniques. This oil is left behind after an oil field has "gone dry," but it can be extracted by **secondary** and **tertiary recovery** techniques. In one simple secondary process, water is pumped into one well, called the injection well. The water floods the reservoir, driving oil to nearby wells, where both the water and oil are extracted. At the surface, the water is separated from the oil and reused, while the oil is sent to the refinery. One tertiary process forces superheated steam into the injection well. The steam heats the oil and makes it more fluid so that it can flow through the rock to an adjacent well.

[2]R. C. Burruss, *Geotimes*, July, 1995, p. 14.

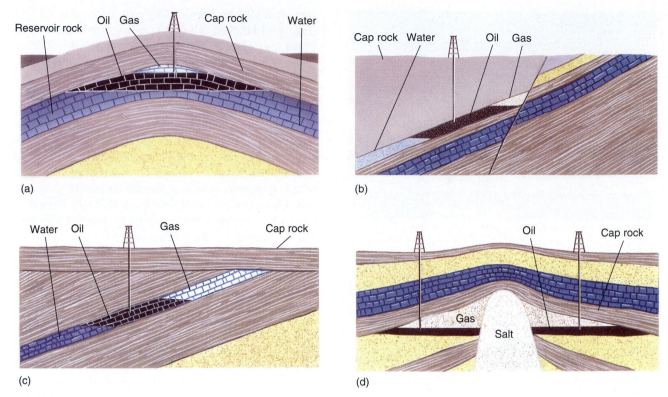

Figure 19–19 Four different types of oil traps. (a) Petroleum rises into permeable limestone in a structural dome. (b) A trap forms where a fault has moved impermeable shale against permeable sandstone. (c) Horizontally bedded shale traps oil in tilted limestone. (d) In a salt dome, a sedimentary salt deposit rises and deforms overlying strata to create a trap.

Because energy is needed to heat the steam, this type of extraction is not always cost effective or energy efficient. Another tertiary process pumps detergent into the reservoir. The detergent dissolves the remaining oil and carries it to an adjacent well, where the petroleum is then recovered and the detergent recycled.

In the United States, more than 300 billion barrels of secondary and tertiary oil remain in oil fields. In 1994, people in the United States consumed 6.4 billion barrels of petroleum. Thus, if all the secondary petroleum could be recovered, it would supply the United States for nearly 50 years. However, not all of the 300 billion barrels can be recovered, and the energy yield from secondary and tertiary extraction is reduced by the energy consumed by the processes.

Because an oil well occupies only a few hundred square meters of land, most cause relatively little environmental damage. However, oil companies have begun to extract petroleum from fragile environments such as the ocean floor and the Arctic tundra. To obtain oil from the sea floor, engineers build platforms on pilings driven into the ocean floor and mount drill rigs on these steel islands (Fig. 19–20). Despite great care, accidents occur during drilling and extraction of oil. When accidents occur at sea, millions of barrels of oil can spread throughout the waters, poisoning marine life and disrupting marine ecosystems. Significant oil spills have occurred in virtually all offshore drilling areas. In addition, tanker accidents have polluted parts of coastal oceans.

Although all oil refineries use expensive pollution-control equipment, these devices are never completely effective. As a result, some toxic and carcinogenic compounds escape into the atmosphere.

TAR SANDS

In some regions, large sand deposits are permeated with heavy oil and an oil-like substance called **bitumen**, which are too thick to be pumped. The richest tar sands exist in Alberta (Canada), Utah, and Venezuela.

In Alberta alone, tar sands contain an estimated 1 trillion barrels of petroleum. About 10 percent of this fuel is shallow enough to be surface mined (Fig. 19–21). Tar sands are dug up and heated with steam to make the

Figure 19–20   An offshore oil drilling platform.   (Sun Oil)

bitumen fluid enough to separate from the sand. The bitumen is then treated chemically and heated to convert it to crude oil. At present, several companies mine tar sands profitably, producing 11 percent of Canada's petroleum. Deeper deposits, composing the remaining 90 percent of the reserve, can be extracted using subsurface techniques similar to those discussed for secondary and tertiary recovery.

Figure 19–21   Mining tar sands in Alberta, Canada. (Syncrude Canada, Limited)

OIL SHALE

Some shales and other sedimentary rocks contain a waxy, solid organic substance called **kerogen**. Kerogen is organic material that has not yet converted to oil. Kerogen-bearing rock is called **oil shale**. If oil shale is mined and heated in the presence of water, the kerogen converts to petroleum. In the United States, oil shales contain the energy equivalent of 2 to 5 trillion barrels of petroleum, enough to fuel the nation for 315 to 775 years at the 1994 consumption rate (Fig. 19–22). However, many oil shales are of such low grade that they require more energy to mine and convert the kerogen to petroleum than is generated by burning the oil, so they will probably never be used for fuel. Oil from higher-grade oil shales in the United States would supply this country for nearly 75 years if consumption rates remained at 1994 levels. Oil shale deposits in most other nations are not as rich, so oil shale is less promising as a global energy source.

Water consumption is a serious problem in oil shale development. Approximately two barrels of water are needed to produce each barrel of oil from shale. Oil shale occurs most abundantly in the semiarid western United States. In this region, the scarce water is also needed for agriculture, domestic use, and industry.

When oil prices rose to $45 per barrel in 1981, major oil companies built experimental oil shale recovery plants. However, when prices plummeted a few years later, most of this activity came to a halt. Today, no large-scale oil shale mining is taking place in the United States.

Figure 19–22 Secondary recovery, tar sands, and oil shale increase our petroleum reserves significantly.

▶ 19.9 NUCLEAR FUELS AND REACTORS

A modern nuclear power plant uses **nuclear fission** to produce heat and generate electricity (Fig. 19–23). One isotope of uranium, U-235, is the major fuel. When a U-235 nucleus is bombarded with a neutron, it breaks apart (the word *fission* means "splitting"). The initial reaction releases two or three neutrons. Each of these neutrons can trigger the fission of additional nuclei; hence, this type of nuclear reaction is called a **branching chain reaction**. Because this fission is initiated by neutron bombardment, it is not a spontaneous process and is different from natural radioactivity.

To fuel a nuclear reactor, concentrated uranium is compressed into small pellets. Each pellet could easily fit into your hand but contains the energy equivalent of 1 ton of coal. A column of pellets is encased in a 2-meter-long pipe called a **fuel rod** (Fig. 19–24). A typical nuclear power plant contains about 50,000 fuel rods bundled into assemblies of 200 rods each. **Control rods** made of neutron-absorbing alloys are spaced among the fuel rods. The control rods fine-tune the reactor. If the reaction speeds up because too many neutrons are striking other uranium atoms, then the power plant operator lowers the control rods to absorb more neutrons and slow down the reaction. If fission slows down because too many neutrons are absorbed, the operator raises the control rods. If an accident occurs and all internal power

systems fail, the control rods fall into the reactor core and quench the fission.

The reactor core produces tremendous amounts of heat. A fluid, usually water, is pumped through the reactor core to cool it. The cooling water (which is now radioactive from exposure to the core) is then passed through a radiator, where it heats another source of water to produce steam. The steam drives a turbine, which in turn generates electricity (Fig. 19–25).

THE NUCLEAR POWER INDUSTRY

Every step in the mining, processing, and use of nuclear fuel produces radioactive wastes. The mine waste discarded during mining is radioactive. Enrichment of the ore produces additional radioactive waste. When a U-235 nucleus undergoes fission in a reactor, it splits into two useless radioactive nuclei that must be discarded. Finally, after several months in a reactor, the U-235 concentration in the fuel rods drops until the fuel pellets are no longer useful. In some countries, these pellets are reprocessed to recover U-235, but in the United States this process is not economical and the pellets are discarded. In Chapter 15 we discussed proposals for storing radioactive wastes. To date, no satisfactory solution has been found.

In recent years, construction of new reactors has become so costly that electricity generated by nuclear power is more expensive than that generated by coal-fired power

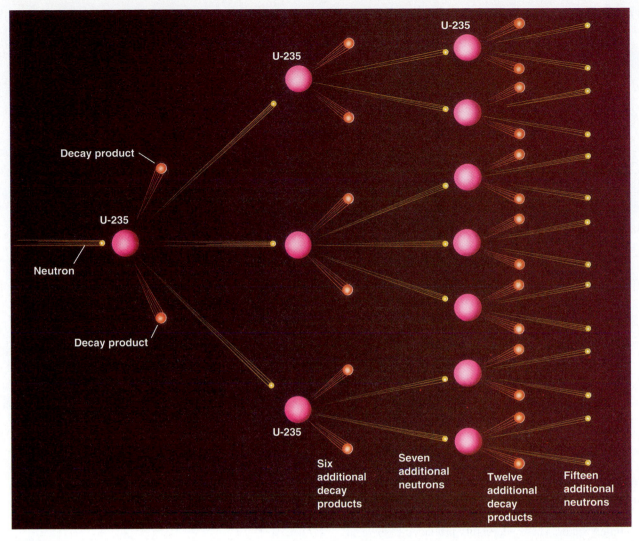

Figure 19–23 When a neutron strikes a uranium-235 nucleus, the nucleus splits into two roughly equal fragments and emits two or three neutrons. These neutrons can then initiate additional reactions, which produce more neutrons. A branching chain reaction accelerates rapidly through a sample of concentrated uranium-235.

Within the diagram:

Decay product

U-235

Neutron

Decay product

U-235

U-235

U-235

Six additional decay products

Seven additional neutrons

Twelve additional decay products

Fifteen additional neutrons

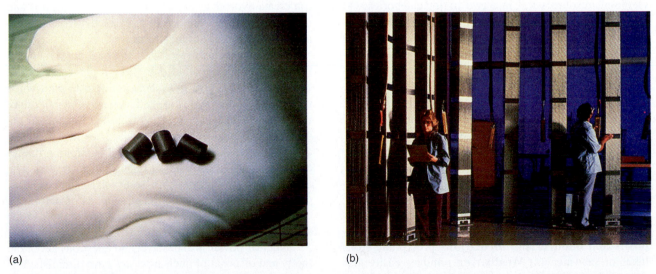

(a)

(b)

Figure 19–24 (a) Fuel pellets containing enriched uranium-235. Each pellet contains the energy equivalent of 1 ton of coal. (b) Fuel pellets are encased into narrow rods that are bundled together and lowered into the reactor core. (Courtesy Westinghouse Electric Corp, Commercial Nuclear Fuel Division)

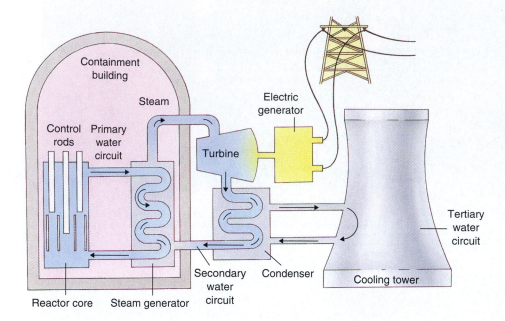

Figure 19–25 In a nuclear power plant, fission energy creates heat, which is used to produce steam. The steam drives a turbine, which generates electricity.

plants. Public concern about accidents and radioactive waste disposal has become acute. The demand for electricity has risen less than expected during the past two decades. As a result, growth of the nuclear power industry has halted. After 1974, many planned nuclear power plants were canceled, and after 1981, no new orders were placed for nuclear power plants in the United States.

In 1994, 109 commercial reactors were operating in the United States. These generators produced 22 percent of the total electricity consumed that year. Those numbers will decline in the coming decade because no new plants have been started and old plants must be decommissioned. *Forbes* business magazine called the United States nuclear power program "the largest managerial disaster in U.S. business history, involving $1 trillion in wasted investment and $10 billion in direct losses to stock holders."

SUMMARY

Geologic resources fall into two major categories: (1) Useful rocks and minerals are called **mineral resources** and include both **nonmetallic** resources and **metals**. All mineral resources are **nonrenewable**. (2) **Energy resources** include **fossil fuels**, **nuclear fuels**, and **alternative energy resources**.

Ore is a rock or other material that can be mined profitably. **Mineral reserves** are the estimated supply of ore in the ground. Five types of geologic processes concentrate elements to form ore. (1) **Magmatic processes**, such as **crystal settling**, form ore as magma solidifies. (2) **Hydrothermal processes** transport and precipitate metals from hot water. (3) Two types of **sedimentary processes** concentrate minerals. Flowing water deposits dense minerals to form **placer deposits**. **Evaporite deposits** and **banded iron formations** precipitate from lakes or seawater. (4) **Weathering** removes easily dissolved elements and minerals, leaving behind **residual deposits** such as **bauxite**. (5) **Metamorphic processes** can create hydrothermal solutions and form asbestos and marble. Metal ores and coal are extracted from **underground mines** and **surface mines**.

Fossil fuels include oil, gas, and coal. If oxygen and flowing water are excluded by burial, plant matter decays partially to form **peat**. Peat converts to **coal** when it is buried further and subjected to elevated temperature and pressure. **Petroleum** forms from the remains of organisms that settle to the ocean floor or lake bed and are incorporated into **source rock**. The organic matter converts to liquid oil when it is buried and heated. The petroleum then migrates to a **reservoir**, where it is retained by an **oil trap**. Additional supplies of petroleum can be recovered by secondary extraction from old wells and from **tar sands** and **oil shale**.

Nuclear power is expensive, and questions about the safety and disposal of nuclear wastes have diminished its future. Inexpensive uranium ore will be available for a century or more.

KEY WORDS

mineral resource *336*	**mineral deposit** *336*	**banded iron formation** *341*	**cap rock** *347*
nonmetallic resources *336*	**ore** *336*	**residual deposit** *341*	**reservoir** *347*
metal *336*	**mineral reserve** *336*	**bauxite** *341*	**secondary recovery** *347*
nonrenewable resource *336*	**crystal settling** *337*	**supergene ore** *342*	**bitumen** *348*
energy resource *336*	**hydrothermal vein** **deposit** *339*	**peat** *344*	**kerogen** *349*
fossil fuel *336*	**disseminated ore deposit** *339*	**coal** *344*	**oil shale** *349*
nuclear fuel *336*	**porphyry copper deposit** *339*	**underground mine** *344*	**branching chain** **reaction** *350*
alternative energy **resource** *336*	**placer deposit** *340*	**surface mine** *344*	**nuclear fission** *350*
quarry *336*	**evaporite deposit** *341*	**petroleum** *347*	**fuel rod** *350*
		source rock *347*	**control rod** *350*
		oil trap *347*	

REVIEW QUESTIONS

1. Describe the two categories of geologic resources.

2. Describe the differences between nonrenewable and renewable resources.

3. What is ore? What are mineral reserves? Describe three factors that can cause changes in estimates of mineral reserves.

4. If most elements are widely distributed in ordinary rocks, why should we worry about running short?

5. Explain crystal settling.

6. Discuss the formation of hydrothermal ore deposits.

7. Discuss the formation of marine evaporites and banded iron formations.

8. Explain why the availability of mineral resources depends on the availability of energy, on other environmental issues, and on political considerations.

9. Explain how coal forms. Why does it form in some environments but not in others?

10. Explain the importance of source rock, reservoir rock, cap rock, and oil traps in the formation of petroleum reserves.

11. Discuss two sources of petroleum that will be available after conventional wells go dry.

12. List the relative advantages and disadvantages of using coal, petroleum, and natural gas as fuels.

13. Explain how a nuclear reactor works. Discuss the behavior of neutrons, the importance of control rods, and how the heat from the reaction is harnessed to produce useful energy.

14. Discuss the status of the nuclear power industry in the United States.

DISCUSSION QUESTIONS

1. What factors can make our metal reserves last longer? What factors can deplete them rapidly?

2. It is common for a single mine to contain ores of two or more metals. Discuss how geologic processes might favor concentration of two metals in a single deposit.

3. List ten objects that you own. What resources are they made of? How long will each of the objects be used before it is discarded? Will the materials eventually be recycled or deposited in the trash? Discuss ways of conserving resources in your own life.

4. If you were searching for petroleum, would you search primarily in sedimentary rock, metamorphic rock, or igneous rock? Explain.

5. If you were a space traveler abandoned on an unknown planet in a distant solar system, what clues would you look for if you were searching for fossil fuels?

6. Is an impermeable cap rock necessary to preserve coal deposits? Why or why not?

7. Discuss problems in predicting the future availability of fossil fuel reserves. What is the value of the predictions?

8. Compare the depletion of mineral reserves with the depletion of fossil fuels. How are the two problems similar, and how are they different?

The Geological Evolution of North America

T he oldest known rocks on Earth are those of the 3.96-billion-year-old Arcasta gneiss in Canada's North-west Territories. They are of granitic composition and are similar to modern continental crust. Thus, a portion of the North American continent formed early in Earth history.

Sometime in the Archean Era, the Earth's outer layers had cooled sufficiently that tectonic plates formed and be-gan gliding over the asthenosphere. Those plates were not like modern plates, however. They may have been thinner and may have moved at different rates than modern plates. However, their movements must have built mountains and caused earthquakes and magmatic activity. Over the past 3.96 billion years, tectonic processes have battered, broken, but ultimately created the North American continent.

As you read this chapter, bear in mind that we are dealing with models and hypotheses. They are based on data—facts that can be observed in the rocks of North America. But the models are interpretations of how the rocks formed. The models may change in the future be-cause geologists will discover new facts, or because they will reinterpret old data. It is the nature of geology that our understanding of the Earth changes as we learn more.

Tectonic processes have built the North American continent over
the past 3.96 billion years.   (Tom Van Sant/Geosphere)

▶ 20.1 THE NORTH AMERICAN CONTINENT

Three major geologic regions make up North America today (Fig. 20–1). The **craton** is the continental interior. It is a tectonically stable region that has seen little or no tectonic activity—no deformation, metamorphism, or magmatic activity—for more than a billion years. The craton consists of two subdivisions: the **shield**, where very old igneous and metamorphic basement rocks are exposed at the surface, and the **platform**, where the same types of basement rocks are covered with a veneer of much younger sedimentary rocks.

The second region consists of the **mountain chains** bordering the craton to the east and west. All of the mountains are young relative to the craton, although some are hundreds of millions of years old.

The third region consists of the **continental shelves** and the **coastal plain**. They compose the region where young sediments eroded from the continent have accumulated on the continental margin. Geologically, the shelves and coastal plain are continuous; they differ only in that the shelf lies below sea level whereas the coastal plain lies above sea level, between the shore and the mountains.

▶ 20.2 THE CRATON

The old igneous and metamorphic basement rocks of the craton consist of several distinct geologic **provinces** (Fig. 20–2). Each province contrasts with adjacent provinces in the following ways:

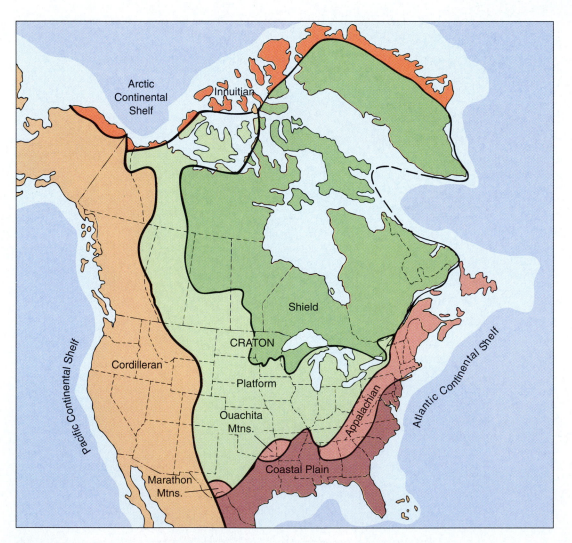

Figure 20–1 Three major geologic regions make up North America. The craton is the tectonically stable continental interior. The mountain ranges surround the craton and are younger. The coastal plain and continental shelves lie between the mountains and the sea.

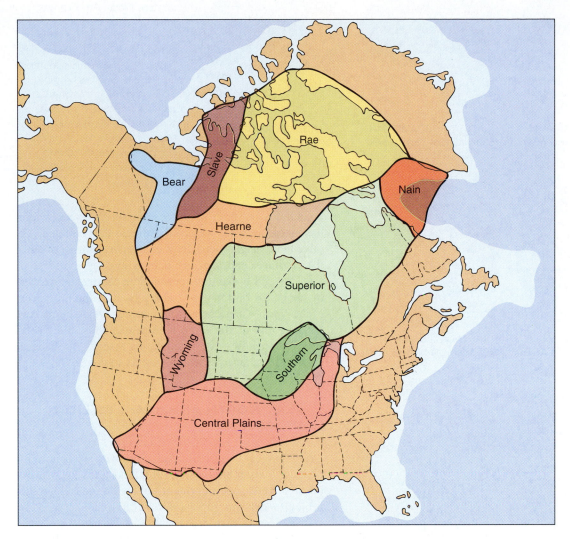

Figure 20–2   The craton consists of several distinct geologic provinces that differ from each other in rock type and age.

1. Geologic relationships seem normal and continuous within a single province but abrupt and discontinuous at a boundary between two provinces. For example, within a single province, low-grade metamorphic rocks may pass gradually to medium and then to high metamorphic grade over a distance of several kilometers. Across a province boundary, however, rocks of low metamorphic grade may lie directly against high-grade rocks, with no intervening medium-grade rocks.

2. The rocks of each province commonly show different radiometric ages from those of adjacent provinces.

3. Rocks in the boundary zone between two provinces are often intensely deformed and metamorphosed—like those now found in the suture zone between India and Asia.

What caused these abrupt geologic changes and discontinuities at the province boundaries?

▶ 20.3   NORTH AMERICA: 2 BILLION YEARS AGO

Prior to 2 billion years ago, large continents as we know them today may not have existed. Instead, many—perhaps hundreds—of small masses of continental crust and island arcs dotted a global ocean basin. They may have been similar to Japan, New Zealand, and the modern island arcs of the southwest Pacific Ocean. Then, between 2 billion and 1.8 billion years ago, tectonic plate movements swept these **microcontinents** together, forming the first supercontinent, which we call **Pangea I**. The North American craton was part of Pangea I.

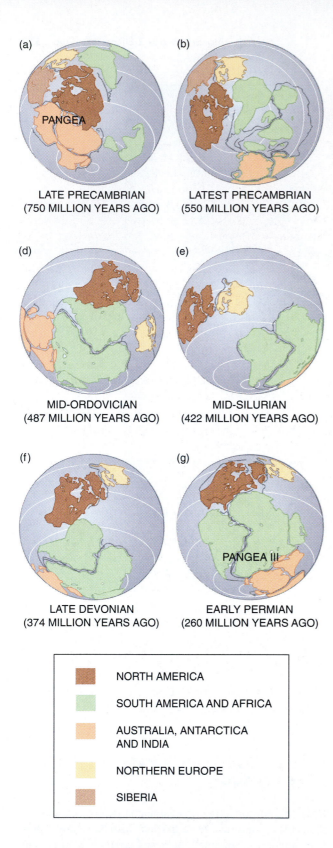

(a)

PANGEA

LATE PRECAMBRIAN
(750 MILLION YEARS AGO)

(b)

LATEST PRECAMBRIAN
(550 MILLION YEARS AGO)

(c)

MIDDLE CAMBRIAN
(530 MILLION YEARS AGO)

(d)

MID-ORDOVICIAN
(487 MILLION YEARS AGO)

(e)

MID-SILURIAN
(422 MILLION YEARS AGO)

(f)

LATE DEVONIAN
(374 MILLION YEARS AGO)

(g)

PANGEA III

EARLY PERMIAN
(260 MILLION YEARS AGO)

NORTH AMERICA

SOUTH AMERICA AND AFRICA

AUSTRALIA, ANTARCTICA
AND INDIA

NORTHERN EUROPE

SIBERIA

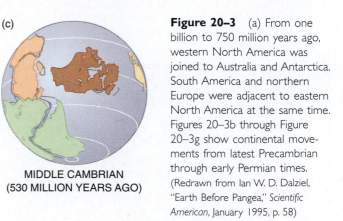

Figure 20–3 (a) From one billion to 750 million years ago, western North America was joined to Australia and Antarctica. South America and northern Europe were adjacent to eastern North America at the same time. Figures 20–3b through Figure 20–3g show continental movements from latest Precambrian through early Permian times. (Redrawn from Ian W. D. Dalziel, "Earth Before Pangea," *Scientific American*, January 1995, p. 58)

boundaries between provinces are suture zones where the microcontinents welded together as they collided. All other modern continents have cratons that consist of similar provinces and province boundaries.

Pangea I broke into several large continental masses by about 1.3 billion years ago. However, the provinces of the North American craton remained welded together, forming one of the large continents. Thus, the North American craton was created during assembly of Pangea I and has remained essentially intact to this day.

▶ 20.4 NORTH AMERICA: 1 BILLION YEARS AGO

After Pangea I split up, the fragments of continental crust reassembled about 1.0 billion years ago, forming a second supercontinent called **Pangea II** (some geologists call this supercontinent "Rodinia").

Geologists have drawn a map of Pangea II by comparing rocks now found on different continents. For example, 1-billion-year-old sedimentary and metamorphic rocks in western North America are similar to rocks of the same age in both Australia and East Antarctica. Ian W. D. Dalziel of The University of Texas at Austin recently suggested that western North America was joined to both Australia and East Antarctica from about 1 billion years ago to 750 million years ago (Fig. 20–3a). Portions of South America and northern Europe lay adjacent to eastern North America at the same time.

Dalziel's maps show that Pangea II then broke apart in late Precambrian time, about 750 million years ago. Australia and Antarctica rifted away from western North America, leaving a shoreline at the western margin of North America. This region, however, did not appear as it does today. Parts of Alaska and western Canada and much of Washington, Oregon, western Idaho, and western California had not yet become part of western North America, as shown in Figure 20–4.

Northern Europe and South America also had rifted away from North America by 550 million years ago, as an ocean basin opened along North America's eastern

The discontinuities between the provinces of the North American craton are ancient boundaries between the microcontinents. The rocks of each province give different radiometric dates because each microcontinent formed at a different time. The sheared and faulted

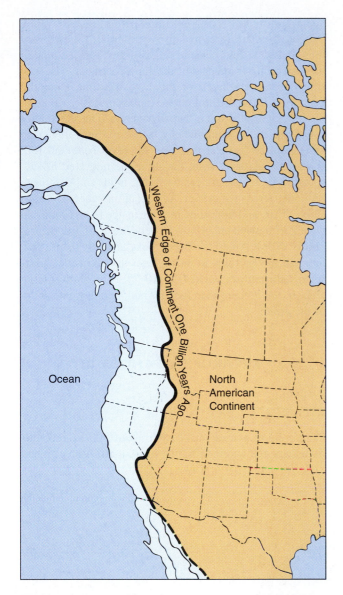

Figure 20–4 Western North America as it appeared following rifting in latest Precambrian time.

margin. Thus, by the end of Precambrian time, North America had become isolated from other continents and was surrounded by oceans (Fig. 20–3b).

► 20.5 NORTH AMERICA: A HALF BILLION YEARS AGO

THE APPALACHIAN MOUNTAINS

Figure 20–3c shows that eastern North America continued to separate from South America as the intervening ocean widened through Middle Cambrian time. Then the two continents began converging again and collided in

mid-Ordovician time (Fig. 20–3d). The convergence of the continents caused subduction of oceanic crust near the east coast of North America. Volcanoes erupted, granite plutons intruded the crust, mountains rose, and great belts of metamorphic rocks formed along the east coast. This first phase of building of the Appalachian mountain chain is called the **Taconic orogeny** after the Taconic Range on the border of New York State with Connecticut and Massachusetts, where rocks deformed by that orogeny are exposed (the term *orogeny* refers to the processes by which mountain ranges are built).

Following the mid-Ordovician collision, North America separated from South America a second time and then collided with it again in late Devonian time (Figs. 20–3e and 20–3f). The collision shoved sedimentary rocks westward from the continental shelf onto the craton, forming tremendous thrust faults and folds along the east coast (Fig. 20–5). This second phase of mountain building in the Appalachians is called the **Acadian orogeny**. It affected the northeastern corner of North America from Newfoundland to Pennsylvania, and it is named for Acadia, the name early settlers gave to that part of North America.

Figure 20–6 is a composite map showing that North America moved along the coast of South America as the two continents separated and collided twice. Then about 265 million years ago, North America slid around the upper end of South America and collided with western Africa. This collision built the central and southern Appalachians in the **Allegheny orogeny**, named for the Allegheny Plateau of the central Appalachian region.

Figure 20–5 Eastern North America collided with South America in late Devonian time, about 374 million years ago, forming great thrust faults and folds such as these in Nova Scotia. (Geological Survey of Canada)

All of these events, beginning with subduction in mid-Ordovician time followed by two collisions with South America and finally one with Africa, are collectively called the **Appalachian orogeny**. They built the Appalachian mountain chain as well as the Ouachita and Marathon mountains in Arkansas, Oklahoma, and eastern Texas. Related events built the Innuitian mountain chain along the northern continental margin (Fig. 20–1).

Note that the Appalachian orogeny was similar to the events that built the Himalayas—subduction of oceanic crust beneath the continent, followed by a continental collision. At one time, the Appalachians must have been immense mountains similar to the Himalayas. Today, however, erosion has worn the Appalachians down to maximum elevations of less than 2000 meters.

As the Appalachians rose over a period of more than 200 million years, all other continents joined the growing landmass. Thus, **Pangea III** had assembled by about 265 million years ago (Fig. 20–3g). North America formed the northwestern portion of Pangea III. This latest supercontinent was Alfred Wegener's Pangea, described in Chapter 2.

FLOODING OF NORTH AMERICA: PLATFORM SEDIMENTARY ROCKS

Recall that a long, rapidly spreading mid-oceanic ridge displaces a large volume of seawater, raising sea level and flooding low-lying portions of continents. When Pangea II rifted apart in late Precambrian time, a very long and wide mid-oceanic ridge system formed among the separating fragments of the supercontinent. The new ridge raised sea level for much of the time from the Cambrian through the Pennsylvanian periods, flooding the low, central craton of North America. The sea did not cover the craton continuously, however. Seas advanced and withdrew several times as sea level rose and fell. Each time sea level rose, the shoreline migrated inland, spreading a blanket of marine sand, mud, and carbonate sediment across the craton. As a result, a veneer of sandstone, shale, and limestone ranging in age from Cambrian through Pennsylvanian now covers much of the central part of North America. These rocks are less than a few hundred meters thick in most places and are called **platform sedimentary rocks** (Fig. 20–7). They blanket large areas of the Precambrian igneous and metamorphic rocks of the craton.

▶ 20.6 BREAKUP OF PANGEA III

OPENING OF THE MODERN ATLANTIC OCEAN

Pangea III remained intact from about 300 to 180 million years ago and then began to rift apart. As North and

Figure 20–6 A composite map showing the movements of North America from 750 million years ago to 265 million years ago. The other continents are shown in their positions of 260 million years ago. (Redrawn from Ian W. D. Dalziel, "Earth Before Pangea," *Scientific American*, January 1995, p. 58)

Figure 20–7 Flat-lying platform sedimentary rocks are exposed along the banks of the Iowa River near Bluffton, Iowa. (D. Cavagnaro/ Visuals Unlimited)

South America separated from Eurasia and Africa, the modern Atlantic Ocean began to open. The Atlantic Ocean was born, and continues to grow today, as a result of sea-floor spreading along the mid-Atlantic ridge. Passive continental margins developed on both sides of the newly opening ocean basin. Tectonic activity on the eastern margin of North America ceased, and the lofty Appalachians began to wear away.

The new eastern margin of North America developed in about the same place it had been before Pangea III assembled. Thus, rifting followed the sutures where continents had welded together 120 million years previously (Fig. 20–8). Suture zones may be lines of weakness within a supercontinent, like the perforations in tear-out advertisements bound into magazines. The rifting did not perfectly follow the old sutures, however. As the supercontinent broke up, small pieces of Europe and Africa remained stuck to the east coast of North America, and parts of North America rode off with Africa and Europe.

SEDIMENTARY ROCKS OF THE COASTAL PLAIN AND CONTINENTAL SHELF

The newly formed mid-Atlantic ridge displaced a great volume of seawater, raising sea level by a hundred meters or more and flooding the continent once again. The seas deposited a new sequence of platform sediments. They also flooded the continental margin, depositing sediment on the continental shelf and coastal plain along the eastern and southeastern margins of North America (Fig. 20–1).

▶ 20.7 BUILDING OF THE WESTERN MOUNTAINS

The **Cordilleran mountain chain** is the long, broad mountain chain of western North America. It reaches from the western edge of the Great Plains to the west coast, and from Alaska to Mexico (Fig. 20–9). The name *Cordillera* is taken from the Spanish word for "chain of mountains." It includes the Rocky Mountains, the Coast Ranges, the Sierra Nevada, and all other mountains and intermountain regions in the western part of our continent. The Cordillera formed as Pangea III rifted apart.

As the Atlantic Ocean opened, the lithospheric plate carrying North America began moving westward. To accommodate this new movement, tectonic plates of the Pacific Ocean sank beneath the western edge of the continent, creating a subduction zone and an Andean-type continental margin. From about 180 to 80 million years ago, great granite batholiths rose into the crust over this subduction zone (Fig. 20–10). Later, tectonic forces raised the batholiths and erosion exposed the rocks, creating the beautiful granite mountains of the Sierra Nevada and many other parts of the Cordillera.

ACCRETED TERRAINS

The land that was to become western Alaska, the Yukon, British Columbia, Washington, Oregon, western Idaho, and western California started to join the continent shortly after Pangea III broke up. At that time, numerous island arcs and microcontinents dotted the Pacific Ocean—much like the southwestern Pacific today. As

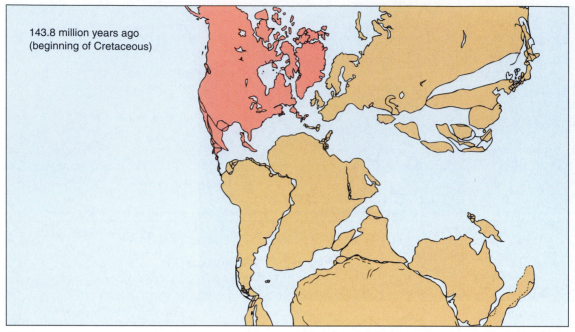

143.8 million years ago
(beginning of Cretaceous)

(a)

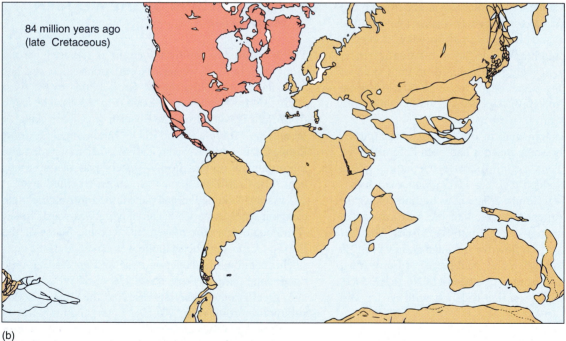

84 million years ago
(late Cretaceous)

(b)

Figure 20–8 Pangea began to break up about 180 million years ago. The rifting followed the lines of the earlier collision zone that had formed the Appalachians. However, bits and pieces of Africa or Eurasia remained stuck to North America, and fragments of North America went off with Africa and Eurasia. (a, b) Positions of continents at two times from early Cretaceous through early Oligocene. (From Bally et al., *The Geology of North America—An Overview*, GSA, 1989)

the tectonic plates of the Pacific sank beneath western North America, these island arcs and microcontinents arrived one by one at the subduction zone. Because they were too buoyant to sink into the mantle, they collided with the western margin of the continent and became parts of the continent. This process is called

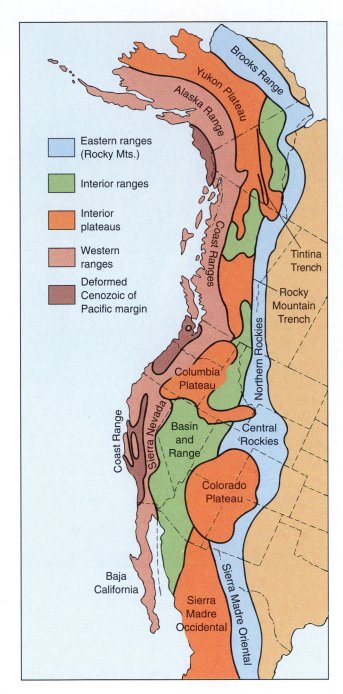

Figure 20–9 The Cordilleran mountain chain includes several subdivisions forming the vast mountainous region of western North America.

the craton are recognized. Geologic relationships within each terrain are continuous, but relations across terrain boundaries are discontinuous. The terrain boundaries are intensely deformed as a result of the collisions, and radiometric ages change abruptly across the boundaries. The terrains originated in many different parts of the Pacific Ocean.

The subduction that brought the accreted terrains to western North America also continued to form granitic magma that rose into the continental margin. As a result of this process, many of the accreted terrains contain granite batholiths and related volcanic rocks that formed as the terrains docked.

Some of the collisions between the accreted terrains and the continent were not direct head-on collisions. Instead, the Pacific plates carrying some terrains were moving northward while the continent was moving west. As a result, many of the accreted terrains were smeared northward for several hundred kilometers along huge strike–slip faults as they docked.

FOLDING AND THRUST FAULTING IN THE CORDILLERA

As terrains crashed into the continent, they created compressive forces like those of a continent–continent collision. The resulting zone of folded and thrust-faulted rocks is called the **Cordilleran fold and thrust belt** (Fig. 20–11). It is only a few hundred kilometers wide but extends north–south for the entire length of the Cordillera (Fig. 20–12).

THE TECTONIC FORCES CHANGE

The western margin of the continent was compressed while convergence of the two tectonic plates and docking of the accreted terrains continued at high speed. However, the rate of convergence suddenly slowed about 45 million years ago for an unknown reason. Then the compressive forces weakened and the warm, thick crust of the Cordillera began to spread out like a mound of honey on a tabletop.

Only the deeper, hotter part of the crust could spread plastically, however. The upper crust was cold and brittle. It fractured and faulted as the spreading deeper rocks pulled it apart. The brittle upper rocks can be likened to a layer of frosting coating the mound of honey. As the honey flows outward, the frosting breaks into segments, which separate as spreading continues.

As the shallow rocks rifted and faulted in this manner, a second magmatic episode began about 45 million years ago, and it continues today. Plutonic and volcanic rocks formed during the last 45 million years are abundant in western North America (Fig. 20–13).

docking, because it is similar to a ship tying up alongside a dock.

In this way, North America grew westward from about 180 to 80 million years ago. Geologists have identified about 40 accreted island arcs and microcontinents, called **accreted terrains**, in the Cordillera. Accreted terrains are identified in the same way as the provinces of

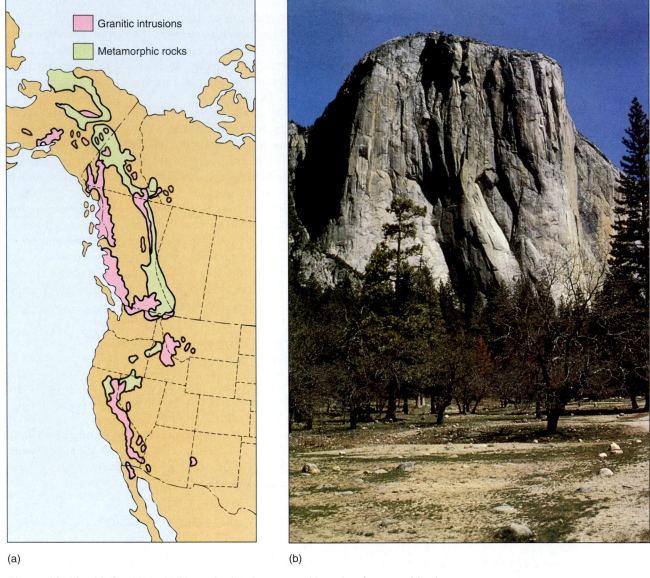

(a)   (b)

Granitic intrusions

Metamorphic rocks

Figure 20–10 (a) Granitic batholiths and related metamorphic rocks of western North America formed as a result of subduction along the western margin of the continent, beginning about 180 million years ago. (b) The granite of El Capitan in Yosemite Valley is part of the Sierra Nevada batholith, which formed in this way.

THE TECTONIC FORCES CHANGE AGAIN: THE SAN ANDREAS FAULT

As North America moved westward, the west coast drew closer to a portion of the mid-oceanic ridge called the **East Pacific rise** (Fig. 20–14). By 30 million years ago, southern California had arrived at the East Pacific rise. At the time, the Pacific plate was moving northwest— nearly parallel to the westward movement of North America. Since these two plates were moving in nearly the same direction, they were not converging; therefore, subduction stopped where California touched the Pacific plate.

But the Pacific plate was moving in a *slightly* different direction from the continent. A new fault developed to accommodate the small difference in direction. The Pacific plate began sliding northwestward against the California coast along the new strike-slip (transform) fault (Fig. 20–14b). The new fault was the beginning of the San Andreas fault. As North America continued to move westward and reached greater lengths of the East

Figure 20–11 Folded limestone in the Lizard Range, British Columbia.

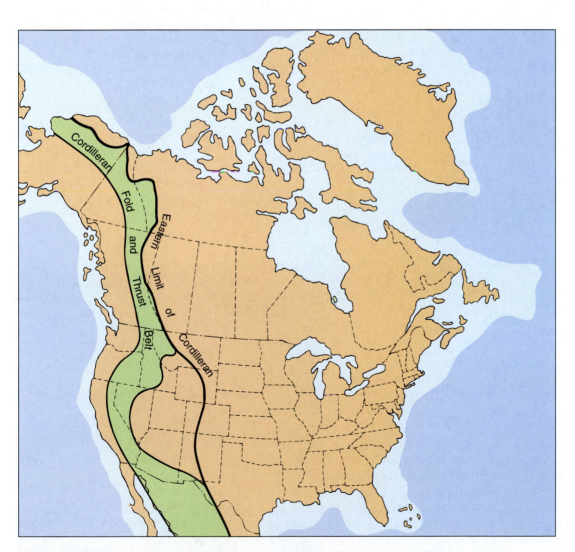

Figure 20–12 The Cordilleran fold and thrust belt is a zone of thrust faults and folded rocks that extends for the entire length of the Cordilleran chain.

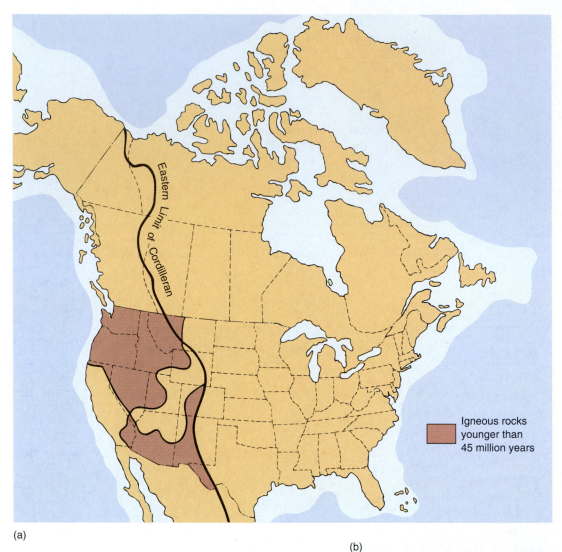

(a)

Eastern Limit of Cordilleran

Igneous rocks
younger than
45 million years

(b)

Figure 20–13 (a) Volcanic and plutonic rocks in the United States formed from 45 million years ago through the present. They are shown as if none had eroded away. (b) Lizard Head Peak is a volcanic plug in the San Juan Mountains, Colorado.

Pacific rise, the San Andreas fault grew longer and migrated inland (Fig. 20–14c).

THE MODERN CASCADE VOLCANOES

Figure 20–14c shows that the San Andreas fault now veers westward in northern California, where it runs out into the Pacific Ocean to connect with the remnants of the East Pacific rise spreading center. This spreading center extends along the coasts of northern California, Oregon, Washington, and southwestern British Columbia. The Juan de Fuca oceanic plate, east of the spreading center, is sinking beneath the westward-moving continent. The subduction creates an active volcanic zone

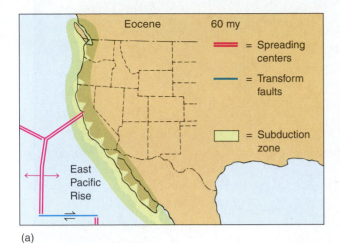

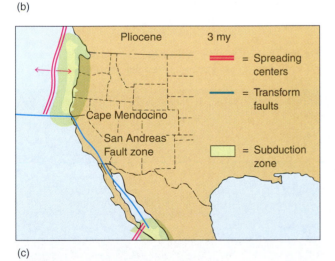

(a)

(b)

(c)

Figure 20–14 The San Andreas fault developed where western North America overran the East Pacific rise, beginning 30 million years ago. The fault has grown longer as more of California has hit the rise. The subduction zone, shown in light green, once extended the entire length of the coast, but it has become inactive in the region of the San Andreas fault.

from northern California to southern British Columbia. Several of the volcanoes, including Mounts Lassen, Shasta, Rainier, St. Helens, and Baker, have erupted recently, some within the past 100 years.

THE BASIN AND RANGE PROVINCE

The San Andreas fault exerts frictional drag against the western margin of North America. Figure 20–15 shows how such a force stretches and fractures the brittle upper crust. Large blocks of crust have dropped as grabens along the faults, leaving other blocks elevated as mountain ranges between the grabens.

For the past 30 million years, this faulting has created the northeast–southwest-oriented mountain ranges and valleys of the **Basin and Range** (Fig. 20–16). Igneous activity accompanied the faulting. Many geologists who work in the Basin and Range believe that the stretching and faulting have at least doubled the width of the region. The tectonic forces associated with the San Andreas fault continue today, and Basin and Range faulting and magmatism are still active.

THE COLORADO PLATEAU

A large block of western North America, known as the **Colorado Plateau**, remained strangely immune to the faulting and igneous activity, although it is surrounded on three sides by the Basin and Range (Fig. 20–9). Perhaps because it has a thicker and stronger crust, the

Figure 20–15 Friction along the San Andreas fault stretches Nevada and nearby regions in a northwest–southeast direction (arrows), forming normal faults (red lines) in the Basin and Range.

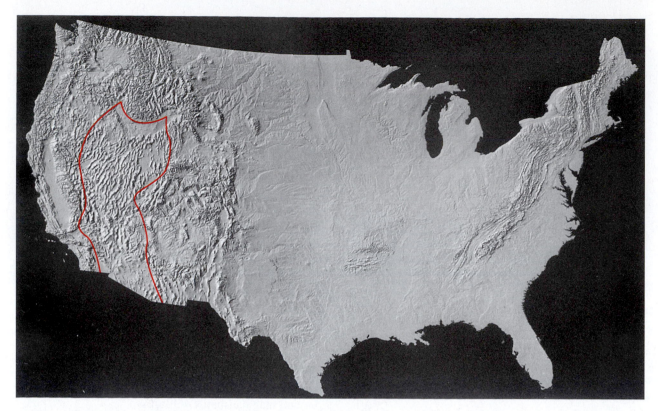

Figure 20–16 A radar image of the United States with the Basin and Range outlined in red. Notice the parallel valleys and mountain ranges caused by normal faulting.

entire Colorado Plateau simply rotated clockwise in response to the tectonic forces that created the Basin and Range. Then between 5 and 10 million years ago, the Colorado Plateau rose without much internal deformation to become a high, nearly circular topographic feature (Fig. 20–17). As the Colorado Plateau rose, the Colorado River cut the Grand Canyon into the rising bedrock.

THE COLUMBIA PLATEAU

The Columbia Plateau is one of the largest basalt plateaus in the world (Fig. 20–9). It formed by rapid extrusion of flood basalt magma about 15 million years ago (Fig. 20–18). Volcanic activity then migrated eastward along the Snake River plain and occurred as recently as a few thousand years ago in Yellowstone National Park at the eastern end of the Snake River plain. Here the story may not be over, because active magma still underlies portions of the Park. The vast Snake River plain and Yellowstone magmatic systems are described in Chapter 5.

▶ **20.8   THE PLEISTOCENE ICE AGE AND THE ARRIVAL OF HUMANS IN NORTH AMERICA**

The aforementioned sequence of tectonic events had created most of the Cordillera by about 5 million years ago.

Later, one additional event sculpted the northern parts of our continent.

At least five major episodes of glaciation have occurred in the Earth's history. The most recent is the

Figure 20–17 Sedimentary rocks of the Colorado Plateau in Grand Canyon have been uplifted but show little folding.

Figure 20–18 Basalt rises above the Columbia River to form the Columbia Plateau. Each layer is a separate basalt flow. (Donald Hyndman)

Pleistocene Ice Age. By 2 million years ago, large ice sheets as much as 3000 meters thick had formed in both the Northern and Southern Hemispheres and were rapidly spreading outward. At its greatest extent, the ice covered about one third of North America. The flowing ice scoured rock and soil from the land, and when the glaciers melted, they deposited huge piles of sand and gravel. Thus, glacial erosion and deposition greatly modified the landscape of North America.

The Bering Strait is part of a shallow-water shelf lying between Alaska and Siberia. When glaciers were at their maximum size during the Pleistocene Ice Age, so much water was stored in continental glaciers that global sea level fell about 130 meters. As a result, the shallow shelf beneath the Bering Strait was exposed as a low-lying, swampy isthmus connecting Asia to North America. Known as **Beringia**, this region was above sea level at least twice during the past 150,000 years and an unknown number of times previously during the Pleistocene Ice Age (Fig. 20–19). Mammoths, bison, caribou, moose, musk ox, mountain sheep, and several other mammals that we now associate with North

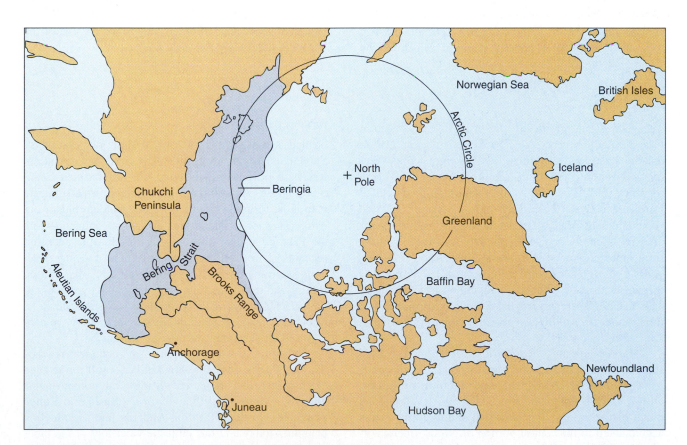

Figure 20–19 Beringia was a landmass measuring 1200 by 2300 kilometers and connecting Siberia to Alaska when Pleistocene glaciers were at their greatest size.

America migrated across Beringia to this continent during Pleistocene time.

Humans also migrated across Beringia to North America, although it is difficult to determine precisely when they first arrived. Coastal areas where the earliest migrants probably lived are now submerged beneath seas that rose as the glaciers melted. Thus, traces of their villages and camps may never be found. However, people probably migrated to North America in several waves. The oldest uncontested remains of humans on this continent are dated between 13,000 and 14,000 years ago. The most recent may have happened only 10,000 years ago, shortly before melting glaciers inundated Beringia.

As they arrived, humans and other animals spread out from the Alaska coast, following several paths (Fig. 20–20). Note how far north several of the pathways are. Why would humans and other mammals migrate northward during times when glaciers were at their maximum growth? And why would they cross to Alaska, a place known for its cold climate and glaciers even now during a nonglacial interval? Some geologists suggest that local climatic patterns created ice-free corridors that directed the migration paths of humans and animals. One corridor extended along the northern edge of the continent, while another ran along the west coast. A third may have extended along the eastern side of the Cordillera, between the alpine glaciers of the mountains and the great continental ice sheet. If these ice-free corridors existed, they must have been mostly treeless, vegetated by

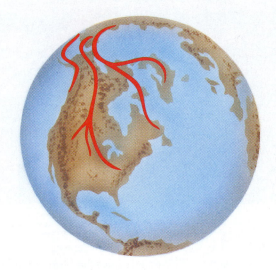

Figure 20–20   Humans and other animals migrated across Beringia and spread throughout the new world by several routes in Pleistocene time.

grasses, heather, and sedge. Small groves of aspen and larch grew in sheltered areas. The terrain was perfect for grazing animals. The flat plains and great animal populations made for easy hunting. When people who continued southward arrived in the vicinity of the modern United States–Canada border, they found the end of the ice and a fabulously rich new land.

SUMMARY

The geologic development of North America began at least 3.96 billion years ago. The North American **craton** consists of several **provinces**, which originated as separate **microcontinents** and island arcs that were swept together during assembly of a supercontinent called **Pangea I**, between 2 and 1.8 billion years ago. This first supercontinent rifted apart a few hundred million years after its assembly, separating the North American craton from other masses of continental crust. About 1 billion years ago, a second supercontinent called **Pangea II** formed, with the North American craton near its center. This second supercontinent broke up in late Precambrian time, about 750 million years ago, isolating the North American craton again.

During assembly of a third supercontinent, called **Pangea III**, subduction followed by a series of continental collisions created the **Appalachian mountain chain**. During Paleozoic time, high sea level flooded much of the craton, and **platform sedimentary rocks** accumulated in the central portion of North America.

When Pangea III broke up, beginning about 180 million years ago, North America once again became iso-

lated from other continents. As the Atlantic Ocean opened, a passive margin formed along our east coast. Subduction began along the west coast in response to westward movement of the continent, forming an Andean-type margin. As subduction continued, great granitic batholiths formed, mountains rose, metamorphism occurred, and many island arcs and microcontinents from the Pacific Ocean basin docked, forming the **accreted terrains** of western North America. The docking of accreted terrains created compressive forces that caused folding and thrust faulting. From 45 million years ago to the present, much of the Cordillera was stretched in an east–west direction, resulting in normal faulting, volcanism, and emplacement of granitic plutons.

When the western margin of the continent reached the **East Pacific rise**, the **San Andreas fault** formed and extensional stress increased, forming the normal faults, grabens, and magmatic activity of the **Basin and Range province**. North of the San Andreas fault, subduction continues today and has built the volcanoes of the modern Cascades. The **Columbia Plateau** formed by massive flood basalt flows about 15 million years ago. The

Colorado Plateau escaped most of the folding and faulting that affected the Cordillera but was uplifted between 5 and 10 million years ago.

In Pleistocene time, glaciers covered one third of North America and modified landforms by both erosion and deposition. During times of maximum size of the glaciers, sea level fell by 130 meters to expose **Beringia**, a large landmass connecting Siberia to Alaska. Many species of animals, including humans, migrated to North America across this land bridge.

KEY WORDS

craton   *356*
shield   *356*
platform   *356*
coastal plain   *356*
province   *356*
microcontinent   *357*

Pangea I, II, and
   III   *357–360*
Appalachian
   orogeny   *360*
platform sedimentary
   rocks   *360*

docking   *363*
accreted terrain   *363*
Cordilleran fold and
   thrust belt   *363*
East Pacific Rise   *364*

Basin and Range   *367*
Colorado Plateau   *367*
Beringia   *369*

REVIEW QUESTIONS

1. What is the approximate age of the oldest rocks in North America? In what part of the continent are they found?

2. Draw a simple map showing the outline of North America and the locations of the three main types of geologic regions that make up the continent.

3. Describe the North American craton.

4. Why are the mountain chains of North America located near the continental margins?

5. Explain the origins of the provinces of the North American craton. What are the main kinds of differences among the provinces?

6. Why are boundaries between the provinces of the craton commonly intensely sheared and faulted?

7. What is a microcontinent? What is a supercontinent?

8. What were the most important tectonic events that built the Appalachian mountain chain?

9. Platform sediments overlie the craton in much of the central portion of North America. How and when did they form?

10. What is the Cordillera?

11. Describe or sketch on a map the locations of the large granitic batholiths of the Cordillera. How and when did these great bodies of granite form?

12. What is an accreted terrain? Where are accreted terrains found in North America?

13. Where is the Cordilleran fold and thrust belt located? Sketch it on a map of North America.

14. Why does recent and modern volcanic activity along the west coast of North America occur only north of the San Andreas fault?

15. What are the main geologic structures of the Basin and Range province? Why are they the most common structures?

16. What is the importance of Beringia in the history of North America? How did Beringia form?

DISCUSSION QUESTIONS

1. The oldest known rocks in North America are gneisses. What does the fact that they are metamorphic rocks tell you about the maximum age of rocks of the craton?

2. Explain the relationship between the continental shelves and the coastal plain of North America.

3. Describe and discuss the model of supercontinent cycles. Does the model seem plausible in light of the data presented in this chapter?

4. Discuss how microcontinents might have formed.

5. Discuss how and why supercontinents form.

6. The breakup of the first supercontinent was accompanied by intrusion of many granite plutons into continental crust. Most granites form in continental crust above subduction zones. Develop and discuss a model in which granite magma forms during rifting of a supercontinent.

7. Discuss the origin of the San Andreas fault and the relationships among the San Andreas fault, the Basin and Range province, and the Colorado Plateau.

8. Using a map of tectonic plate movements, give a plausible scenario for the movement of continents over the next 200 million years. Which oceans will grow larger? Which ones will shrink?

APPENDIX A
Systems of Measurement

I THE SI SYSTEM

In the past, scientists from different parts of the world have used different systems of measurement. However, global cooperation and communication make it essential to adopt a standard system. The International System of Units (SI) defines various units of measurement as well as prefixes for multiplying or dividing the units by decimal factors. Some primary and derived units important to geologists are listed below.

Time The SI unit is the **second**, s or sec, which used to be based on the rotation of the Earth but is now related to the vibration of atoms of cesium-133. SI prefixes are used for fractions of a second (such as milliseconds or microseconds), but the common words **minutes**, **hours**, and **days** are still used to express multiples of seconds.

Length The SI unit is the **meter**, m, which used to be based on a standard platinum bar but is now defined in terms of wavelengths of light. The closest English equivalent is the **yard** (0.914 m). A **mile** is 1.61 kilometers (km). An inch is exactly 2.54 centimeters (cm).

Area Area is length squared, as in **square meter**, **square foot**, and so on. The SI unit of area is the **are**, a, which is 100 sq m. More commonly used is the **hectare**, ha, which is 100 ares, or a square that is 100 m on each side. (The length of a U.S. football field plus one end zone is just about 100 m.) A hectare is 2.47 acres. An **acre** is 43,560 sq ft, which is a plot of 220 ft by 198 ft, for example.

Volume Volume is length cubed, as in **cubic centimeter**, cm^3, **cubic foot**, ft^3, and so on. The SI unit is the **liter**, L, which is 1000 cm^3. A **quart** is 0.946 L; a U.S. liquid **gallon** (gal) is 3.785 L. A **barrel** of petroleum (U.S.) is 42 gal, or 159 L.

Mass Mass is the amount of matter in an object. **Weight** is the force of gravity on an object. To illustrate the difference, an astronaut in space has no weight but still has mass. On Earth, the two terms are directly proportional and often used interchangeably. The SI unit of mass is the **kilogram**, kg, which is based on a standard platinum mass. A **pound** (avdp), lb, is a unit of weight. On the surface of the Earth, 1 lb is equal to 0.454 kg. A **metric ton**, also written as **tonne**, is 1000 kg, or about 2205 lb.

Temperature The Celsius scale is used in most laboratories to measure temperature. On the Celsius scale the freezing point of water is 0°C and the boiling point of water is 100°C.

The SI unit of temperature is the **Kelvin**. The coldest possible temperature, which is −273°C, is zero on the Kelvin scale. The size of 1 degree Kelvin is equal to 1 degree Celsius.

$$\text{Celsius temperature (°C)} = \text{Kelvin temperature (K)} - 273 \text{ K}$$

Fahrenheit temperature (°F) is not used in scientific writing, although it is still popular in English-speaking countries. Conversion between Fahrenheit and Celsius is shown below.

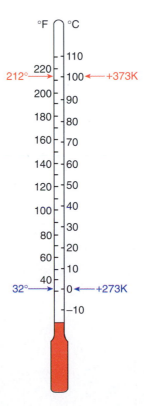

Energy   Energy is a measure of work or heat, which were once thought to be different quantities. Hence, two different sets of units were adopted and still persist, although we now know that work and heat are both forms of energy.

The SI unit of energy is the **joule**, J, the work required to exert a force of 1 newton through a distance of 1 m. In turn, a newton is the force that gives a mass of 1 kg an acceleration of 1 m/sec^2. In human terms, a joule is not much—it is about the amount of work required to lift a 100-g weight to a height of 1 m. Therefore, joule units are too small for discussions of machines, power plants, or energy policy. Larger units are

megajoule, MJ = 10^6 J (a day's work by one person)

gigajoule, GJ = 10^9 J (energy in half a tank of gasoline)

The energy unit used for heat is the **calorie**, cal, which is exactly 4.184 J. One calorie is just enough energy to warm 1 g of water 1°C. The more common unit used in measuring food energy is the **kilocalorie**, kcal, which is 1000 cal. When **Calorie** is spelled with a capital C, it means kcal. If a cookbook says that a jelly doughnut has 185 calories, that is an error—it should say 185 Calories (capital C), or 185 kcal. A value of 185 calories (small c) would be the energy in about one quarter of a thin slice of cucumber.

The unit of energy in the British system is the **British thermal unit**, or Btu, which is the energy needed to warm 1 lb of water 1°F.

$$1 \text{ Btu} = 1054 \text{ J} = 1.054 \text{ kJ} = 252 \text{ cal}$$

The unit often referred to in discussions of national energy policies is the **quad**, which is 1 quadrillion Btu, or 10^{15} Btu.

Some approximate energy values are

1 barrel (42 gal) of petroleum = 5900 MJ
1 ton of coal = 29,000 MJ
1 quad = 170 million barrels of oil, or 34 million tons of coal

II   PREFIXES FOR USE WITH BASIC UNITS OF THE METRIC SYSTEM

PREFIX	SYMBOL†	POWER		EQUIVALENT
geo*		10^{20}		
tera	T	10^{12}	= 1,000,000,000,000	Trillion
giga	G	10^{9}	= 1,000,000,000	Billion
mega	M	10^{6}	= 1,000,000	Million
kilo	k	10^{3}	= 1,000	Thousand
hecto	h	10^{2}	= 100	Hundred
deca	da	10^{1}	= 10	Ten
— — —	—	10^{0}	= 1	One
deci	d	10^{-1}	= .1	Tenth
centi	c	10^{-2}	= .01	Hundredth
milli	m	10^{-3}	= .001	Thousandth
micro	μ	10^{-6}	= .000001	Millionth
nano	n	10^{-9}	= .000000001	Billionth
pico	p	10^{-12}	= .000000000001	Trillionth

* Not an official SI prefix but commonly used to describe very large quantities such as the mass of water in the oceans.

† The SI rules specify that SI symbols are not followed by periods, nor are they changed in the plural. Thus, it is correct to write "The tree is 10 m high," not "10 m. high" or "10 ms high."

III   HANDY CONVERSION FACTORS

TO CONVERT FROM	TO	MULTIPLY BY	
Centimeters	Feet	0.0328 ft/cm	
	Inches	0.394 in/cm	
	Meters	0.01 m/cm	(exactly)
	Micrometers (Microns)	1000 μm/cm	(")
	Miles (statute)	6.214×10^{-6} mi/cm	
	Millimeters	10 mm/cm	(exactly)
Feet	Centimeters	30.48 cm/ft	(exactly)
	Inches	12 in/ft	(")
	Meters	0.3048 m/ft	(")
	Micrometers (Microns)	304800 μm/ft	(")
	Miles (statute)	0.000189 mi/ft	
Grams	Kilograms	0.01 kg/g	(exactly)
	Micrograms	1×10^{6} μg/g	(")
	Ounces (avdp.)	0.03527 oz/g	
	Pounds (avdp.)	0.002205 lb/g	
Hectares	Acres	2.47 acres/ha	
Inches	Centimeters	2.54 cm/in	(exactly)
	Feet	0.0833 ft/in	
	Meters	0.0254 m/in	(exactly)
	Yards	0.0278 yd/in	
Kilograms	Ounces (avdp.)	35.27 oz/kg	
	Pounds (avdp.)	2.205 lb/kg	
Kilometers	Miles	0.6214 mi/km	
Meters	Centimeters	100 cm/m	(exactly)
	Feet	3.2808 ft/m	
	Inches	39.37 in/m	
	Kilometers	0.001 km/m	(exactly)
	Miles (statute)	0.0006214 mi/m	
	Millimeters	1000 mm/m	(exactly)
	Yards	1.0936 yd/m	

III HANDY CONVERSION FACTORS (CONTINUED)

TO CONVERT FROM	TO	MULTIPLY BY	
Miles (statute)	Centimeters	160934 cm/mi	
	Feet	5280 ft/mi	(exactly)
	Inches	63360 in/mi	(exactly)
	Kilometers	1.609 km/mi	
	Meters	1609 m/mi	
	Yards	1760 yd/mi	(exactly)
Ounces (avdp.)	Grams	28.35 g/oz	
	Pounds (avdp.)	0.0625 lb/oz	(exactly)
Pounds (avdp.)	Grams	453.6 g/lb	
	Kilograms	0.454 kg/lb	
	Ounces (avdp.)	16 oz/lb	(exactly)

IV EXPONENTIAL OR SCIENTIFIC NOTATION

Exponential or scientific notation is used by scientists all over the world. This system is based on exponents of 10, which are shorthand notations for repeated multiplications or divisions.

A positive exponent is a symbol for a number that is to be multiplied by itself a given number of times. Thus, the number 10^2 (read "ten squared" or "ten to the second power") is exponential notation for $10 \cdot 10 = 100$. Similarly, $3^4 = 3 \cdot 3 \cdot 3 \cdot 3 = 81$. The reciprocals of these numbers are expressed by negative exponents. Thus $10^{-2} = 1/10^2 = 1/(10 \cdot 10) = 1/100 = 0.01$.

To write 10^4 in longhand form you simply start with the number 1 and move the decimal four places to the right: 10000. Similarly, to write 10^{-4} you start with the number 1 and move the decimal four places to the left: 0.0001.

It is just as easy to go the other way—that is, to convert a number written in longhand form to an exponential expression. Thus, the decimal place of the number 1,000,000 is six places to the right of 1:

$$1\,000\,000 = 10^6$$
$$\text{6 places}$$

Similarly, the decimal place of the number 0.000001 is six places to the left of 1 and

$$0.000001 = 10^{-6}$$
$$\text{6 places}$$

What about a number like 3,000,000? If you write it $3 \cdot 1,000,000$, the exponential expression is simply $3 \cdot 10^6$. Thus, the mass of the Earth, which, expressed in long numerical form is 3,120,000,000,000,000,000,000,000 kg, can be written more conveniently as 3.12×10^{24} kg.

APPENDIX B
Rock Symbols

The symbols used in this book for types of rocks are shown below:

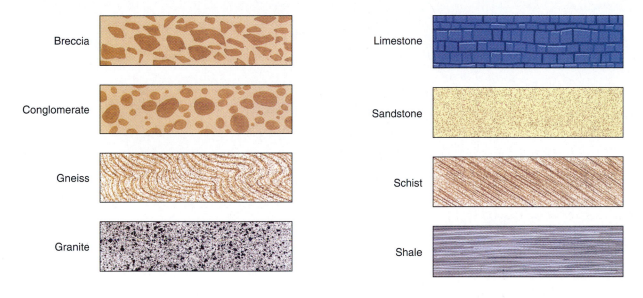

Breccia

Conglomerate

Gneiss

Granite

Limestone

Sandstone

Schist

Shale

In this book we have adopted consistent colors and style for depicting magma and layers in the upper mantle and crust.

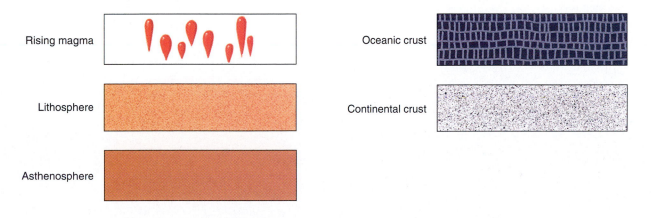

Rising magma

Lithosphere

Asthenosphere

Oceanic crust

Continental crust

GLOSSARY

A horizon   The uppermost layer of soil composed of a mixture of humus and leached and weathered minerals. (*syn*: topsoil)

aa   A lava flow that has a jagged, rubbly, broken surface.

ablation area   (See zone of ablation.)

abrasion   The mechanical wearing and grinding of rock surfaces by friction and impact.

absolute age   Age, or time measured in years.

abyssal fan   A large, fan-shaped accumulation of sediment deposited at the bases of many submarine canyons adjacent to the deep-sea floor. (*syn*: submarine fan)

abyssal plain   A flat, level, largely featureless part of the ocean floor between the mid-oceanic ridge and the continental rise.

accessory mineral   A mineral that is common, but usually found only in small amounts.

accreted terrain   A landmass that originated as an island arc or a microcontinent and was later added onto a continent.

accumulation area   (See zone of accumulation.)

acid precipitation   A condition in which natural precipitation becomes acidic after reacting with air pollutants. Often called acid rain.

active continental margin   A continental margin characterized by subduction of an oceanic lithospheric plate beneath a continental plate. (*syn*: Andean margin)

active volcano   A volcano that is erupting or is expected to erupt.

albedo   The reflectivity of a surface. A mirror or bright snowy surface reflects most of the incoming light and has a high albedo, whereas a rough flat road surface has a low albedo.

alluvial fan   A fan-like accumulation of sediment created where a steep stream slows down rapidly as it reaches a relatively flat valley floor.

alpine glacier   A glacier that forms in mountainous terrain.

amphibole   A group of double chain silicate minerals. Hornblende is a common amphibole.

Andean margin   A continental margin characterized by subduction of an oceanic lithospheric plate beneath a continental plate. (*syn*: active continental margin)

andesite   A fine-grained gray or green volcanic rock intermediate in composition between basalt and granite, consisting of about equal amounts of plagioclase feldspar and mafic minerals.

angle of repose   The maximum slope or angle at which loose material remains stable.

angular unconformity   An unconformity in which younger sediment or sedimentary rocks rest on the eroded surface of tilted or folded older rocks.

anion   An ion that has a negative charge.

antecedent stream   A stream that was established before local uplift started and cut its channel at the same rate the land was rising.

anticline   A fold in rock that resembles an arch; the fold is convex upward, and the oldest rocks are in the middle.

aquifer   A porous and permeable body of rock that can yield economically significant quantities of ground water.

Archean Eon   A division of geologic time 3.8 to 2.5 billion years ago.

arête   A sharp, narrow ridge between adjacent valleys formed by glacial erosion.

arkose   A feldspar-rich sandstone formed adjacent to granite cliffs.

artesian aquifer   An inclined aquifer that is bounded top and bottom by layers of impermeable rock so the water is under pressure. (*syn*: confined aquifer)

artesian well   A well drilled into an artesian aquifer in which the water rises without pumping and in some cases spurts to the surface.

asbestos   An industrial name for a group of minerals that crystallize as thin fibers. The two most common types are fibrous varieties of the minerals chrysotile and amphibole.

asbestosis   An often lethal lung disease most commonly found among asbestos miners and others who work with asbestos.

aseismic ridge   A submarine mountain chain with little or no earthquake activity.

ash (volcanic)   Fine pyroclastic material less than 2 mm in diameter.

ash flow   A mixture of volcanic ash, larger pyroclastic particles, and gas that flows rapidly along the Earth's surface as a result of an explosive volcanic eruption. (*syn*: nuée ardente)

asteroid   One of the many small celestial bodies in orbit around the Sun. Most asteroids orbit between Mars and Jupiter.

asthenosphere   The portion of the upper mantle beneath the lithosphere. It consists of weak, plastic rock and extends from a depth of about 100 kilometers to about 350 kilometers below the surface of the Earth.

atmosphere   A mixture of gases, mostly nitrogen and oxygen, that envelops the Earth.

atoll   A circular coral reef that surrounds a lagoon and is bounded on the outside by the deep water of the open sea.

atom   The fundamental unit of elements, consisting of a small, dense, positively charged center called a nucleus surrounded by a diffuse cloud of negatively charged electrons.

aulacogen   A tectonic trough on a craton bounded by normal faults and commonly filled with sediment. An aulacogen forms when one limb of a continental rift becomes inactive shortly after it forms.

axial plane   An imaginary plane that runs through the axis and divides a fold as symmetrically as possible into two halves.

B horizon   The soil layer just below the A horizon, called the subsoil, where ions leached from the A horizon accumulate.

back arc basin   A sedimentary basin on the opposite side of the magmatic arc from the trench, either in an island arc or in an Andean continental margin.

backshore   The upper zone of a beach that is usually dry but is washed by waves during storms.

bajada   A broad depositional surface extending outward from a mountain front and formed by merging alluvial fans.

banks   The rising slopes bordering the two sides of a stream channel.

bar   An elongate mound of sediment, usually composed of sand or gravel, in a stream channel or along a coastline.

barchan dune   A crescent-shaped dune, highest in the center, with the tips facing downwind.

barrier island   A long, narrow, low-lying island that extends parallel to the shoreline and is separated from the mainland by a lagoon.

basal slip   Movement of the entire mass of a glacier along the bedrock.

basalt   A dark-colored, very fine-grained, mafic, volcanic rock composed of about half calcium-rich plagioclase feldspar and half pyroxene.

basalt plateau   A sequence of horizontal basalt lava flows that were extruded rapidly to cover a large region of the Earth's surface. (*syn*: flood basalt, lava plateau)

base level   The deepest level to which a stream can erode its bed. The ultimate base level is usually sea level, but this is seldom attained.

basement rocks   The older granitic and related metamorphic rocks of the Earth's crust that make up the foundations of continents.

basin   A circular or elliptical synclinal structure, commonly filled with sediment.

batholith   A large plutonic mass of intrusive rock with more than 100 square kilometers of surface exposed.

bauxite   A gray, yellow, or reddish brown rock composed of a mixture of aluminum oxides and hydroxides. It is the principal ore of aluminum.

baymouth bar   A spit that extends partially or completely across the entrance to a bay.

beach   Any strip of shoreline washed by waves or tides. Most beaches are covered by sediment.

beach drift   The concerted movement of sediment along a beach caused by waves striking the shore at an angle.

beach terrace   A level portion of old beach elevated above the modern beach by uplift of the shoreline or fall of sea level.

bed   The floor of a stream channel. Also the thinnest layer in sedimentary rocks, commonly ranging in thickness from a centimeter to a meter or two.

bed load   That portion of a stream's load that is transported on or immediately above the stream bed.

bedding   Layering that develops as sediment is deposited.

bedrock   The solid rock that underlies soil or regolith.

Benioff zone   An inclined zone of earthquake activity that traces the upper portion of a subducting plate in a subduction zone.

bioclastic sediment   Clastic sediment composed of fragments of organisms such as clams, oysters, coral, etc.

biogeochemical cycle   The movement of nutrients through the atmosphere, biosphere, hydrosphere, and solid Earth in response to physical, biological, and chemical processes.

biomass energy   Electricity or other forms of energy produced by combustion of plant fuels.

bioremediation   Use of microorganisms to decompose a ground-water contaminant.

biosphere   The thin zone near the Earth's surface that is inhabited by life.

biotite   Black, rock-forming mineral of the mica group.

bitumen   A general term for solid and semi-solid hydrocarbons that are fusible and soluble in carbon bisulfide. The term includes petroleum, asphalt, natural mineral waxes, and asphaltites.

blowout   A saucer- or trough-shaped depression created by wind erosion.

body waves   Seismic waves that travel through the interior of the Earth.

boulder   A rounded rock fragment larger than a cobble (diameter greater than 256 cm).

Bowen's reaction series   A series of minerals in which any early-formed mineral crystallizing from a cooling magma reacts with the magma to form minerals lower in the series.

braided stream   A stream that divides into a network of branching and reuniting shallow channels separated by mid-channel bars.

breccia   A coarse-grained sedimentary rock composed of angular, broken fragments larger than 2 mm in diameter cemented in a fine-grained matrix of sand or silt.

brittle fracture   The rupture that occurs when a rock breaks sharply.

butte   A flat-topped mountain, with several steep cliff faces. A butte is smaller and more tower-like than a mesa.

C horizon   The lowest soil layer, composed mainly of partly weathered bedrock grading downward into unweathered parent rock.

calcite   A common rock-forming mineral, $CaCO_3$.

caldera   A large circular depression caused by an explosive volcanic eruption.

caliche   A hard soil layer formed when calcium carbonate precipitates and cements the soil.

calving   A process in which large chunks of ice break off from tidewater glaciers to form icebergs.

cap rock   An impermeable rock, usually shale, that prevents oil or gas from escaping upward from a reservoir.

capacity   The maximum quantity of sediment that a stream can carry.

capillary action   The action by which water is pulled upward through small pores by electrical attraction to the pore walls.

capillary fringe   A zone above the water table in which the pores are filled with water due to capillary action.

carbonate rocks   Rocks such as limestone and dolomite, made up primarily of carbonate minerals.

carbonatite   A carbonate rock of magmatic origin composed mostly of calcite or dolomite.

carbonization   A process in which a fossil forms when the volatile components of the soft tissues are driven off, leaving behind a thin film of carbon.

cast   A fossil formed when sedimentary rock or mineral matter fills a natural mold.

catastrophism   A principle that states that catastrophic events have been important in Earth history and modify the path of slow change.

cation   A positively charged ion.

cavern   An underground cavity or series of chambers created when ground water dissolves large amounts of rock, usually limestone. (*syn*: cave)

cementation   The process by which clastic sediment is lithified by precipitation of a mineral cement among the grains of the sediment.

Cenozoic era   The most recent era; 65 million years ago to the present.

chalk   A very fine-grained, soft, white to gray bioclastic limestone made of the shells and skeletons of marine microorganisms.

chemical bond   The linkage between atoms in molecules and between molecules and ions in crystals.

chemical weathering   The chemical decomposition of rocks and minerals by exposure to air, water, and other chemicals in the environment.

chert   A hard, dense, sedimentary rock composed of microcrystalline quartz. (*syn*: flint)

chondrule   A small grain, composed largely of olivine and pyroxene, found in stony meteorites.

cinder cone   A small volcano, as high as 300 meters, made up of loose pyroclastic fragments blasted out of a central vent.

cinders (volcanic)   Glassy pyroclastic volcanic fragments 4 to 32 mm in size.

cirque   A steep-walled semicircular depression eroded into a mountain peak by a glacier.

clastic sediment   Sediment composed of fragments of weathered rock that have been transported and deposited at the Earth's surface.

clastic sedimentary rocks   Rocks composed of lithified clastic sediment.

clay   Any clastic mineral particle less than 1/256 millimeter in diameter. Also a group of layer silicate minerals.

claystone   A fine-grained clastic sedimentary rock composed predominantly of clay minerals and small amounts of quartz and other minerals of clay size.

cleavage   The tendency of some minerals to break along certain crystallographic planes.

climate   The composite pattern of long-term weather conditions that can be expected in a given region.

coal   A flammable organic sedimentary rock formed from partially decomposed plant material and composed mainly of carbon.

cobbles   Rounded rock fragments in the 64- to 256-mm size range, larger than pebbles and smaller than boulders.

column   A dripstone or speleothem formed when a stalactite and a stalagmite meet and fuse together.

columnar joints   The regularly spaced cracks that commonly develop in lava flows, forming five- or six-sided columns.

comet   An interplanetary body, composed of loosely bound rock and ice, that forms a bright head and an extended fuzzy tail when it enters the inner portion of the solar system.

compaction   A process whereby the weight of overlying sediment compresses deeper sediment, decreasing pore space and causing weak lithification.

competence   A measure of the largest particles that a stream can transport.

composite volcano   A volcano that consists of alternate layers of unconsolidated pyroclastic material and lava flows. (*syn:* stratovolcano)

compressive stress   Stress that acts to shorten an object by squeezing it.

concordant   Pertaining to an igneous intrusion that is parallel to the layering of country rock.

cone of depression   A cone-like depression in the water table formed when water is pumped out of a well more rapidly than it can flow through the aquifer.

confining stress   Stress produced when rock or sediment is buried.

conformable   The condition in which sedimentary layers were deposited continuously without interruption.

conglomerate   A coarse-grained clastic sedimentary rock, composed of rounded fragments larger than 2 mm in diameter, cemented in a fine-grained matrix of sand or silt.

contact   A boundary between two different rock types or between rocks of different ages.

contact metamorphic ore deposit   An ore deposit formed by contact metamorphism.

contact metamorphism   Metamorphism caused by heating of country rock, and/or addition of fluids, from a nearby igneous intrusion.

continental crust   The predominantly granitic portion of the crust, 20 to 80 kilometers thick, that makes up the continents.

continental drift   The theory proposed by Alfred Wegener that continents were once joined together and later split and drifted apart. The continental drift theory has been replaced by the more complete plate tectonics theory.

continental glacier   A glacier that forms a continuous cover of ice over areas of 50,000 square kilometers or more and spreads outward under the influence of its own weight. (*syn:* ice sheet)

continental margin   The region between the shoreline of a continent and the deep ocean basins, including the continental shelf, continental slope, and continental rise. Also the region where thick, granitic continental crust joins thinner, basaltic oceanic crust.

continental margin basin   A sediment-filled depression or other thick accumulation of sediment and sedimentary rocks near the margin of a continent.

continental rifting   The process by which a continent is pulled apart at a divergent plate boundary.

continental rise   An apron of sediment between the continental slope and the deep sea floor.

continental shelf   A shallow, nearly level area of continental crust covered by sediment and sedimentary rocks that is submerged below sea level at the edge of a continent between the shoreline and the continental slope.

continental slope   The relatively steep (3° to 6°) underwater slope between the continental shelf and the continental rise.

continental suture   The junction created where two continents collide and weld into a single mass of continental crust.

control rod   A column of neutron-absorbing alloys that is placed among fuel rods to control nuclear fission in a reactor.

convection current   A current in a fluid or plastic material, formed when heated materials rise and cooler materials sink.

convergent plate boundary   A boundary where two lithospheric plates collide head-on.

coquina   A bioclastic limestone consisting of coarse shell fragments cemented together.

core   The innermost region of the Earth, probably consisting of iron and nickel.

correlation   Demonstration of the age equivalence of rocks or geologic features from different locations.

cost-benefit analysis   A system of analysis that attempts to weigh the cost of an act or policy, such as pollution control, directly against the economic benefits.

country rock   The older rock intruded by a younger igneous intrusion or mineral deposit.

covalent bond   A chemical bond in which two or more atoms share electrons to produce the effect of filled outer electron shells.

crater   A bowl-like depression at the summit of the volcano.

craton   A segment of continental crust, usually in the interior of a continent, that has been tectonically stable for a long time, commonly a billion years or longer.

creep   The slow movement of unconsolidated material downslope under the influence of gravity.

crest (of a wave)   The highest part of a wave.

crevasse   A fracture or crack in the upper 40 to 50 meters of a glacier.

cross-bedding   An arrangement of small beds lying at an angle to the main sedimentary layering.

cross-cutting relationship   (See principle of cross-cutting relationships.)

crust   The Earth's outermost layer, about 5 to 80 kilometers thick, composed of relatively low-density silicate rocks.

crystal A solid element or compound whose atoms are arranged in a regular, orderly, periodically repeated array.

crystal face A planar surface that develops if a crystal grows freely in an uncrowded environment.

crystal habit The shape in which individual crystals grow and the manner in which crystals grow together in aggregates.

crystal settling A process in which the crystals that solidify first from a cooling magma settle to the bottom of a magma chamber because the solid minerals are more dense than liquid magma.

Curie point The temperature below which rocks can retain magnetism.

current A continuous flow of water in a concerted direction.

daughter isotope An isotope formed by radioactive decay of another isotope.

debris flow A type of mass wasting in which particles move as a fluid and more than half of the particles are larger than sand.

deflation Erosion by wind.

deformation Folding, faulting and other changes in shape of rocks or minerals in response to mechanical forces, such as those that occur in tectonically active regions.

delta The nearly flat, alluvial, fan-shaped tract of land at the mouth of a stream.

dendritic drainage pattern A pattern of stream tributaries which branches like the veins in a leaf. It often indicates uniform underlying bedrock.

deposition The laying-down of sediment by any natural agent.

depositional environment Any setting in which sediment is deposited.

depositional remanent magnetism Remanent magnetism resulting from mechanical orientation of magnetic mineral grains during sedimentation.

desert A region with less than 25 cm of rainfall a year. Also defined as a region that supports only a sparse plant cover.

desert pavement A continuous cover of stones created as wind erodes fine sediment, leaving larger rocks behind.

desertification A process by which semiarid land is converted to desert, often by improper farming or by climate change.

differential weathering The process by which certain rocks weather more rapidly than adjacent rocks, usually resulting in an uneven surface.

dike A sheet-like igneous rock that cuts across the structure of country rock.

dike swarm A group of dikes that form in parallel or radial sets.

diorite A rock that is the medium- to coarse-grained plutonic equivalent of andesite.

dip The angle of inclination of bedding, measured from the horizontal.

directed stress Stress that acts most strongly in one direction.

discharge The volume of water flowing downstream per unit time. It is measured in units of m^3/sec.

disconformity A type of unconformity in which the sedimentary layers above and below the unconformity are parallel.

discordant Pertaining to a dike or other feature that cuts across sedimentary layers or other kinds of layering in country rock.

disseminated ore deposit A large low-grade ore deposit in which generally fine-grained metal-bearing minerals are widely scattered throughout a rock body in sufficient concentration to make the deposit economical to mine.

dissolution The process by which soluble rocks and minerals dissolve in water or water solutions.

dissolved load The portion of a stream's sediment load that is carried in solution.

distributary A channel that flows outward from the main stream channel, such as is commonly found in deltas.

divergent plate boundary The boundary or zone where lithospheric plates separate from each other. (*syn:* spreading center, rift zone)

docking The accretion of island arcs or microcontinents onto a continental margin.

dolomite A common rock-forming mineral, $CaMg(CO_3)_2$.

dome A circular or elliptical anticlinal structure.

dormant volcano a volcano that is not now erupting but has erupted in the past and will probably do so again.

downcutting Downward erosion by a stream.

drainage basin The region that is ultimately drained by a single river.

drainage divide A ridge or other topographically higher region that separates adjacent drainage basins.

drift (glacial) All rock or sediment transported and deposited by a glacier or by glacial meltwater.

dripstone A deposit formed in a cavern when calcite precipitates from dripping water.

drumlin An elongate hill formed when a glacier flows over and reshapes a mound of till or stratified drift.

dune A mound or ridge of wind-deposited sand.

earthflow A flowing mass of fine-grained soil particles mixed with water. Earthflows are less fluid than mudflows.

earthquake   A sudden motion or trembling of the Earth caused by the abrupt release of slowly accumulated elastic energy in rocks.

echo sounder   An instrument that emits sound waves and then records them after they reflect off the sea floor. The data is then used to record the topography of the sea floor.

effluent stream   A stream that receives water from ground water because its channel lies below the water table. (*syn*: gaining stream)

elastic deformation   A type of deformation in which an object returns to its original size and shape when stress is removed.

elastic limit   The maximum stress that an object can withstand without permanent deformation.

electron   A fundamental particle which forms a diffuse cloud of negative charge around an atom.

element   A substance that cannot be broken down into other substances by ordinary chemical means. An element is made up of the same kind of atoms.

emergent coastline   A coastline that was recently under water but has been exposed either because the land has risen or sea level has fallen.

end moraine   A moraine that forms at the end, or terminus, of a glacier.

eon   The longest unit of geologic time. The most recent, the Phanerozoic Eon, is further subdivided into eras and periods.

epicenter   The point on the Earth's surface directly above the focus of an earthquake.

epidemiology   The study of the distribution of sickness in a population.

epoch   The smallest unit of geologic time. Periods are divided into epochs.

era   A geologic time unit. Eons are divided into eras, and in turn eras are subdivided into periods.

erosion   The removal of weathered rocks and minerals by moving water, wind, ice, and gravity.

erratic   A boulder that was transported to its present location by a glacier, deposited at some distance from its original outcrop, and generally resting on a different type of bedrock.

esker   A long snake-like ridge formed by deposition in a stream that flowed on, within, or beneath a glacier.

estuary   A shallow bay that formed when a broad river valley was submerged by rising sea level or a sinking coast.

eutrophic lake   A lake characterized by abundant dissolved nitrates, phosphates, and other plant nutrients, and by a seasonal deficiency of oxygen in bottom water. Such lakes are commonly shallow.

evaporation   The transformation of a liquid into a gas.

evaporite deposit   A chemically precipitated sedimentary rock that formed when dissolved ions were concentrated by evaporation of water.

evolution   The change in the physical and genetic characteristics of a species over time.

exfoliation   Weathering in which concentric plates or shells split from the main rock mass like the layers of an onion.

extensional stress   Tectonic stress in which rocks are pulled apart.

external mold   A fossil cavity created in sediment by a shell or other hard body part that bears the impression of the exterior of the original.

externality   An environmental cost not directly associated with manufacturing. Examples include the costs of acid rain and purifying polluted water.

extinct volcano   A volcano that is expected never to erupt again.

extrusive rock   An igneous rock formed from material that has erupted onto the surface of the Earth.

eustatic sea level change   Global sea level change caused by changes in water temperature, changes in the volume of the mid-oceanic ridge, or growth or melting of glaciers.

fall   A type of mass wasting in which rock or regolith falls freely or bounces down the face of a cliff.

fault   A fracture in rock along which one rock has moved relative to rock on the other side.

fault creep   A continuous, slow movement of solid rock along a fault, resulting from a constant stress acting over a long time.

fault zone   An area of numerous closely spaced faults.

faunal succession   (See principle of faunal succession.)

feldspar   A common group of aluminum silicate rock-forming minerals that contain potassium, sodium, or calcium.

fetch   The distance that the wind has travelled over the ocean without interruption.

firn   Hard, dense snow that has survived through one summer melt season. Firn is transitional between snow and glacial ice.

fissility   Fine layering along which a rock splits easily.

fission (nuclear)   The spontaneous or induced splitting by particle collision of a heavy nucleus into a pair of nearly equal fission fragments plus some neutrons. Fission releases large amounts of energy (see fusion).

fjord   A long, deep, narrow arm of the sea bounded by steep walls, generally formed by submergence of a glacially eroded valley. (Also spelled fiord.)

flash flood   A rapid, intense, local flood of short duration, commonly occurring in deserts.

flood basalt   Basaltic lava that erupts gently in great volume to cover large areas of land and form a basalt plateau.

flood plain   That portion of a river valley adjacent to the channel; it is built by sediment deposited during floods and is covered by water during a flood.

flow   Mass wasting in which individual particles move downslope as a semi-fluid, not as a consolidated mass.

focus   The initial rupture point of an earthquake.

fold   A bend in rock.

foliation   Layering in rock created by metamorphism.

footwall   The rock beneath an inclined fault.

forearc basin   A sedimentary basin between the subduction complex and the magmatic arc in either an island arc or the Andean continental margin.

foreshock   Small earthquakes that precede a large quake by a few seconds to a few weeks.

foreshore   The zone that lies between the high and low tides; the intertidal region.

formation   A lithologically distinct body of sedimentary, igneous, or metamorphic rock that can be recognized in the field and can be mapped.

fossil   Any preserved trace, imprint, or remains of a plant or animal.

fossil fuel   Fuels formed from the partially decayed remains of plants and animals. The most commonly used fossil fuels are petroleum, coal, and natural gas.

fractional crystallization   Crystallization from a magma in which early-formed crystals are prevented from reacting with the magma, resulting in the evolution of a final magma that is enriched in silica and other components of granite.

fracture   (a) The manner in which minerals break other than along planes of cleavage. (b) A crack, joint, or fault in bedrock.

frost wedging   A process in which water freezes in a crack in rock and the expansion wedges the rock apart.

fuel rod   A 2-meter-long column of fuel-grade uranium pellets used to fuel a nuclear reactor.

fusion (of atomic nuclei)   The combination of two light nuclei to form a heavier nucleus. Fusion releases large amounts of energy. (See fission.)

gabbro   Igneous rock that is mineralogically identical to basalt but that has a medium- to coarse-grained texture because of its plutonic origin.

gaining stream   A stream that receives water from ground water because its channel lies below the water table. (*syn*: effluent stream)

gem   a mineral that is prized primarily for its beauty. Any precious or semiprecious stone, especially when cut or polished for ornamental use.

geologic column   A composite columnar diagram that shows the sequence of rocks at a given place or region arranged to show their position in the geologic time scale.

geologic structure   Any feature formed by rock deformation, such as a fold or a fault. Also, the combination of all such features of an area or region.

geologic time scale   A chronological arrangement of geologic time subdivided into units.

geology   The study of the Earth, including the materials that it is made of, the physical and chemical changes that occur on its surface and in its interior, and the history of the planet and its life forms.

geothermal energy   Energy derived from the heat of the Earth.

geothermal gradient   The rate at which temperature increases with depth in the Earth.

geyser   A type of hot spring that intermittently erupts jets of hot water and steam. Geysers occur when ground water comes in contact with hot rock.

glacial polish   A smooth polish on bedrock created when fine particles transported at the base of a glacier abrade the bedrock.

glacial striations   Parallel grooves and scratches in bedrock that form as rocks are dragged along at the base of a glacier.

glacier   A massive, long-lasting accumulation of compacted snow and ice that forms on land and moves downslope or outward under its own weight.

gneiss   A foliated rock with banded appearance formed by regional metamorphism.

Gondwanaland   The southern part of Wegener's Pangea, which was the late Paleozoic supercontinent. (*syn*: Gondwana)

graben   A wedge-shaped block of rock that has dropped downward between two normal faults.

graded bedding   A type of bedding in which larger particles are at the bottom of each bed, and the particle size decreases towards the top.

graded stream   A stream with a smooth concave profile. A graded stream is in equilibrium with its sediment supply; once a stream becomes graded, the rate of channel erosion becomes equal to the rate at which the stream deposits sediment in its channel. Thus, there is no net erosion or deposition, and the stream profile no longer changes.

gradient   The vertical drop of a stream over a specific distance.

granite   A medium- to coarse-grained felsic, plutonic rock made predominantly of potassium feldspar and quartz.

gravel   Unconsolidated sediment consisting of rounded particles larger than 2 millimeters in diameter.

graywacke   A poorly sorted sandstone, commonly dark in color and consisting mainly of quartz, feldspar, and rock fragments with considerable quantities of silt and clay in its pores.

greenhouse effect   An increase in the temperature of a planet's atmosphere caused by infrared-absorbing gases in the atmosphere.

groin   A narrow wall built perpendicular to the shore to trap sand transported by currents and waves.

ground moraine   A moraine formed when a melting glacier deposits till in a relatively thin layer over a broad area.

ground water   Water contained in soil and bedrock. All subsurface water.

guyot   A flat-topped seamount.

gypsum   A mineral with the formula ($CaSO_4 \cdot 2H_2O$). It commonly forms in evaporite deposits.

Hadean Eon   The earliest time in the Earth's history, from about 4.6 billion years ago to 3.8 billion years ago.

half-life   The time it takes for half of the nuclei of a radioactive isotope in a sample to decompose.

halite   A mineral, NaCl. (*syn*: common salt)

hanging valley   A tributary glacial valley whose mouth lies high above the floor of the main valley.

hanging wall   The rock above an inclined fault.

hardness   The resistance of the surface of a mineral to scratching.

headward erosion   The lengthening of a valley in an upstream direction.

heat flow   The amount of heat energy leaving the Earth per cm^2/sec, measured in calories/cm^2/sec.

horn   A sharp, pyramid-shaped rock summit formed by glacial erosion of three or more cirques into a mountain peak.

hornblende   A rock-forming mineral. The most common member of the amphibole group.

hornfels   A fine-grained rock formed by contact metamorphism.

horst   A block of rock that has moved relatively upward and is bounded by two faults.

hot spot   A persistent volcanic center thought to be located directly above a rising plume of hot mantle rock.

hot spring   A spring formed where hot ground water flows to the surface.

humus   The dark organic component of soil composed of litter that has decomposed sufficiently so that the origin of the individual pieces cannot be determined.

hydraulic action   The mechanical loosening and removal of material by flowing water.

hydride   A compound of hydrogen and one or more metals. Hydrides can be heated to release hydrogen gas for use as a fuel.

hydroelectric energy   Electricity produced by turbines that harness the energy of water dropping downward through a dam.

hydrogeologist   A scientist who studies ground water and related aspects of surface water.

hydrologic cycle   The constant circulation of water among the sea, the atmosphere, and the land.

hydrolysis   A weathering process in which water reacts with a mineral to form a new mineral with water incorporated into its crystal structure.

hydrosphere   The collection of all water at or near the Earth's surface.

hydrothermal metamorphism   Changes in rock that are primarily caused by migrating hot water and by ions dissolved in the hot water. (*syn*: hydrothermal alteration)

hydrothermal vein   A sheet-like mineral deposit that fills a fault or other fracture, precipitated from hot water solutions.

ice age   A time of extensive glacial activity, when alpine glaciers descended into lowland valleys and continental glaciers spread over the higher latitudes.

ice sheet   A glacier that forms a continuous cover of ice over areas of 50,000 square kilometers or more and spreads outward under the influence of its own weight. (*syn*: continental glacier)

iceberg   A large chunk of ice that breaks from a glacier into a body of water.

igneous rock   Rock that solidified from magma.

incised meander   A stream meander that is cut below the level at which it originally formed, usually caused by rejuvenation.

index fossil   A fossil that dates the layers in which it is found. Index fossils are abundantly preserved in rocks, widespread geographically, and existed as a species or genus for only a relatively short time.

industrial mineral   Any rock or mineral of economic value exclusive of metal ores, fuels, and gems.

influent stream   A stream that lies above the water table. Water percolates from the stream channel downward into the saturated zone. (*syn*: losing stream)

intermediate rocks   Igneous rocks with chemical and mineral compositions between those of granite and basalt.

internal mold   A fossil that forms when the inside of a shell fills with sediment or precipitated minerals.

internal processes   Earth processes and movements that are initiated within the Earth—for example, formation of magma, earthquakes, mountain building, and tectonic plate movement.

intertidal zone   The part of a beach that lies between the high and low tide lines.

intracratonic basin   A sedimentary basin located within a craton.

intrusive rock   A rock formed when magma solidifies within bodies of preexisting rock.

ion An atom with an electrical charge.

ionic bond A chemical bond in which cations and anions are attracted by their opposite electronic charges, and thus bond together.

ionic substitution The replacement of one ion by another in a mineral; usually the two ions are of similar size and charge.

island arc A gently curving chain of volcanic islands in the ocean formed by convergence of two plates, each bearing ocean crust, and the resulting subduction of one plate beneath the other.

isostasy The condition in which the lithosphere floats on the asthenosphere as an iceberg floats on water.

isostatic adjustment The rising and settling of portions of the lithosphere to maintain equilibrium as they float on the plastic asthenosphere.

isotopes Atoms of the same element that have the same number of protons but different numbers of neutrons.

joint A fracture that occurs without movement of rock on either side of the break.

Jovian planets The outer planets—Jupiter, Saturn, Uranus, and Neptune—which are massive and are composed of a high proportion of the lighter elements.

kame A small mound or ridge of layered sediment deposited by a stream at the margin of a melting glacier or in a low place on the surface of a glacier.

kaolinite A common clay mineral, $Al_2Si_2O_5(OH)_4$.

karst topography A type of topography formed over limestone or other soluble rock and characterized by caverns, sinkholes, and underground drainage.

kerogen The solid bituminous mineraloid substance in oil shales that yields oil when the shales are distilled.

kettle A depression in outwash created by melting of a large chunk of ice left buried in the drift by a receding glacier.

key bed A thin, widespread, easily recognized sedimentary layer that can be used for correlation.

kimberlite An alkalic peridotite containing phenocrysts of olivine and phlogopite in a groundmass of calcite, olivine, and phlogopite. The name is derived from Kimberley, South Africa, where the rock contains diamonds.

L wave An earthquake wave that travels along the surface of the Earth, or along a boundary between layers within the Earth. (*syn:* surface wave)

lagoon A protected body of water separated from the sea by a reef or barrier island.

lake a large, inland body of standing water that occupies a depression in the land surface.

landslide A general term for the downslope movement of rock and regolith under the influence of gravity.

lateral moraine A moraine that forms on or adjacent to the sides of a mountain glacier.

laterite A highly weathered soil rich in oxides of iron and aluminum that usually develops in warm, moist tropical or temperate regions.

Laurasia The northern part of Wegener's Pangea, which was the late Paleozoic supercontinent.

lava Fluid magma that flows onto the Earth's surface from a volcano or fissure. Also, the rock formed by solidification of the same material.

lava plateau A sequence of horizontal basalt lava flows that were extruded rapidly to cover a large region of the Earth's surface. (*syn:* flood basalt, basalt plateau)

leaching The dissolution and downward movement of soluble components of rock and soil by percolating water.

limb The side of a fold in rock.

limestone A sedimentary rock consisting chiefly of calcium carbonate.

lithification The conversion of loose sediment to solid rock.

lithosphere The cool, rigid, outer layer of the Earth, about 100 kilometers thick, which includes the crust and part of the upper mantle.

litter Leaves, twigs, and other plant or animal material that has fallen to the surface of the soil but is still recognizable.

loam Soil that contains a mixture of sand, clay, and silt and a generous amount of organic matter.

loess A homogenous, unlayered deposit of windblown silt, usually of glacial origin.

longitudinal dune A long, symmetrical dune oriented parallel with the direction of the prevailing wind.

longshore current A current flowing parallel and close to the coast that is generated when waves strike a shore at an angle.

losing stream A stream that lies above the water table. Water percolates from the stream channel downward into the saturated zone. (*syn:* influent stream)

Love wave a surface seismic wave that produces side-to-side motion.

luster The quality and intensity of light reflected from the surface of a mineral.

mafic rock Dark-colored igneous rock with high magnesium and iron content, and composed chiefly of iron- and magnesium-rich minerals.

magma Molten rock generated within the Earth.

magmatic arc A narrow, elongate band of intrusive and volcanic activity associated with subduction.

magnetic reversal A change in the Earth's magnetic field in which the north magnetic pole becomes the south magnetic pole, and vice versa.

magnetometer An instrument that measures the Earth's magnetic field.

manganese nodule A manganese-rich, potato-shaped rock found on the ocean floor.

mantle A mostly solid layer of the Earth lying beneath the crust and above the core. The mantle extends from the base of the crust to a depth of about 2900 kilometers.

mantle convection The convective flow of solid rock in the mantle.

mantle plume A rising vertical column of mantle rock.

marble A metamorphic rock consisting of fine- to coarse-grained recrystallized calcite and/or dolomite.

maria Dry, barren, flat expanses of volcanic rock on the Moon, first thought to be seas.

mass wasting The movement of earth material down-slope primarily under the influence of gravity.

meander One of a series of sinuous curves or loops in the course of a stream.

mechanical weathering The disintegration of rock into smaller pieces by physical processes.

medial moraine A moraine formed in or on the mid-dle of a glacier by the merging of lateral moraines as two glaciers flow together.

mesa A flat-topped mountain or a tableland that is smaller than a plateau and larger than a butte.

Mesozoic era The portion of geologic time roughly 245 to 65 million years ago. Dinosaurs rose to promi-nence during this era. The end of the Mesozoic era is marked by the extinction of the dinosaurs.

metallic bond A chemical bond in which the metal atoms are surrounded by a matrix of outer-level elec-trons that are free to move from one atom to another.

metamorphic facies A set of all metamorphic rock types that formed under similar temperature and pressure conditions.

metamorphic grade The intensity of metamorphism that formed a rock; the maximum temperature and pressure attained during metamorphism.

metamorphic rock A rock formed when igneous, sed-imentary, or other metamorphic rocks recrystallize in response to elevated temperature, increased pres-sure, chemical change, and/or deformation.

metamorphism The process by which rocks and min-erals change in response to changes in temperature, pressure, chemical conditions, and/or deformation.

metasomatism Metamorphism accompanied by the introduction of ions from an external source.

meteorite A fallen meteoroid.

meteoroid A small interplanetary body in an irregular orbit. Many meteoroids are asteroids or comet frag-ments.

mica A layer silicate mineral with a distinctive platy crystal habit and perfect cleavage. Muscovite and bi-otite are common micas.

mid-channel bar An elongate lobe of sand and gravel formed in a stream channel.

mid-oceanic ridge A continuous submarine mountain chain that forms at the boundary between divergent tectonic plates within oceanic crust.

migmatite A rock composed of both igneous and metamorphic-looking materials. It forms at very high metamorphic grades when rock begins to partially melt to form magma.

mineral A naturally occurring inorganic solid with a characteristic chemical composition and a crystalline structure.

mineral deposit A local enrichment of one or more minerals.

mineral reserve The known supply of ore in the ground.

mineralization A process of fossilization in which the organic components of an organism are replaced by minerals.

Mohorovičić discontinuity (Moho) The boundary be-tween the crust and the mantle, identified by a change in the velocity of seismic waves.

Mohs hardness scale A standard numbered from 1 to 10, to measure and express the hardness of minerals based on a series of ten fairly common minerals, each of which is harder than those lower on the scale.

moment magnitude scale A scale used to measure and express the energy released during an earth-quake.

monocline A fold with only one limb.

moraine A mound or ridge of till deposited directly by glacial ice.

mountain chain A number of mountain ranges grouped together in an elongate zone.

mountain range A series of mountains or mountain ridges that are closely related in position, direction, age, and mode of formation.

mud Wet silt and clay.

mud cracks Irregular, usually polygonal fractures that develop when mud dries. The patterns may be pre-served when the mud is lithified.

mudflow Mass wasting of fine-grained soil particles mixed with a large amount of water.

mudstone A non-fissile rock composed of clay and silt.

mummification A process in which the remains of an animal are preserved by dehydration.

natural gas A mixture of naturally occurring light hydrocarbons composed mainly of methane (CH_4).

natural levee A ridge or embankment of flood-deposited sediment along both banks of a stream channel.

neutron A subatomic particle with the mass of a proton but no electrical charge.

nonconformity A type of unconformity in which layered sedimentary rocks lie on igneous or metamorphic rocks.

nonfoliated The lack of layering in metamorphic rock.

non-point source pollution Pollution that is generated over a broad area, such as that originating from fertilizers and pesticides spread over fields.

nonrenewable resource A resource in which formation of new deposits occurs much more slowly than consumption.

normal fault A fault in which the hanging wall has moved downward relative to the footwall.

normal polarity A magnetic orientation the same as that of the Earth's modern magnetic field.

nucleus The small, dense, central portion of an atom composed of protons and neutrons. Nearly all of the mass of an atom is concentrated in the nucleus.

nuée ardente A swiftly flowing, often red-hot cloud of gas, volcanic ash, and other pyroclastics formed by an explosive volcanic eruption. (*syn*: ash flow)

O horizon The uppermost soil layer, consisting mostly of litter and humus with a small proportion of minerals.

obsidian A black or dark-colored glassy volcanic rock, usually of rhyolitic composition.

oceanic crust The 7- to 10-kilometer-thick layer of sediment and basalt that underlies the ocean basins.

oceanic island A seamount, usually of volcanic origin, that rises above sea level.

Ogallala aquifer The aquifer that extends for almost 1000 kilometers from the Rocky Mountains eastward beneath portions of the Great Plains.

oil A naturally occurring liquid or gas composed of a complex mixture of hydrocarbons. (*syn*: petroleum)

oil shale A kerogen-bearing sedimentary rock that yields liquid or gaseous hydrocarbons when heated.

oil trap Any rock barrier that accumulates oil or gas by preventing its upward movement.

oligotrophic lake A lake characterized by nearly pure water with low concentrations of nitrates, phosphates, and other plant nutrients. Oligotrophic lakes have low productivities and sustain relatively few organisms, although lakes of this type typically contain a few huge trout or similar game fish and are commonly deep.

olivine A common rock-forming mineral in mafic and ultramafic rocks with a composition that varies between Mg_2SiO_4 and Fe_2SiO_4.

ooid A small rounded accretionary body in sedimentary rock, generally formed of concentric layers of calcium carbonate around a nucleus such as a sand grain.

ore A natural material that is sufficiently enriched in one or more minerals to be mined profitably.

original horizontality (principle of) (See principle of original horizontality.)

orogeny The process of mountain building; all tectonic processes associated with mountain building.

orographic lifting Lifting of air that occurs when air flows over a mountain.

orthoclase A common rock-forming mineral; a variety of potassium feldspar, $(KAlSi_3O_8)$.

outwash Sediment deposited by streams beyond the glacial terminus.

outwash plain A broad, level surface composed of outwash.

oxbow lake A crescent-shaped lake formed where a meander is cut off from a stream and the ends of the cut-off meander become plugged with sediment.

oxidation The loss of electrons from a compound or element during a chemical reaction. In the weathering of common minerals, oxidation usually occurs when a mineral reacts with molecular oxygen.

ozone hole The unusually low concentration of ozone in the upper atmosphere, first discovered in 1985.

P wave (Also called a compressional wave.) A seismic wave that causes alternate compression and expansion of rock.

pahoehoe A basaltic lava flow with a smooth, billowy, or "ropy" surface.

paleoclimatology The study of ancient climates.

paleomagnetism The study of natural remnant magnetism in rocks and of the history of the Earth's magnetic field.

paleontology The study of life that existed in the past.

Paleozoic era The part of geologic time 538 to 245 million years ago. During this era invertebrates, fishes, amphibians, reptiles, ferns, and cone-bearing trees were dominant.

Pangea A supercontinent, identified and named by Alfred Wegener, that existed from about 300 to 200 million years ago and included most of the continental crust of the Earth. In this book we refer to three supercontinents: Pangea I (about 2.0 billion to 1.3 billion years ago), Pangea II (1 billion to 700 million years ago), and Pangea III (300 million to 200 million years ago).

parabolic dune A crescent-shaped dune with tips pointing into the wind.

parent rock Any original rock before it is changed by metamorphism or other geological processes.

partial melting The process in which a silicate rock only partly melts as it is heated, to form magma that is more silica-rich than the original rock.

passive continental margin A margin characterized by a firm connection between continental and oceanic crust, where little tectonic activity occurs.

paternoster lake One of a series of lakes, strung out like beads and connected by short streams and waterfalls, created by glacial erosion.

peat A loose, unconsolidated, brownish mass of partially decayed plant matter; a precursor to coal.

pebble A sedimentary particle between 2 and 64 millimeters in diameter, larger than sand and smaller than a cobble.

pedalfer A soil type that forms in humid environments, characterized by abundant iron and aluminum oxides and a concentration of clay in the B horizon.

pediment A gently sloping erosional surface that forms along a mountain front uphill from a bajada, usually covered by a patchy veneer of gravel only a few meters thick.

pedocal A soil formed in arid and semiarid climates characterized by an accumulation of calcium carbonate and other minerals in the B horizon.

pegmatite An exceptionally coarse-grained igneous rock, usually with the same mineral content as granite.

pelagic sediment Muddy ocean sediment that consists of a mixture of clay and the skeletons of microscopic marine organisms.

peneplain According to a model popular in the first half of this century, streams erode mountain ranges ultimately to form a large, low, nearly featureless surface called a peneplain. However, the theory fails to consider tectonic rejuvenation, and peneplains do not actually exist.

perched water table The top of a localized lens of ground water that lies above the main water table, formed by a layer of impermeable rock or clay.

peridotite A coarse-grained plutonic rock composed mainly of olivine; it may also contain pyroxene, amphibole, or mica but little or no feldspar. The upper part of the mantle is thought to be composed mostly of peridotite.

period A geologic time unit longer than an epoch and shorter than an era.

permafrost A layer of permanently frozen soil or subsoil which lies from about a half meter to a few meters beneath the surface in arctic environments.

permeability A measure of the ease with which fluid can travel through a porous material.

permineralization Fossilization that occurs when mineral matter is deposited in cavities or pores.

petroleum A naturally occurring liquid composed of a complex mixture of hydrocarbons.

Phanerozoic Eon The most recent 538 million years of geologic time, represented by rocks that contain evident and abundant fossils.

phenocryst A large, early-formed crystal in a finer matrix in igneous rock.

phyllite A metamorphic rock with a silky appearance and commonly wrinkled surface, intermediate in grade between slate and schist.

phytoplankton All floating plants, such as diatoms.

pillow lava Lava that solidified under water, forming spheroidal lumps like a stack of pillows.

pipe A vertical conduit below a volcano, through which magmatic materials passed. It is usually filled with solidified magma and/or brecciated rock.

placer deposit A surface mineral deposit formed by the mechanical concentration of mineral particles (usually by water) from weathered debris.

planetesimal One of many small rocky spheres that formed early in the history of the solar system and later coalesced to form the planets.

plankton Floating and drifting aquatic organisms.

plastic deformation A type of deformation in which the material changes shape permanently without fracture.

plate A relatively rigid independent segment of the lithosphere that can move independently of other plates.

plate boundary A boundary between two lithospheric plates.

plate tectonics theory A theory of global tectonics in which the lithosphere is segmented into several plates that move about relative to one another by floating on and gliding over the plastic asthenosphere. Seismic and tectonic activity occur mainly at the plate boundaries.

plateau A large elevated area of comparatively flat land.

platform The part of a continent covered by a thin layer of nearly horizontal sedimentary rocks overlying older igneous and metamorphic rocks of the craton.

playa A dry desert lake bed.

playa lake An intermittent desert lake.

Pleistocene epoch A span of time from roughly 2 million to 8000 years ago, characterized by several advances and retreats of glaciers.

plucking A process in which glacial ice erodes rock by loosening particles and then lifting and carrying them downslope.

plunging fold A fold with a dipping or plunging axis.

pluton An igneous intrusion.

plutonic rock An igneous rock that forms deep (a kilometer or more) beneath the Earth's surface.

pluvial lake A lake formed during a time of abundant precipitation. Many pluvial lakes formed as continental ice sheets melted.

point bar A stream deposit located on the inside of a growing meander.

point source pollution Pollution which arises from a specific site such as a septic tank or a factory.

polarity The magnetically positive (north) or negative (south) character of a magnetic pole.

polymorph A mineral that crystallizes with more than one crystal structure.

pore space The open space between grains in rock, sediment, or soil.

porosity The proportion of the volume of a material that consists of open spaces.

porphyry Any igneous rock containing larger crystals (phenocrysts) in a relatively fine-grained matrix.

porphyry copper deposit A large body of porphyritic igneous rock that contains disseminated copper sulfide minerals, usually mined by surface mining methods.

pothole A smooth, rounded depression in bedrock in a stream bed, caused by abrasion when currents circulate stones or coarse sediment.

Precambrian All of geologic time before the Paleozoic era, encompassing approximately the first 4 billion years of Earth's history. Also, all rocks formed during that time.

precautionary principle A guideline that recommends that environmental precautions be taken without absolute proof that the perturbation is harmful.

precipitation (a) A chemical reaction that produces a solid salt, or precipitate, from a solution. (b) Any form in which atmospheric moisture returns to the Earth's surface—rain, snow, hail, and sleet.

preservation A process in which an entire organism or a part of an organism is preserved with very little chemical or physical change.

pressure-release fracturing The process by which rock fractures as overlying rock erodes away and the pressure diminishes.

pressure release melting The melting of rock and the resulting formation of magma caused by a drop in pressure at constant temperature.

primary (P) wave A seismic wave formed by alternate compression and expansion of rock. P waves travel faster than any other seismic waves.

principle of crosscutting relationships The principle that a dike or other feature cutting through rock must be younger than the rock.

principle of faunal succession The principle that fossil organisms succeed one another in a definite and recognizable sequence, so that sedimentary rocks of different ages contain different fossils, and rocks of the same age contain identical fossils. Therefore, the relative ages of rocks can be identified from their fossils.

principle of original horizontality The principle that most sediment is deposited as nearly horizontal beds, and therefore most sedimentary rocks started out with nearly horizontal layering.

principle of superposition The principle that states that in any undisturbed sequence of sediment or sedimentary rocks, the age becomes progressively younger from bottom to top.

Proterozoic Eon The portion of geological time from 2.5 billion to 538 million years ago.

proton A dense, massive, positively charged particle found in the nucleus of an atom.

pumice Frothy, usually rhyolitic magma solidified into a rock so full of gas bubbles that it can float on water.

pyroclastic rock Any rock made up of material ejected explosively from a volcanic vent.

pyroxene A rock-forming silicate mineral group that consists of many similar minerals. Members of the pyroxene group are major constituents of basalt and gabbro.

quartz A rock-forming silicate mineral, SiO_2. Quartz is a widespread and abundant component of continental rocks but is rare in the oceanic crust and mantle.

quartz sandstone Sandstone containing more than 90 percent quartz.

quartzite A metamorphic rock composed mostly of quartz, formed by recrystallization of sandstone.

radial drainage pattern A drainage pattern formed when a number of streams originate on a mountain and flow outward like the spokes on a wheel.

radioactivity The natural spontaneous decay of unstable nuclei.

radiometric age dating The process of measuring the absolute age of geologic material by measuring the concentrations of radioactive isotopes and their decay products.

radon A radioactive gas formed by radioactive decay of uranium that commonly accumulates in some igneous and sedimentary rocks.

rain shadow desert A desert formed on the lee side of a mountain range.

Rayleigh wave A surface seismic wave with an up-and-down rolling motion.

recessional moraine A moraine that forms at the terminus of a glacier as the glacier stabilizes temporarily during retreat.

recharge The replenishment of an aquifer by the addition of water.

rectangular drainage pattern A drainage pattern in which the main stream and its tributaries are of approximately the same length and intersect at right angles.

reef A wave-resistant ridge or mound built by corals or other marine organisms.

reflection The return of a wave that strikes a surface.

refraction The bending of a wave that occurs when the wave changes velocity as it passes from one medium to another.

regional burial metamorphism   Metamorphism of a broad area of the Earth's crust caused by elevated temperatures and pressures resulting from simple burial.

regional dynamothermal metamorphism   Metamorphism accompanied by deformation affecting an extensive region of the Earth's crust.

regional metamorphism   Metamorphism that is broadly regional in extent, involving very large areas and volumes of rock. Includes both regional dynamothermal and regional burial metamorphism.

regolith   The loose, unconsolidated, weathered material that overlies bedrock.

rejuvenated stream   A stream that has had its gradient steepened and its erosive ability renewed by tectonic uplift or a drop of sea level.

relative age   Age expressed as the order in which rocks formed and geological events occurred, but not measured in years.

relief   The vertical distance between a high and a low point on the Earth's surface.

remediation (of a contaminated aquifer)   The treatment of a contaminated aquifer to remove or decompose a pollutant.

remote sensing   The collection of information about an object by instruments that are not in direct contact with it.

replacement   Fossilization in which the original organic material is replaced by new minerals.

reserves   Known geological deposits that can be extracted profitably under current conditions.

reservoir rock   Porous and permeable rock in which liquid petroleum or gas accumulates.

reverse fault   A fault in which the hanging wall has moved up relative to the footwall.

reversed polarity   Magnetic orientations in rock which are opposite to the present orientation of the Earth's field. Also, the condition in which the Earth's magnetic field is opposite to its present orientation.

rhyolite   A fine-grained extrusive igneous rock compositionally equivalent to granite.

Richter scale   A numerical scale of earthquake magnitude measured by the amplitude of the largest wave on a standardized seismograph.

rift   A zone of separation of tectonic plates at a divergent plate boundary.

rift valley   An elongate depression that develops at a divergent plate boundary. Examples include continental rift valleys and the rift valley along the center of the mid-oceanic ridge system.

rift zone   The boundary or zone where lithospheric plates rift or separate from each other. (*syn*: divergent plate boundary, spreading center)

ring of fire   The belt of subduction zones and major tectonic activity including extensive volcanism that borders the Pacific Ocean along the continental margins of Asia and the Americas.

rip current   A current created when water flows back toward the sea after a wave breaks against the shore. (*syn*: undertow)

ripple marks   Small, nearly parallel ridges and troughs formed in loose sediment by wind or water currents and waves. They may then be preserved when the sediment is lithified.

risk assessment   The analysis of risk and the implementation of policy based on that analysis.

roche moutonnée   An elongate, streamlined bedrock hill sculpted by a glacier.

rock   A naturally formed solid that is an aggregate of one or more different minerals.

rock avalanche   A type of mass wasting in which a segment of bedrock slides over a tilted bedding plane or fracture. The moving mass usually breaks into fragments. (*syn*: rockslide)

rock cycle   The sequence of events in which rocks are formed, destroyed, altered, and reformed by geological processes.

rock flour   Finely ground, silt-sized rock fragments formed by glacial abrasion.

rockslide   A type of slide in which a segment of bedrock slides along a tilted bedding plane or fracture. The moving mass usually breaks into fragments. (*syn*: rock avalanche)

rounding   The sedimentary process in which sharp, angular edges and corners of grains are smoothed.

rubble   Angular particles with diameters greater than 2 millimeters.

runoff   Water that flows back to the oceans in surface streams.

S wave   A seismic wave consisting of a shearing motion in which the oscillation is perpendicular to the direction of wave travel. S waves travel more slowly than P waves.

salinization   A process whereby salts accumulate in soil when water, especially irrigation water, evaporates from the soil.

salt cracking   A weathering process in which salty water migrates into the pores in rock. When the water evaporates, the salts crystallize, pushing grains apart.

saltation   Sediment transport in which particles bounce and hop along the surface.

sand   Sedimentary grains that range from 1/16 to 2 millimeters in diameter.

sandstone   Clastic sedimentary rock composed primarily of lithified sand.

saturated zone   The region below the water table where all the pores in rock or regolith are filled with water.

scarp A line of cliffs created by faulting or by erosion.

schist A strongly foliated metamorphic rock that has a well developed parallelism of minerals such as micas.

sea arch An opening created when a cave is eroded all the way through a narrow headland.

sea stack A pillar of rock left when a sea arch collapses or when the inshore portion of a headland erodes faster than the tip.

sea-floor spreading The hypothesis that segments of oceanic crust are separating at the mid-oceanic ridge.

seamount A submarine mountain, usually of volcanic origin, that rises 1 kilometer or more above the surrounding sea floor.

secondary recovery Production of oil or gas as a result of artificially augmenting the reservoir energy by injection of water or other fluids. Secondary recovery methods are usually applied after substantial depletion of the reservoir.

sediment Solid rock or mineral fragments transported and deposited by wind, water, gravity, or ice, precipitated by chemical reactions, or secreted by organisms, and that accumulate as layers in loose, unconsolidated form.

sedimentary rock A rock formed when sediment is lithified.

sedimentary structure Any structure formed in sedimentary rock during deposition or by later sedimentary processes; for example, bedding.

seismic gap An immobile region of a fault bounded by moving segments.

seismic profiler A device used to construct a topographic profile of the ocean floor and to reveal layering in sediment and rock beneath the sea floor.

seismic tomography A technique whereby seismic data from many earthquakes and recording stations are analyzed to provide a three-dimensional view of the Earth's interior.

seismic wave All elastic waves that travel through rock, produced by an earthquake or explosion.

seismogram The record made by a seismograph.

seismograph An instrument that records seismic waves.

seismology The study of earthquake waves and the interpretation of these data to elucidate the structure of the interior of the Earth.

semiarid Any zone that receives between 25 and 50 centimeters of rainfall annually. Semiarid zones surround most deserts.

serpentinite A rock composed largely of serpentine-group minerals, usually chrysotile and antigorite, commonly derived from alteration of peridotite or sea floor basalt.

shale A fine-grained clastic sedimentary rock with finely layered structure composed predominantly of clay minerals.

shear stress Stress that acts in parallel but opposite directions.

shear wave (See S wave.)

sheet flood A broad, thin sheet of flowing water that is not concentrated into channels, typically in arid regions.

shield A large region of exposed basement rocks that are commonly of Precambrian age.

shield volcano A large, gently sloping volcanic mountain formed by successive flows of basaltic magma.

sialic rock A rock such as granite and rhyolite that contains large proportions of silicon and aluminum.

silica Silicon dioxide, SiO_2. Includes quartz, opal, chert, and many other varieties.

silicate A mineral whose crystal structure contains silicate tetrahedra. All rocks composed principally of silicate minerals.

silicate tetrahedron A pyramid-shaped structure of a silicon ion bonded to four oxygen ions, $(SiO_4)^{4-}$.

sill A tabular or sheetlike igneous intrusion that lies parallel to the grain or layering of country rock.

silt All sedimentary particles from 1/256 to 1/16 millimeter in size.

siltstone A rock composed of lithified silt.

sinkhole A circular depression in karst topography caused by the collapse of a cavern roof or by dissolution of surface rocks.

slate A compact, fine-grained, low-grade metamorphic rock with slaty cleavage that can be split into slabs and thin plates, intermediate in grade between shale and phyllite.

slaty cleavage Metamorphic foliation aligned in a plane perpendicular to the direction of maximum tectonic compressive stress.

slide Any type of mass wasting in which the rock or regolith initially moves as coherent blocks over a fracture surface.

slip The distance that rocks on opposite sides of a fault have moved.

slip face The steep lee side of a dune that is at the angle of repose for loose sand so that the sand slides or slips.

slump A type of mass wasting in which the rock and regolith move as a consolidated unit with a backward rotation along a concave fracture.

smectite A type of clay mineral that contains the abundant elements weathered from feldspar and silicate rocks.

snowline The boundary on a glacier between permanent glacial ice and seasonal snow. Above the snowline, winter snow does not melt completely during summer, while below the snowline it does.

soil The upper layers of regolith that support plant growth.

soil horizon A layer of soil that is distinguishable from other horizons because of differences in appearance and in physical and chemical properties.

soil-moisture belt The relatively thin, moist surface layer of soil above the unsaturated zone beneath it.

solar cell A device that produces electricity directly from sunlight.

solar energy Energy derived from the Sun. Current technologies allow us to use solar energy in three ways: passive solar heating, active solar heating, and electricity production by solar cells.

solar wind A stream of ions and electrons shot into space by violent storms occurring in the outer regions of the Sun's atmosphere.

solifluction The slow mass wasting of water-saturated soil that commonly occurs over permafrost.

sorting A process in which flowing water or wind separates sediment according to particle size, shape, or density.

source rock The geologic formation in which oil or gas originates.

specific gravity The weight of a substance relative to the weight of an equal volume of water.

speleothems Any mineral deposit formed in caves by the action of water.

spheroidal weathering Weathering in which the edges and corners of a rock weather more rapidly than the flat faces, giving rise to a rounded shape.

spit A long ridge of sand or gravel extending from shore into a body of water.

spreading center The boundary or zone where lithospheric plates rift or separate from each other. (*syn*: divergent plate boundary, rift zone)

spring A place where ground water flows out of the Earth to form a small stream or pool.

stalactite An icicle-like dripstone deposited from drops of water that hang from the ceiling of a cavern.

stalagmite A deposit of mineral matter that forms on the floor of a cavern by the action of dripping water.

stock An igneous intrusion with an exposed surface area of less than 100 square kilometers.

strain The deformation (change in size or shape) that results from stress.

stratification The arrangement of sedimentary rocks in strata or beds.

stratified drift Sediment that was transported by a glacier and then transported, sorted, and deposited by glacial meltwater.

stratovolcano A steep-sided volcano formed by an alternating series of lava flows and pyroclastic eruptions. (*syn*: composite volcano)

streak The color of a fine powder of a mineral usually obtained by rubbing the mineral on an unglazed porcelain streak plate.

stream A moving body of water confined in a channel and flowing downslope.

stream piracy The natural diversion of the headwaters of one stream into the channel of another.

stream terrace An abandoned flood plain above the level of the present stream.

stress The force per unit area exerted against an object.

striations Parallel scratches in bedrock caused by rocks embedded in the base of a flowing glacier.

strike The compass direction of the line produced by the intersection of a tilted rock or structure with a horizontal plane.

strike-slip fault A fault on which the motion is parallel with its strike and is primarily horizontal.

subduction The process in which a lithospheric plate descends beneath another plate and dives into the asthenosphere.

subduction complex Rock and sediment scraped onto an island arc or continental margin during convergence and subduction.

subduction zone (or subduction boundary) The region or boundary where a lithospheric plate descends into the asthenosphere.

sublimation The process by which a solid transforms directly into a vapor or a vapor transforms directly into a solid without passing through the liquid phase.

submarine canyon A deep, V-shaped, steep-walled trough eroded into a continental shelf and slope.

submarine fan A large, fan-shaped accumulation of sediment deposited at the bases of many submarine canyons adjacent to the deep sea floor. (*syn*: abyssal fan)

submergent coastline A coastline that was recently above sea level but has been drowned either because the land has sunk or sea level has risen.

subsidence Settling of the Earth's surface which can occur as either petroleum or ground water is removed by natural processes.

supergene (ore) An ore deposit that has been enriched by weathering processes that leach metals from a metal deposit, carry them downward, and reprecipitate them to form more highly concentrated ore.

superposed stream A stream that has downcut through several rock units and maintained its course as it encountered older geologic structures and rocks.

superposition (principle of) (See principle of superposition.)

surf The chaotic turbulence created when a wave breaks near the beach.

surface mine   A hole excavated into the Earth's surface for the purpose of recovering mineral or fuel resources.

surface processes   All processes that sculpt the Earth's surface, such as erosion, transport, and deposition.

surface wave   An earthquake wave that travels along the surface of the Earth or along a boundary between layers within the Earth. (*syn*: L wave)

suspended load   That portion of a stream's load that is carried for a considerable time in suspension, free from contact with the stream bed.

suture   The junction created when two continents or other masses of crust collide and weld into a single mass of continental crust.

syncline   A fold that arches downward and whose center contains the youngest rocks.

tactite   A rock formed by contact metamorphism of carbonate rocks. It is typically coarse-grained and rich in garnet.

talus slope   An accumulation of loose angular rocks at the base of a cliff that has fallen mainly as a result of frost wedging.

tarn   A small lake at the base of a cirque.

tectonics   A branch of geology dealing with the broad architecture of the outer part of the Earth; specifically the relationships, origins, and histories of major structural and deformational features.

terminal moraine   An end moraine that forms when a glacier is at its greatest advance.

terminus   The end or foot of a glacier.

terrestrial planets   The four Earth-like planets closest to the sun—Mercury, Venus, Earth, and Mars—which are composed primarily of rocky and metallic materials.

terrigenous sediment   Sea-floor sediment derived directly from land.

tertiary recovery   Production of oil or gas by artificially augmenting the reservoir energy, as by injection of steam or detergents. Tertiary recovery methods are usually applied after secondary recovery methods have been used.

thermoremanent magnetism   The permanent magnetism of rocks and minerals that results from cooling through the Curie point.

thrust fault   A type of reverse fault with a dip of 45° or less over most of its extent.

tidal current   A current caused by the tides.

tide   The cyclic rise and fall of ocean water caused by the gravitational force of the Moon and, to a lesser extent, of the Sun.

tidewater glacier   A glacier that flows directly into the sea.

till   Sediment deposited directly by glacial ice and that has not been resorted by a stream.

tillite   A sedimentary rock formed of lithified till.

trace fossil   A sedimentary structure consisting of tracks, burrows, or other marks made by an organism.

traction   Sediment transport in which particles are dragged or rolled along a stream bed, beach, or desert surface.

transform fault   A strike-slip fault between two offset segments of a mid-oceanic ridge.

transform plate boundary   A boundary between two lithospheric plates where the plates are sliding horizontally past one another.

transpiration   Direct evaporation from the leaf surfaces of plants.

transport   The movement of sediment by flowing water, ice, wind, or gravity.

transverse dune   A relatively long, straight dune that is oriented perpendicular to the prevailing wind.

trellis drainage pattern   A drainage pattern characterized by a series of fairly straight parallel streams joined at right angles by tributaries.

trench   A long, narrow depression of the sea floor formed where a subducting plate sinks into the mantle.

tributary   Any stream that contributes water to another stream.

trough   The lowest part of a wave.

truncated spur   A triangular-shaped rock face that forms when a valley glacier cuts off the lower portion of an arête.

tsunami   A large sea wave produced by a submarine earthquake or a volcano, characterized by long wavelength and great speed.

tuff   A general term for all consolidated pyroclastic rocks.

turbidity current   A rapidly flowing submarine current laden with suspended sediment, that results from mass wasting on the continental shelf or slope.

turbulent flow   A pattern in which water flows in an irregular and chaotic manner. It is typical of stream flow.

U-shaped valley   A glacially eroded valley with a characteristic U-shaped cross section.

ultimate base level   The lowest possible level of downcutting of a stream, usually sea level.

ultramafic rock   Rock composed mostly of minerals containing iron and magnesium— for example, peridotite.

unconformity   A gap in the geological record, such as an interruption of deposition of sediments, or a break between eroded igneous and overlying sedimentary strata, usually of long duration.

underground mine   A mine consisting of subterranean passages that commonly follow ore veins or coal seams.

undertow A current created by water flowing back toward the sea after a wave breaks. (*syn*: rip current)

uniformitarianism The principle that states that geological change occurs over long periods of time, by a sequence of almost imperceptible events. In addition, processes and scientific laws operating today also operated in the past and thus past geologic events can be explained by forces observable today.

unit cell The smallest group of atoms that perfectly describes the arrangement of all atoms in a crystal, and repeats itself to form the crystal structure.

unsaturated zone A subsurface zone above the water table that may be moist but is not saturated; it lies above the zone of saturation. (*syn*: zone of aeration)

upper mantle The part of the mantle that extends from the base of the crust downward to about 670 kilometers beneath the surface.

upwelling A rising ocean current that transports water from the depths to the surface.

valley train A long and relatively narrow strip of outwash deposited in a mountain valley by the streams flowing from an alpine glacier.

Van der Waals forces Weak electrical forces that bond molecules together. They result from an uneven distribution of electrons around individual molecules, so that one portion of a molecule may have a greater density of negative charge while another portion has a partial positive charge.

varve A pair of light and dark layers that was deposited in a year's time as sediment settled out of a body of still water. Most commonly formed in sediment deposited in a glacial lake.

vent A volcanic opening through which lava and rock fragments erupt.

ventifact Cobbles and boulders found in desert environments which have one or more faces flattened and polished by windblown sand.

vesicle A bubble formed by expanding gases in volcanic rocks.

viscosity The property of a substance that offers internal resistance to flow.

volcanic bomb A small blob of molten lava hurled out of a volcanic vent that acquired a rounded shape while in flight.

volcanic neck A vertical pipe-like intrusion formed by the solidification of magma in the vent of a volcano.

volcanic rock A rock that formed when magma erupted, cooled, and solidified within a kilometer or less of the Earth's surface.

volcano A hill or mountain formed from lava and rock fragments ejected through a volcanic vent.

wash An intermittent stream channel found in a desert.

water table The upper surface of a body of ground water at the top of the zone of saturation and below the zone of aeration.

wave height The vertical distance from the crest to the trough of a wave.

wave period The time interval between two crests (or two troughs) as a wave passes a stationary observer.

wave-cut cliff A cliff created when a rocky coast is eroded by waves.

wave-cut platform A flat or gently sloping platform created by erosion of a rocky shoreline.

wavelength The distance between successive wave crests (or troughs).

weather The condition of the atmosphere, including temperature, precipitation, cloudiness, humidity, and wind, at any given time and place.

weathering The decomposition and disintegration of rocks and minerals at the Earth's surface by mechanical and chemical processes.

welded tuff A hard, tough glass-rich pyroclastic rock formed by cooling of an ash flow that was hot enough to deform plastically and partly melt after it stopped moving; it often appears layered or streaky.

wetlands Known as swamps, bogs, marshes, sloughs, mud flats, and flood plains, wetlands develop where the water table intersects the land surface. Some are water soaked or flooded throughout the entire year; others are dry for much of the year and wet only during times of high water. Still others are wet only during exceptionally wet years and may be dry for several years at a time.

X-ray diffraction A powerful technique for the study of crystal structure in which the regular, periodic arrangement of atoms in a crystal splits an x-ray beam into many separate beams, the pattern of which reflects the crystal structure.

zone of ablation The lower portion of a glacier where more snow melts in summer than accumulates in winter so that there is a net loss of glacial ice.

zone of accumulation The upper portion of a glacier where more snow accumulates in winter than melts in summer, and snow accumulates from year to year.

zone of aeration A subsurface zone above the water table that may be moist but is not saturated; it lies above the zone of saturation. (*syn*: unsaturated zone)

zone of saturation A subsurface zone below the water table in which the soil and bedrock are completely saturated with water.

zooplankton Animal forms of plankton, e.g., jellyfish. They consume phytoplankton.

INDEX

Boldface page numbers indicate where key terms are highlighted; an italicized page number indicates a figure; t indicates a table; n indicates a footnote.

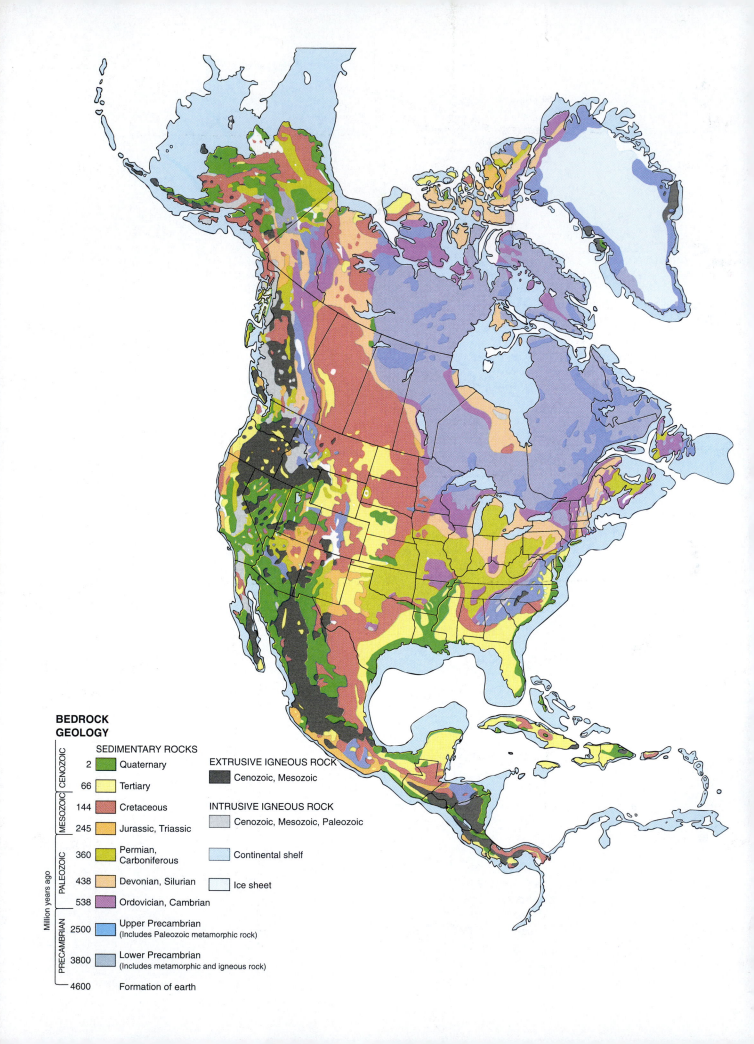

BEDROCK GEOLOGY

SEDIMENTARY ROCKS

Million years ago			
CENOZOIC	2	🟩	Quaternary
	66	🟨	Tertiary
MESOZOIC	144	🟥	Cretaceous
	245	🟧	Jurassic, Triassic
PALEOZOIC	360	🟩	Permian, Carboniferous
	438	🟧	Devonian, Silurian
	538	🟪	Ordovician, Cambrian
PRECAMBRIAN	2500	🟦	Upper Precambrian (Includes Paleozoic metamorphic rock)
	3800	⬜	Lower Precambrian (Includes metamorphic and igneous rock)
	4600		Formation of earth

EXTRUSIVE IGNEOUS ROCK

⬛ Cenozoic, Mesozoic

INTRUSIVE IGNEOUS ROCK

⬜ Cenozoic, Mesozoic, Paleozoic

🟦 Continental shelf

🟦 Ice sheet